LATTICE ENGINEERING

edited by Shumin Wang

TECHNOLOGY AND APPLICATIONS

LATTICE ENGINEERING

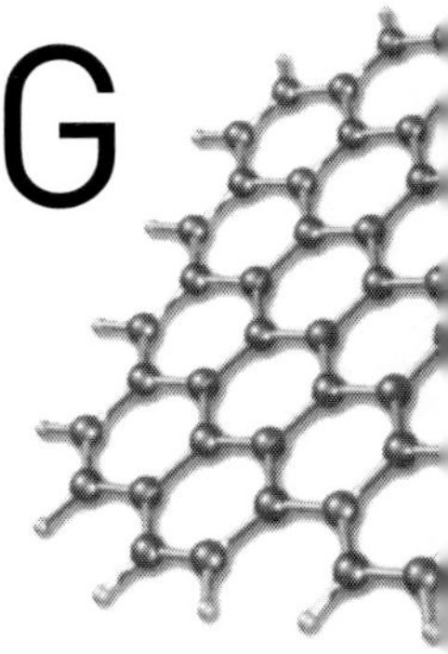

PAN STANFORD PUBLISHING

Published by

Pan Stanford Publishing Pte. Ltd.
Penthouse Level, Suntec Tower 3
8 Temasek Boulevard
Singapore 038988

Email: editorial@panstanford.com
Web: www.panstanford.com

British Library Cataloguing-in-Publication Data
A catalogue record for this book is available from the British Library.

Lattice Engineering: Technology and Applications

ISBN 978-981-4316-29-3 (Hardcover)
ISBN 978-981-4364-25-6 (eBook)

Printed in the USA

Contents

Preface

Semiconductor heterostructures have revolutionized modern information and communication technology (ICT) and are continuously finding new device applications in ICT, energy and our daily life. This is realized by the integration of dissimilar nanoscale thin films using modern epitaxy. One of the biggest technological challenges is that a device-quality heterostructure has to be grown with its lattice constant equal or close to that of a substrate underneath to avoid deleterious structural defects. This lattice constraint restricts heterostructure designs on a few commercially available substrates, often in small size, containing less abundant elements and expensive, thus limiting the flexibility for band engineering. Although the technology to bridge two different semiconductors can date back to 1954–1957, when Herbert Kroemer proposed the concept of heterostructures and used SiGe on Ge substrates through diffusion to fabricate graded gap transistors (*Rev. Mod. Phys.* 73, 783 (2001)), research to solve the lattice mismatch problem has been pursued progressively for over 30 years when epitaxial tools became mature and is still a very active research topic nowadays due to ever-increasing demands for faster, smaller, and energy-efficient devices and new applications on cheap and environmental-friendly substrates.

This book contains reviews and updates of several most important technologies to harness lattice mismatch in several examples of semiconductor heterostructures, including III–Vs, Ge/Si, III–V/Si, and III–N, and their device applications. The first four chapters focus on technologies, including interfacial misfit method (Chapter 1), compliant substrates (Chapter 2), patterned substrate epitaxy (Chapter 3), and low-temperature direct wafer bonding (Chapter 4). Chapter 5 is dedicated to heterostructure nanowires where strain relaxation and lattice engineering are very different when the feature size is down to the nanometer scale. The last five chapters present successful device demonstrations (also a part of Chapter 1), and also some technologies such as conventional metamorphic growth (Chapters 6, 7, 8, and 9), dislocation filtering by nanostructures (Chapter 9), and mediation of strain relaxation by dopants (Chapter 8). Chapter 10 focuses on III–N grown by plasma-assisted molecular beam epitaxy and demonstration of blue and green light-emitting diodes and laser diodes (LDs), a hot topic in the research field of III–N. Other demonstrated electronic and optoelectronic devices include metamorphic high-electron-mobility transistors and their integration with CMOS and with PIN photodiodes to form monolithic optoelectronic integrated circuits, metamorphic heterojunction bipolar transistors, metamorphic infrared quantum well and quantum dot LDs on GaAs and Si, quantum dot solar cells, and type II superlattice infrared photodetectors. Successful lattice engineering technologies have also been implemented in other important materials such as II–VIs, ZnO, and oxides and used to demonstrate multiple junction solar cells with a record high efficiency under concentrated sunlight where the metamorphic growth is a crucial step. We anticipate more innovative concepts and methods to come in the future such that high-performance devices can be eventually fabricated on cheap and sustainable substrates such as Si.

Shumin Wang
Winter 2012

Chapter 1

Interfacial Misfit Dislocation Arrays

Diana L. Huffaker and Kalyan Nunna

California NanoSystems Institute, University of California, Los Angeles, CA 90095 USA

huffaker@ee.ucla.edu, kalyan.nunna@rfmd.com

Lattice mismatch epitaxy is vital to synthesizing a wide range of materials on a limited number of available substrates. To realize compound semiconductor devices for any desired application, a variety of material systems and alloy combinations must be tuned to match the rather small choice of existing substrates. Considerable efforts have been made over the last few decades to epitaxially grow dissimilar materials with larger or smaller lattice size than that of the substrate. Strain buildup in such epilayers results in defects that are detrimental to reliable device operation. With advances in growth techniques combined with careful lattice engineering, significant progress has been made to reduce the defect density. The applications for semiconductor devices are numerous and the interest in *heteroepitaxy* has been increasing steadily to implement and integrate different material systems with the already existing and well-established technology. Other factors, such as cost, size, and quality of the available wafers, electrical/optical/thermal properties, compatibility of device integration, make mismatch lattice engineering even more significant. This chapter focuses on

Lattice Engineering: Technology and Applications
Edited by Shumin Wang

ISBN 978-981-4316-29-3 (Hardcover), 978-981-4364-25-6 (eBook)
www.panstanford.com

antimonide-based materials and devices that can be grown on two very well-studied substrates, GaAs and Si. A very unique scheme that can accommodate the strain mismatch, called the interfacial-misfit (IMF)-dislocation-array technique, is presented. In the later sections of this chapter, the growth and formation of misfit dislocation arrays, structural characterizations, strain energy calculations, and modeling of these arrays will be presented.

1.1 Introduction

An epilayer is considered to be compressive or tensile stained, depending on its lattice constant being larger or smaller, respectively, relative to the substrate. As the deposition of the strained material continues, the strain energy either due to the lattice mismatch or a very thick epilayer increases and eventually is relieved through forming a network of defects or dislocations. These dislocations are initially in the form of misfits, sites where atoms have been skipped, becoming sources for threading defects. The quality and performance of devices are degraded drastically as the dislocations propagate vertically into the active regions, acting as non-radiative recombination centers and current leakage paths. Hence, the lattice mismatch between the epilayer/substrate has to be mitigated and the dislocations need to be suppressed. Over the years researchers have developed different thick metamorphic buffer layer techniques to accommodate the strain mismatch: compositional graded [1, 2] and step-graded layers [3, 4], strained layer superlattices (SLSs) [5], wafer bonding [6], and IMF dislocation arrays [7, 8]. In this chapter, we will discuss the misfit dislocation array method, which is a very effective strain relieving and "buffer-free" technique, with emphasis on technologically important GaSb epilayers on GaAs and Si substrates.

Antimony related compound semiconductors, such as GaSb, AlSb, and InAs or the "6.1 Å family", are an attractive material systems for several infrared (IR) wavelength optoelectronic devices [9–24] and high-speed, low-power consumption electronic applications [25–27]. The III-Sb material combinations with lattice constant close to 6.1 Å offer a wide range of IR bandgaps, band offsets, and electronic barriers along with high carrier mobility. Recent research advances have shown the potential advantages of GaSb-based structures in a wide range of devices: tunability to cover wavelengths from near-IR to far-IR [10–12, 14] and improved quantum efficiency in IR detectors [12], lower threshold current densities in near-IR lasers [16], high-frequency operation at minimal drain voltages in high electron mobility transistors (HEMTs) [27], improved efficiency in tandem solar cells [28], and thermo-photovoltaic (TPV) cells [23]. These key properties and increase in the uses of IR photonics generated enormous interest over the last decade in GaSb material research as well as in novel device fabrications. The III-Sb-based devices are highly promising candidates for numerous applications spanning from military to civil sectors: surveillance and IR imaging, sensors for toxic and hazardous gas detection, pollution monitoring, photovoltaics, near-IR spectroscopy, medical diagnostics, and next generation optical fiber communications. While recent technical advances have enabled good quality GaSb epitaxy on native GaSb substrates, for many applications GaAs substrates are highly desirable. GaSb substrates are not suitable due to several factors:

- Lack of semi-insulating option: Important for device isolation and high-frequency operation in high-speed transistors [27, 29, 30] and detectors.
- Optical loss: GaSb has higher absorption losses at long wavelengths above 5 μm due to which the GaSb substrate has to be removed for developing IR focal plane arrays. On the other hand, GaAs has two orders lower absorption coefficients in the same wavelength range [24, 31]. Deep level defects found in bulk GaSb can also act as loss sites around 2–2.5 μm wavelengths [32].
- Electrical contacts: GaAs technology is very well established and provides highly ohmic *n*-type contacts with specific contact resistances in the range of $\sim 10^{-6}-10^{-7} \Omega$ cm^{-2}. On the other hand ohmic *n*-GaSb metallization has been challenging and is very sensitive to alloying conditions and/or metal schemes with specific contact resistances $\sim 10^{-4}-10^{-5} \Omega$ cm^{-2}.

- Thermal: The GaAs/AlGaAs DBRs offer lower thermal impedance than the GaSb/AlGaSb mirrors, which is very useful in vertical cavity high-power lasers [32].
- Quality: The commercially available GaSb substrates are of poor quality with relatively high defect density, are more expensive, and offer only smaller wafer size (up to 3" diameter).

Hence, commercialization of the Sb-based devices can be greatly enhanced and the problems presented by GaSb substrate can be circumvented by developing the devices on a well-established platform such as GaAs.

Integrating III–V materials and devices on Si substrates offers a different set of advantages and potential applications. Monolithic coupling of Si complementary metal-oxide semiconductor (CMOS) and integrated circuit (IC) technology with III–V photonic devices is of considerable interest with applications in optical fiber communication systems, optical interconnects between ICs, evanescent coupled lasers on silicon-on-insulator (SOI) substrates [33]. Successful epitaxial growth on Si may further reduce the cost per device by eliminating the relatively expensive III–V substrates and allow growth over large Si wafers. Several optoelectronic devices have already been shown based by monolithic growth of GaAs/Si and III-Sb/Si systems.

There are several major issues to growing good quality crystals on Si, as the III–V material properties are substantially different than the Si substrate. Large lattice mismatch causes dislocation defects; thermal expansion coefficient mismatch causes strain related micro cracks in the thin films, and very importantly the polar surface of the III–V's on non-polar Si causes defects due to multiple crystal domains. Considerably, research has been done to overcome these problems over the past two decades with reasonable success. Miscut Si substrates, instead of (100) Si surface, can be used to suppress or annihilate the anti-phase domains (ADPs). AlSb has been found to be highly suitable to nucleate on Si as the thermal mismatch is the smallest. The threading defects have been reduced by various buffer layer schemes, which sometimes involve AlSb island nucleation, strained superlattice layers, etc. In later sections (Section 1.3.4) of this chapter, we will discuss the most recent development using misfit dislocation array technique, which addresses all the above problems to achieve device quality GaSb layers grown on Si substrates with AlSb nucleation layers.

1.2 Lattice Mismatch and Dislocations

Frank and van der Merwe established the foundation for mismatched yet crystalline growth in the 1940s. Later, Matthews and Blakeslee [34] pointed out that the careful growth of an epitaxial layer whose lattice constant is close, but not exactly equal to the lattice constant of the substrate, can result in coherent strain, as opposed to polycrystalline or amorphous incoherent growth.

1.2.1 *Tetragonal Distortion*

If an epilayer takes on the morphology or the lattice constant of the underlying substrate, the strained layer undergoes *tetragonal distortion*. As the growth of the strained material continues, a "*critical thickness*" is reached beyond which the accumulated strain energy is relieved by forming dislocations. The critical thickness arises as a result of a competition between strain energy and chemical energy. Below the critical thickness, the minimum energy condition of the epilayer is achieved through strain. This critical thickness depends upon the extent of the mismatch, material parameters, and the properties of the dislocations that form in a particular material. Critical thickness is usually given by the following equation.

$$h_c = \frac{a_0\left(1-\frac{\upsilon_{PR}}{4}\right)\left[\ln\left(\frac{h_c\sqrt{2}}{a_0}\right)+1\right]}{2\sqrt{2}\pi|f|(1+\upsilon_{PR})} \tag{1.1}$$

where

a_o: Lattice constant of substrate

f: Lattice mismatch

υ_{PR}: Poisson's ratio (~1/3 for most semiconductors).

As the growth proceeds beyond the critical thickness, *relaxation* occurs. Relaxation implies that the material that was once strained and distorted has now come back to its original shape. In the

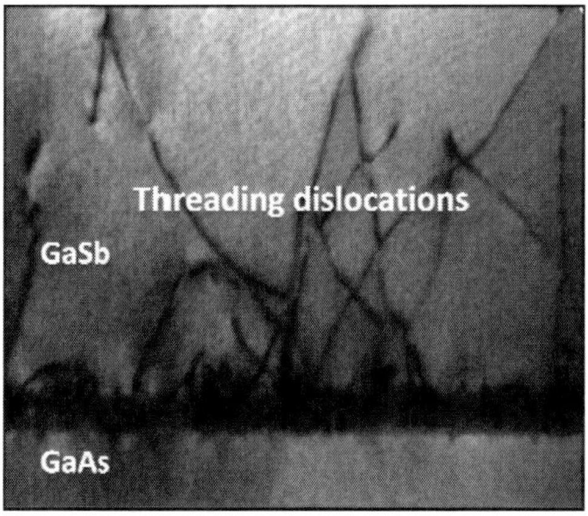

Figure 1.1. A cross-sectional TEM image of detrimental threading dislocations propagating into the GaSb epilayer grown on a GaAs substrate with a lattice mismatch of ~7.8% [7].

strain relieving process the atoms skip the lattice sites and cause misfits dislocations, a source for *threading dislocations*. As the name suggests these defects propagate vertically threading through the crystal by breaking bonds. Figure 1.1 shows a transmission electron microscope (TEM) image of such threading dislocations in a GaSb/GaAs system with a lattice mismatch of ~7.8%. Depending on the material properties, the crystal may take several microns of growth from the onset of dislocations to completely relax. By doing so, the material is said to have moved from a pseudomorphic phase to a metamorphic phase.

Thick metamorphic buffer layers have been investigated as a potential approach that could provide any lattice constant to grow on, thus eliminating the problem of limited substrates. Researchers have used several techniques to reduce the defect densities caused by the lattice mismatch. For instance, linearly graded [1, 2, 35, 36] or step-graded [3] buffer layer compositions and strain layer superlattices (SLS) [1, 5] have been shown in systems such as SiGe/Si, InP/GaAs, and AlSb/Si. To grow GaAs on Si substrates, Fitzgerald *et al.* [4] used step-graded SiGe buffer layers as the lattice constant of Ge is approximately equal to that of GaAs. Despite a number of GaAs/Ge/Si integration issues, GaAs/AlGaAs continuous wave room temperature (RT) lasers were demonstrated, as the surface threading dislocation densities were as low as $\sim 2 \times 10^6$ cm^{-2}.

High quality InP layers on GaAs were grown using compositional grading [37] of a quaternary alloy AlGaInAs to extend the lattice constant from GaAs to InP and $In_{0.63}Ga_{0.37}As$/InP SLS [5] structures. Group III-Sb growth on Si was first demonstrated by van der Ziel *et al.* [15] using AlSb nucleation layer and a 1 μm thick GaSb/AlSb strain-relief superlattice. While the layers had extensive threading dislocations (5×10^7 cm^{-2}), they were among the first III-Sb devices to lase on Si.

Even though the tetragonal distortion and metamorphic buffer layers have enabled device demonstrations, this approach has its limitations. Threading dislocation densities are relatively high, causing significant material/device degradation over time. In spite of very thick buffers (often >5 μm) [18, 38–40], defects manage to propagate through the layers into the active regions. Moreover, the metamorphic-buffer-layer approach is relatively complicated and is often time-intensive. Additionally, the dislocations that do not filter out completely act as non-radiative recombination centers and current leakage paths rendering the material unusable for any devices. Ideally, the excess strain should be relieved at the heterointerface and not result in dislocations that travel through the material in the growth direction. The *IMF-array* technique is close to that ideal method. In this growth mode, misfit dislocations are forced to form in a very well-defined manner and as they are pure-edge type they travel parallel to the growth plane rather than into the epilayers.

1.2.2 *90° Dislocations*

Lomer or edge dislocations are formed by inserting an extra half plane of atoms into a perfect crystal and are marked by an edge of an incomplete plane of atoms. An edge dislocation is represented by the symbol "$\perp$", in which the horizontal line represents the glide plane and vertical line, the tip of inserted half plane. Edge dislocations are further categorized depending on the angle of the dislocation line with respect to the interface. Misfit dislocations that are parallel to the interface with an angle

0° are referred to as screw dislocations. The other commonly observed edge dislocations are 30°, 60°, and 90°. While the 30° dislocations form in systems with a small lattice mismatch, 60° and 90° dislocations are often observed in highly mismatched systems (GaSb/GaAs). A 90° dislocation can sometimes arise from the interaction between two 60° dislocations and can propagate to the surface forming threading dislocations. The 60° misfits are about two times closely packed than the 90° misfits, and therefore the GaSb/GaAs (100) heterointerface has higher elastic energy when relaxed by 60° dislocations [41]. Although, the 90° dislocations are energetically favorable, under unsuitable growth conditions the formation of 90° misfits is disrupted and can be partially replaced by the 60° misfits as an alternate relaxation mechanism.

In the early 1990s, Rocher *et al.* reported that the 90° misfit dislocation networks are easily generated in strain-relaxed GaSb islands grown on GaAs (100) [42–44]. These misfits are very well defined with a certain periodicity, which is attributed to the $\{111\}$ facet formation and flat top shape of the islands. However, as the GaSb deposition continued, forming a thin film, they additionally observed threading defects on the order of $\sim 1 \times 10^{10}$ cm^{-2}. When the GaSb islands coalesce and the misfit networks meet, it is unlikely that all the misfit networks will be in phase. Hence, it was believed that this interaction between various networks led to 60° dislocations that thread to the surface. The presence of 60° misfits was also surprising as they are not often expected in highly mismatched systems such as GaSb/GaAs ($\sim$7.8%).

Qian *et al.* investigated these interfaces to identify the source of threading dislocations observed in the layers in spite of the formation of the 90° misfit network in the islands [45]. They suggest that the misfits are nucleated spontaneously at the edge of the leading edges of the $\{111\}$ faceted islands, where the stress field is the highest. These misfits then glide inward on the (100) plane until an equilibrium position is reached dropping the stress field at the edge. However, the residual compressive strain in the growing layer induces another misfit nucleation. This cycle repeats with further material deposition resulting in a periodicity equivalent to the equilibrium misfit spacing. The choice of 90° misfit nucleation over the 60° one is because the former accommodates the strain twice as efficiently. As far as the threading defects are concerned, Qian *et al.* proposes that they are all associated with the 60° misfit dislocations directly [45]. The 90° pure-edge-misfit dislocations are not likely to introduce any threading segments as they are elastically stable and cannot glide on the close packed {111} planes. In contrast, the 60° dislocations nucleated in the GaSb epilayer can easily glide into the interface along the {111} planes resulting in threading segments.

The conditions for this selection of the type of misfit (60° or 90° or 60° + 90°) formation have not been clearly identified but the amount of lattice mismatch or the growth temperature are considered to be influential parameters. Some researchers have stated that the lattice mismatch is of critical importance in the formation of 90° misfits [46]. Low strain systems ($<2\%$) have resulted in 60° misfits, moderate strain (3–4%) in mixed 90° and 60° misfits and high strain ($>6\%$) in pure 90° misfits. The GaSb growth temperature has also been shown to favor a certain type of misfit to form and thread all the way to the surface [44, 47]. In this chapter, we will discuss the growth mechanism, where a highly periodic array of 90° misfits not only relieve the strain instantaneously but also does not propagate into relaxed GaSb epilayers.

1.3 Interfacial Misfit Dislocation Arrays

Matthews's theory of lattice-mismatched growth using tetragonal distortion states that the excess strain is relieved by forming misfit dislocations when the critical thickness is reached [34]. However, the pure-edge misfits fundamentally do not follow the tetragonal distortion method. The early studies as discussed in Section 1.2.2 were focused on the generation of 90° misfit dislocations at the interface in the {111} plane-bound GaSb islands resulting in threading defects when a continuous film is formed.

The IMF-dislocation-array technique, the main subject of this chapter, is an alternate and unique approach to realize defect-free, strain-relaxed GaSb epilayers on GaAs substrates.

Highly periodic arrays of pure-edge 90° misfit dislocations are initially formed in both [110] and [1–10] directions on the entire GaAs (100) surface, prior to any GaSb material is deposited. After the IMF-array formation, subsequent material deposition proceeds in a strain-free growth mode. The IMF array makes use of the atomic arrangements on the substrate surface to relieve the strain even before the critical thickness is reached. The Sb atoms self-assemble such that each Sb atom forms a single bond with an underlying Ga atom. In order to accommodate the difference in the lattice sizes, the Sb-Ga bonds bend and stretch, and one Sb-Ga bond every 13 atomic sites has to be skipped to relieve the excess strain energy buildup. This skipped bond, a misfit dislocation repeats itself periodically every 5.6 nm, corresponding to quasi-complete relaxation.

Interestingly, regardless of the amount of mismatch, the misfit arrays form under appropriate growth conditions and relieve almost all the excess strain at the heterointerface. Misfit arrays have been reported in several systems with lattice-mismatches ranging from $\sim$0.4% to $\sim$13%. Example systems include GaP/Si [48], GaAs/Si [49], InAs/GaAs [50], InAs/GaP [51], AlSb/Si [8, 52], GaSb/GaAs [7, 44, 53, 54], and InP/GaAs [6]. The IMF array has also been demonstrated under tensile strain conditions, i.e., GaAs on GaSb layer [55]. This can be useful in realizing III-Sb-based materials completely embedded within a GaAs system matrix such as in vertical cavity lasers with GaAs/AlGaAs distributed Bragg reflectors [56]. Type II GaSb quantum dots (QDs) have been demonstrated using the IMF, which are formed differently than the traditional Stranski–Krastanov (SK) growth mode QDs [57, 58]. The particular advantage of GaSb IMF QDs is the type II band alignment and possibility to stack up layers of QDs to increase the absorption volume in solar cell devices [56], without the need for strain compensation layers.

The IMF arrays can be formed using different growth techniques such as metal organic chemical vapor deposition (MOCVD) [48, 49], molecular beam epitaxial MBE [7, 8, 50, 51, 53], atmospheric pressure metal organic vapor phase epitaxy (MOVPE) [54], and wafer bonding [6]. The MBE technique offers important features, such as monolayer precision control and real time *in-situ* monitoring of surface reconstructions by reflection high energy electron diffraction (RHEED).

In the following sections, we will discuss the MBE growth of the 2D IMF array in the GaSb-GaAs and AlSb/Si systems. The

IMF formation is analyzed by the cross-sectional and plan-view transmission electron microscopy (TEM). These TEM observations of the interface are later explained by atomistic strain distribution modeling using molecular mechanics. Further, the quality of these epilayers is assessed by etch pit density (EPD) and x-ray diffraction (XRD) measurements. Finally, examples of successful operation of IMF-based optoelectronic devices are shown with some insight into the effect of the IMF on the electrical properties of *p–n* diodes.

1.3.1 *GaSb/GaAs Compressive IMF Array*

The IMF-array formation does not follow the critical thickness approach and is governed by the surface chemistry and atomic reconstructions. The atomic arrangement of Sb on Ga-rich-GaAs is favored by specific surface chemistry. The Sb atoms form a good bond with the underlying Ga atoms but do not react with the GaAs substrate to displace the As atoms. This effect has been noted by Losurdo *et al.* [59], where they document a positive enthalpy for both reactions of anion exchange with and without isoelectronic compound formation:

$$GaAs + Sb_2 \rightarrow GaSb + AsSb,\ \Delta H^\circ = 13.7\ \text{kJ/mol} \tag{1.2}$$

$$2GaAs + Sb_2 \rightarrow 2GaSb + As_2,\ \Delta H^\circ = 47.6\ \text{kJ/mol} \tag{1.3}$$

Hence, the process can be classified as self-assembly, and RHEED can be a powerful tool to observe the surface during the IMF-array formation in real time.

1.3.1.1 Growth of GaSb/GaAs IMF

The MBE growth of GaSb/GaAs IMF proceeds as follows. The surface oxide is desorbed at 620°C, followed by initial GaAs buffer layer deposition at 580°C, which shows a (2 × 4) diffraction pattern as shown in Fig. 1.2a. The key to forming a long range, highly periodic misfit is to establish a Ga-rich surface. This is accomplished by stopping the GaAs growth at 580°C and the As beam supply. The As atoms desorb from the surface leaving behind a Ga-rich surface. During this step, the RHEED pattern changes from an As-rich (2 × 4) to a Ga-rich (4 × 2) pattern. It is important to allow the As atoms

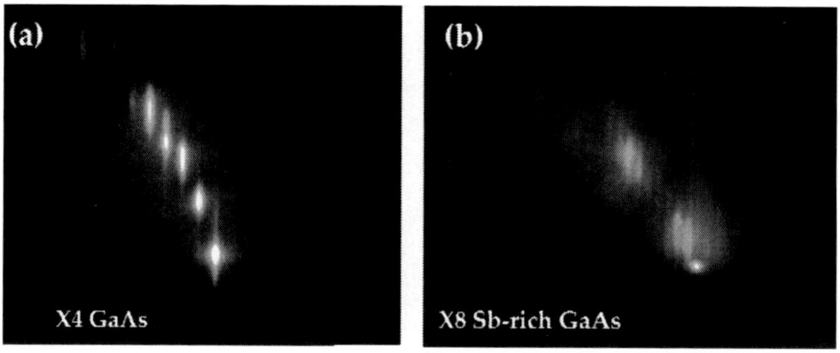

Figure 1.2. RHEED patterns of (a) ×4 pattern of GaAs surface and (b) ×8 pattern of Sb-soaked Ga-rich surface indicating the IMF formation.

from only the surface to desorb and not from underlying layers. Once a stable (4 × 2) pattern is established, Sb beam flux is introduced and the surface transforms into a (2 × 8) reconstruction indicating that the Sb atoms have incorporated and the surface has achieved an equilibrium (Fig. 1.2b) [60]. The (2 × 8) surface is indicative of atomic packing instead of tetragonal distortion resulting in the 90° dislocations. The substrate is cooled down immediately under Sb overpressure while retaining the (2 × 8) pattern. Once the substrate is stabilized at 510°C, GaSb deposition is initiated and the RHEED pattern becomes spotty as the GaSb 2D islands take shape. With continued deposition, the islands coalesce and the surface quickly changes to a clear (1 × 3) pattern after a few monolayers.

1.3.1.2 Structural characterization of GaSb/GaAs IMF by TEM

The structural quality of the GaSb epitaxial layers grown on GaAs and Si using the IMF technique is investigated using low- and high-resolution TEM. We will first inspect the IMF formation by cross-sectional TEM along [1–10] and then confirm the 2D nature of the IMF array in both the [1–10] and [110] directions by high-resolution plan-view TEM. Burgers vector analysis is then used to identify the dislocations as pure 90° edge type (Lomer dislocations). In order to study the initial stages of the IMF formation and how the growth evolves, TEM of very thin layers/islands of GaSb on GaAs are also discussed. The tensile GaAs/GaSb IMF is briefly discussed, followed

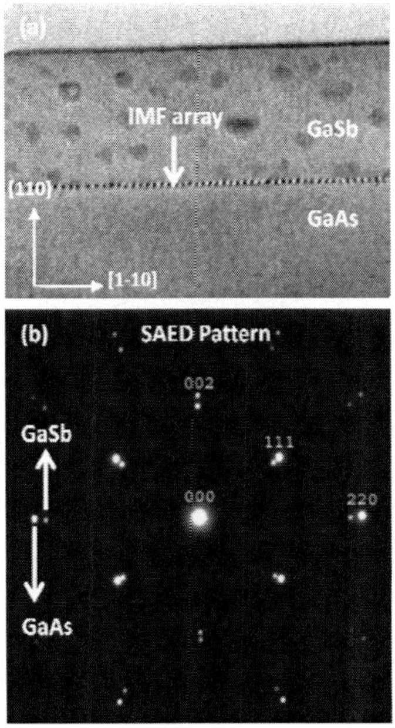

Figure 1.3. Cross-sectional TEM image of 100 nm of GaSb on GaAs. (a) shows a highly periodic array of misfit dislocations at the GaSb/GaAs interface (b) shows SAED pattern of the scan area [7, 61].

by the comparison of the GaSb/GaAs IMF QDs with the SK QDs by XTEM and atomic force microscopy (AFM).

A cross-sectional TEM image in Fig. 1.3a shows a strain-relaxed, defect-free 100 nm thick GaSb buffer layer grown on GaAs substrate. The GaSb/GaAs interface along the [1–10] direction is defined by highly periodic array of misfit dislocations that are localized entirely at the interface. There are no other features suggesting an island growth mode or dislocations in the material associated with the coalescence of large islands. This is further verified by the absence of threading dislocations in this section of the material. The selected-area-electron-diffraction (SAED) pattern along the GaSb/GaAs interface (Fig. 1.3b) indicates that the two sets of cubic diffraction spots are initially overlapped and gradually separated with the increase of the plane indices.

The higher order spots are split into two spots, corresponding to each of the two materials. It reveals that GaSb epilayer has the same crystal growth orientation as that of GaAs layer, and the deposited GaSb is almost completely relaxed with respect to the GaAs substrate. This is an initial confirmation that the IMF array is indeed an efficient growth technique to realize strain-relaxed GaSb epilayer on GaAs without any defects propagating into the layer. To determine if the observed misfits are 90° pure-edge-type dislocations, we use Burgers vector analysis.

The difference between the distorted and a perfect lattice around a dislocation can be quantified by a crystal vector called Burgers vector. The direction of the vector depends on the plane of dislocation, which is usually on the closest-packed plane of the unit cell. The magnitude of the vector is represented by:

$$\|b\| = \frac{a}{2}\sqrt{h^2 + k^2 + l^2} \tag{1.4}$$

where a is the lattice constant of the crystal, and h, k, l are the Miller index components of Burgers vector, **b** =<h k l>. In a 2D primitive square lattice, the Burgers vector of an edge dislocation can be determined by tracing around (red arrow) the end of the dislocation plane to form a closed loop as shown in Fig. 1.4a. In a perfect lattice, a trace in the same path, moving the same number of lattice vectors along each direction as before is shown in Fig. 1.4b. This loop will not be complete, and the closure failure (blue arrow) is the Burgers vector.

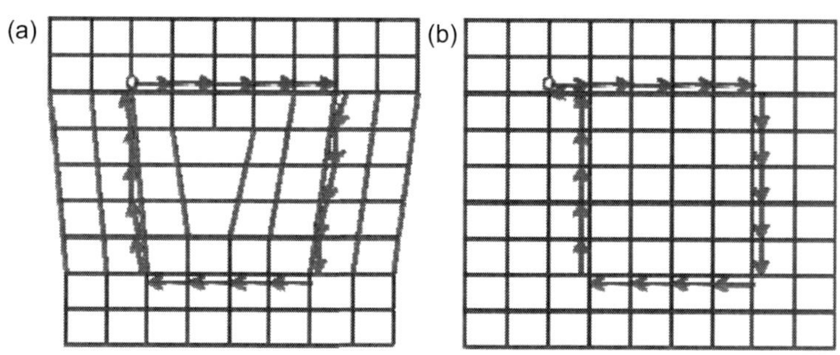

Figure 1.4. Burger vector tracing around a dislocation and a perfect crystal lattice.

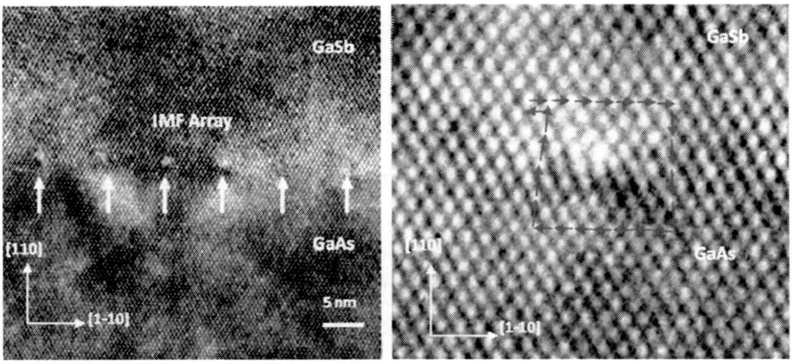

Figure 1.5. (a) Shows a high-resolution cross-sectional TEM image of 100 nm of GaSb epilayer grown on GaAs and (b) shows the Burgers vector trace around the dislocation [61].

The high-resolution cross-sectional TEM images of the GaSb/GaAs IMF are shown in Fig. 1.5. The interfacial misfits in Fig. 1.5a can be identified by the light spots as indicated by the arrows. The Burgers circuit theory around a misfit dislocation, shown in Fig. 1.5b, indicates that the Burgers vector, $a/2[110]$, lies along the interface and identifies this misfit as 90° pure-edge type. The misfit separation, measured to be 5.6 nm, corresponds to exactly 13 GaSb lattice sites grown on 14 GaAs lattice sites. A simple arithmetic shows that the lattice space occupied by 14 atoms of GaAs along the [110] direction is about the same as that occupied by 13 atomic spaces of GaSb.

$$\text{For GaAs, } (14 \times \mathrm{d}_{110}) = 14 \times (5.65/1.414) = 55.94\ \text{Å}$$
$$\text{For GaSb, } (13 \times \mathrm{d}_{110}) = 13 \times (6.09/1.414) = 55.99\ \text{Å}$$

Similar misfit spacing has been obtained in GaSb/GaAs islands by Rocher *et al.* and Qian *et al.*, who suggest the quasi-perfect relaxation of GaSb is indeed possible as the 90° misfits are highly efficient in accommodating the strain mismatch. They also reported that these misfit arrays have been observed along both [110] and [1–10] directions forming a two dimensional net.

1.3.1.3 2D network of IMF arrays

After a few MLs of GaSb growth, the surface transforms into a clear (1 × 3) reconstruction and is observed on the RHEED (Section 1.3.1. The substrate is cooled down after few MLs of deposition during which the (1 × 3) reconstruction transforms into a spotty RHEED indicating island formation. A plan-view scanning electron microscope (SEM) image is shown in Fig. 1.6a for a 9 ML thick GaSb on the IMF array. The islands are found to be elongated along the [1–10] direction indicating that the Sb-terminated {111}B facets grow

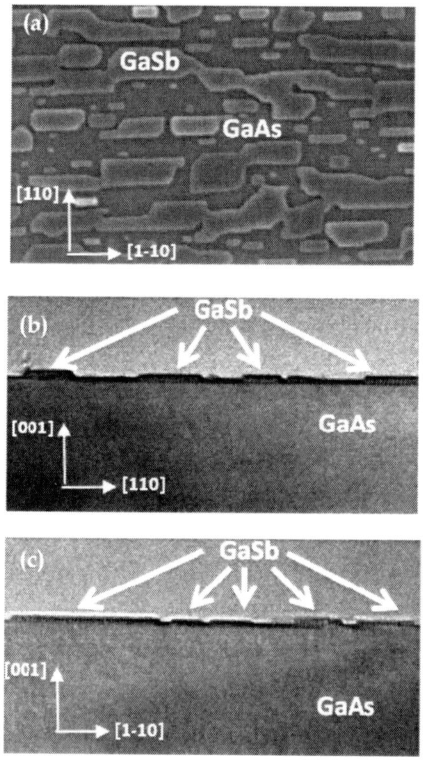

Figure 1.6. Electron microscopic images of 9 ML deposition of GaSb on IMF/GaAs (a) plan-view SEM, showing formation of relaxed islands that are elongated along the [1–10] direction. Cross-sectional TEM along (b) the [110] and (c) the [1–10] directions showing the misfit array formation at the interface and highly periodic under and in between the islands [62].

faster than the Ga-terminated {111}A facets. The average island dimensions are 10 nm in height, 450 nm in length, and 120 nm in width. It is to be noted that the misfit array is present not only under the GaSb islands but also in areas where there is no material coverage. This is considered as the key difference between the IMF technique and the efforts of early researchers [40, 44, 45]. Hence, there is no dependence of the size, shape, nucleation, coalescence of the GaSb islands on the misfit generation, and the film growth proceeds without any threading defects.

In order to verify the 2D nature of the IMF by XTEM, the same sample cleaved along both [110] and [1–10] directions is shown in Fig. 1.6b,c, respectively. These images show the presence of the periodic IMF-dislocation array along both [110] and [1–10] directions. However, there are very few defective islands with no misfit dislocations at the GaSb/GaAs interface, indicating the importance of the misfits in enabling this high-quality growth mode. The origin of these non-IMF islands is a result of perhaps substrate imperfections (steps) and/or growth conditions.

1.3.1.4 Plan-view TEM-Moiré fringe analysis

Cross-sectional TEM only images the IMF array in one particular direction at a time and does not truly depict the 2D network and its periodicity. Hence, plan-view TEM has been employed to examine a ~54 ML thick IMF-based GaSb layer grown on GaAs substrate. A [001] bright field plan-view TEM image in Fig. 1.7 shows Moiré fringes of the IMF array along both (220) and (220) diffraction vectors indicating a 2D Lomer dislocation network. In the IMF-growth mode, the Moiré fringes cover the entire scan area and hence can relieve most of the strain at the interface. This network will be elastically stable without a tendency to attract or repel if the arrays are comprised of only pure-edge dislocations. Therefore it is expected that ideally no threading dislocation will be generated as long as the interface is made up of a perfect 90° dislocation net. Rocher *et al.* analyzed Moiré fringes at the GaSb/GaAs interface [53] where they observe very regular 90° dislocation array networks. However, threading defects associated with 60° dislocations were

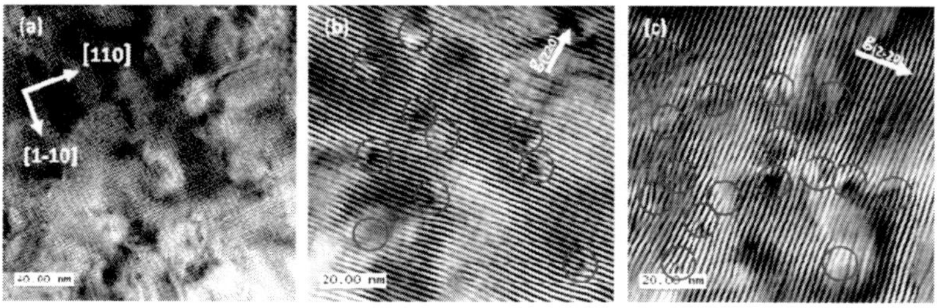

Figure 1.7. Plan-view TEM images of 54 ML GaSb on GaAs. (a) Shows a 2D Lomer dislocation network along both [110] and [1–10] measured using diffraction vectors (220), (2–20), respectively, (b) and (c) shows the same fringes separately along [110] and [1–10], respectively. Consecutive white and dark lines represent Moiré fringes, and the circles represent the edge dislocations [61].

also observed in continuous GaSb films. In Fig. 1.7a, no threading defects are visible in the IMF-growth mode.

To further verify that the IMF array consist of only 90° (not 60° + 90°) and measure the misfit separation, plan-view TEM along both g(220) and g(2–20) diffraction vectors of the same sample are imaged as shown in Fig. 1.7b,c, respectively. The pattern of Moiré fringes in Fig. 1.7 reveals that the IMF array consists mainly of 90° misfit dislocations at the GaSb/GaAs interface along both [110] and [1–10] directions, which supports the XTEM observations, shown in Fig. 1.5. The edge dislocations are indicated by circles in the Fig. 1.7b,c.

Moiré fringes are sensitive to interface distortions such as stacking faults, which are revealed as displacements from these patterns. Since displacement of the fringes is not observed in the IMF samples, no stacking faults or partial dislocations are ascribed to the IMF-growth mode. Kang *et al.* [44] estimated a threading density of $\sim 1 \times 10^{10}$ cm^{-2}, which are related to half-period shifts in the fringes caused by the interaction of 60° misfit dislocations. Threading segments can also arise as the 60° dislocations can easily glide on the closely packed {111} planes [40]. The plan-view TEM images of Fig. 1.8 does not show any half-period shifts (no 60° misfits) in the IMF-growth mode in either direction, highlighting the effect of pure-edge-misfit-dislocation array on quality of the epilayer.

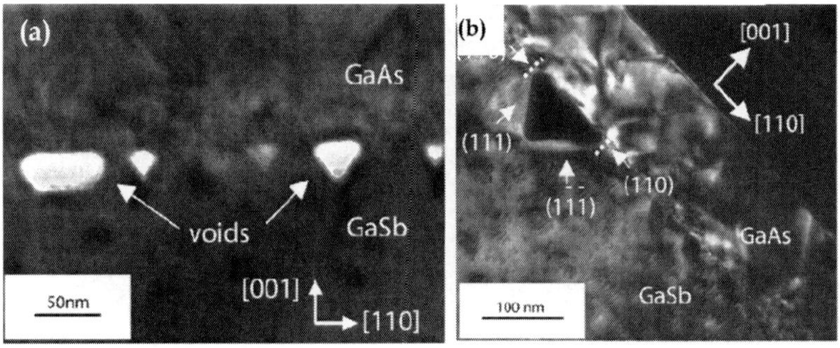

Figure 1.8. XTEM images of the tensile GaAs/GaSb interface after the GaSb surface is exposed to 60 sec of As_2 showing (a) encapsulated "*v*" and "*truncated v*" shaped nanovoids, (b) dark-field image highlighting the "*v*" shaped void in darker contrast and threading dislocations [63].

Moiré fringes are the interference patterns that are formed when two crystals with different orientations or lattice constants overlap, thus providing an excellent indication of amount of relaxation of the epilayer with respect to the substrate. This is particularly relevant in thin-film (<50 nm) experiments where XRD signal is insufficient for assessing the strain. The theoretical parallel Moiré fringe spacing for an epitaxial system, D_m, is given by the general equation

$$D_m = \frac{d_e d_s}{d_e - d_s} \tag{1.5}$$

where d_s and d_e are the inter-planar spacings of parallel sets of planes in the substrate and the epilayer, respectively. Given the lattice constants of GaSb $= 6.096$ Å and GaAs $= 5.653$ Å, the Moiré fringes spacing D_m for diffraction vector $g(220)$ reflections is 2.75 nm. The measured value of $\sim$2.83 nm is in good agreement with the theoretical spacing implying that the GaSb epilayer is almost fully relaxed.

The structural characterization and analysis by TEM hence shows that the IMF-array growth mode of lattice mismatch epitaxy results in very high-quality, threading defect-free, strain-relaxed GaSb layers on GaAs without any thick metamorphic buffer layers.

1.3.2 GaAs/GaSb Tensile IMF Array

As mentioned before, the IMF array can also be formed in the tensile mode when GaAs is grown on a GaSb epilayer/substrate. In contrast, for the As_2/GaSb system, the enthalpy is negative for both reactions of the As-for-Sb anion exchange with and without isoelectronic compound formation

$$GaSb + As_2 \rightarrow GaAs + AsSb,\ \Delta H^\circ = -33.9\ \text{kJ/mol} \quad (1.6)$$

$$2GaSb + As_2 \rightarrow 2GaAs + Sb_2,\ \Delta H^\circ = -47.6\ \text{kJ/mol} \quad (1.7)$$

Initially, the surface oxide of the GaSb substrate is desorbed at 550°C, followed by 250 nm of GaSb buffer layer deposition at 510°C. Conceptually, the growth is similar to the compressive IMF, in that, a Ga-rich surface is established by desorbing the Sb from the surface and supplying As to form the misfit array. However, during the As exposure, the As species react aggressively with the GaSb surface resulting in highly crystallographic pits, making it difficult to establish the reconstructed As layer [63]. The negative enthalpy in Eqs. 1.6, 1.7 indicates that both cases are thermodynamically favorable and can compete with each other. During the IMF formation at a tensile GaAs/GaSb interface, the As atoms try to replace the Sb and result in etching nano voids [55]. Hence, the quality of the IMF was studied as a function of As soak time, and it was found that best misfit array formed without any As soak. The smooth GaSb surface indicated by (1×3) pattern switches to (4×2) Ga rich, and after the As soak, GaAs layer is initiated with a (2×4) pattern.

The excess strain is immediately relieved by skipping an Sb–Ga and an As–Ga bond every 13 lattice sites in the compressive and tensile IMF cases, respectively. In contrast to the non-reactive nature of the Sb atom on the GaAs surface, the GaAs/GaSb interface is more complex due to the aggressive interaction of the As_2 species with the GaSb surface during the As_2 soak. This reactivity leads to the formation of highly crystallographic nanoscale pits at the GaAs/GaSb interface, obstructing the smooth reconstruction of the As layer on the GaSb surface. Samples with different As_2 soak durations are carefully studied by low and high-resolution TEMs [63]. Figure 1.8 shows the XTEM of the GaAs/GaSb interface with an As_2 soak time of 60 seconds. The heterointerface is defined by a high density of nano-pits and threading defects, which are reduced with smaller As_2 soak times. Initial GaAs growth proceeds with nucleation and partial coalescence only on the un-etched portions of the GaSb surface. The subsequent GaAs coverage on these interfaces progresses via lateral overgrowth over the nano-pits resulting in encapsulated "*v*" or "*truncated v*" shaped voids. A closer look at these voids reveals crystal facets and suggests that the observed shapes are due to the different etching rates along specific crystallographic directions. The voids appear to be filled with Sb or AsSb compound and are amorphous in nature as they do not produce any diffraction patterns.

The interface with no As_2 soak as shown in Fig. 1.9a,b does not form any other defects or voids except for the misfit dislocations seen as bright spots arranged in a highly periodic array and localized at the GaAs/GaSb interface. The Burgers vector: $a/2$ [1–10] around a misfit dislocation lies along the interface and identifies the misfit as 90° pure-edge type. The misfit separation is measured to be 5.6 nm, corresponding to exactly 14 GaAs lattice sites grown on 13 GaSb lattice sites. Identical misfit arrays have been observed along both [110] and [1–10] directions confirming a 2D IMF-array network. While the single As atomic layer seems to form on the Ga atomic layer by the ambient As overpressure, it appears to lack long-range uniformity. This leads to relatively higher defect densities (discussed in later sections) than that achieved in the GaSb/GaAs IMF. However, if the strong reaction between the As_2 and GaSb surface is suppressed efficiently, high-quality GaAs layers can be realized using the tensile IMF array.

1.3.3 *GaSb IMF Quantum Dots*

The GaSb IMF QDs are formed differently than the traditional SK growth mode. The growth conditions for the formation of the GaSb/GaAs QDs in both IMF and SK modes are similar with the exception of the V/III ratios during QD formation. In both cases, the QDs ensemble is formed at 500°C with a total coverage of 3 ML at a growth rate of 0.3 ML/s and V/III ratios of 10 and 1 for the IMF and SK QDs, respectively. The corresponding RHEED patterns during the growth of the two types of ensembles are shown in Fig. 1.10. The

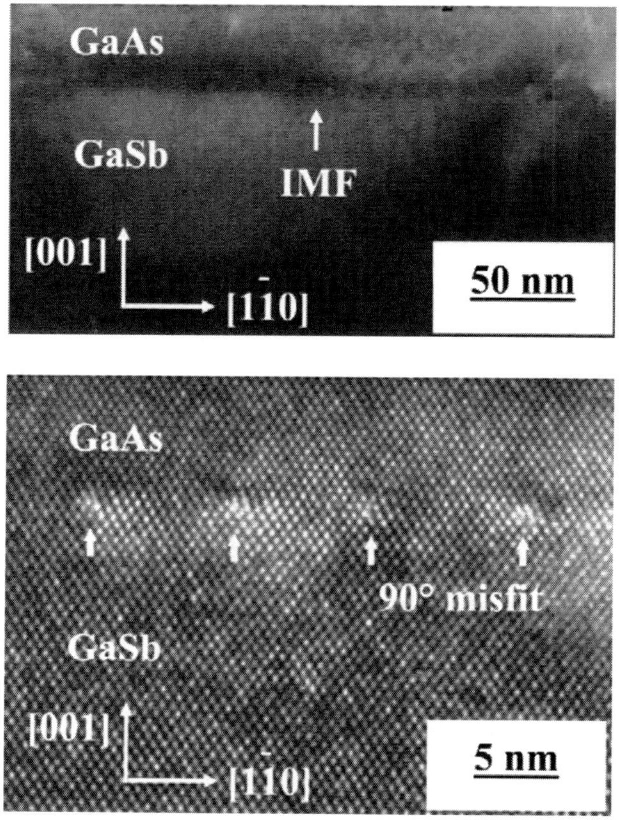

Figure 1.9. XTEM images of the tensile GaAs/GaSb interface with no As_2 soak (a) nanovoids and other defects are not observed except the misfits localized at the interface, (b) high-resolution image showing the misfits as bright spots [55].

QDs are nucleated on a smooth (2 $\times$ 4) reconstructed GaAs surface. Figure 1.10a shows the classical chevrons like features associated with the strained SK QDs. A comparison of the angles involved in the chevron pattern with those obtained from InAs QDs indicates a close match suggesting that the crystallographic planes enclosing the two ensembles may be very similar. Figure 1.10b shows the IMF QD formation with a pattern that does not resemble the traditional chevrons. Components that belong to the $\{111\}$ and $\{100\}$ planes seem to be present indicating a relaxed form.

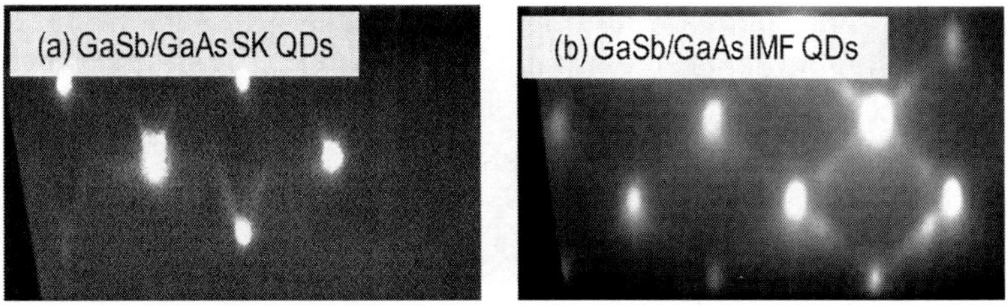

Figure 1.10. RHEED patterns of the GaSb/GaAs (a) SK QDs and (b) IMF QDs. The SK QDs exhibit chevron patterns unlike the IMF QDs due to strained and relaxed natures of the dots, respectively [58].

A V/III ratio in the range of 2 to 10 is employed to produce IMF QDs, while a low ratio of ~1 favors the SK dots. An opposite trend was observed in the InAs/GaAs QD system where, a V/III (As:In) ratio of 6 produces SK QDs and a low ratio of ~1 produces an IMF-based interface. This is because the large atom in the former system belongs to group III (In), and in the latter, it belongs to group V (Sb). From the above scenarios, two primary requirements for the dominance of IMF QD mode over the SK QD mode can be established. First, there must be sufficient size disparity between the constituent adatoms, and second, the larger adatom must be in abundant quantity on the growth surface compared to the small adatom. In contrast, when the larger atom is present in much lower quantity, the SK growth mode is observed.

The structural morphology of the GaSb self-assembled QDs formed on GaAs in IMF and SK modes are examined in this section by AFM and XTEM. As mentioned above, the Sb-rich conditions favor the IMF QD formation over SK type QDs. The AFM images of both GaSb QD ensembles are shown in Fig. 1.11a. The average dot densities for an overall GaSb coverage of 3 MLs in IMF and SK QDs are correspondingly 6×10^{10} cm^{-2} and 3×10^{10} cm^{-2}. The IMF dots are slightly elongated along [1–10] with average dimensions of 50, 30, and 6 nm in length, width, and height, respectively. The SK QD ensemble in Fig. 1.11b displays typical characteristics of the strain driven QDs with average width and height of 10 and 5 nm, respectively.

The presence of IMF array within the QD is confirmed by XTEM as shown in Fig. 1.11a. The misfit periodicity of ~5.6 nm, similar to that

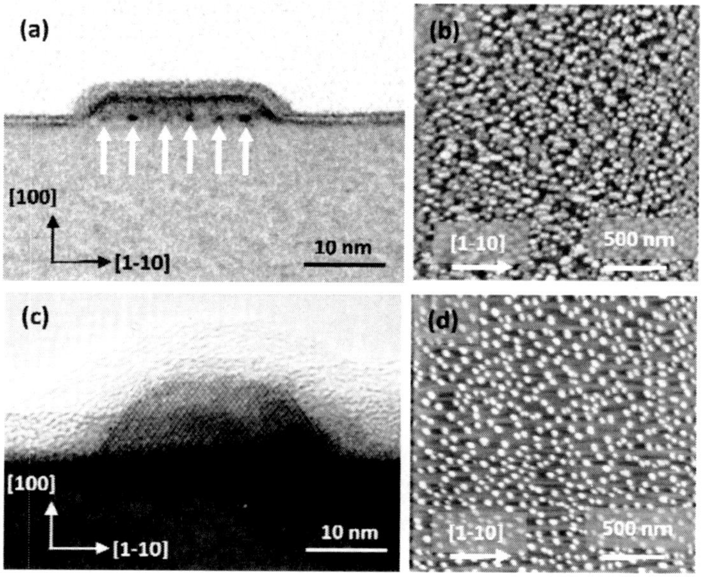

Figure 1.11. XTEM and AFM images of the type II GaSb QDs grown via IMF shown in (a) and (b), respectively. The IMF array is shown with arrows in at the interface with flat top QDs. The SK dots in (c) XTEM and (d) AFM show spherical tops [57].

of the bulk GaSb/GaAs IMF along [1–10] and [110] is observed. The IMF QDs have a flat and broad top, while the SK dots show typical highly crystalline dome shape (Fig. 1.11c). The interface is abrupt in the case of GaSb SK QDs and does not show any indication of the misfit array formation.

Multiple stacks of IMF GaSb QD layers with excellent crystalline quality have been demonstrated [57]. Further, the optical properties of these strain-relaxed IMF QDs have been investigated and successfully implemented into emitters and solar cell devices [56].

1.3.4 *AlSb/Si Mismatch Epitaxy*

There are two major issues preventing growth of high-quality III–V materials on Si substrates other than lattice mismatch: the thermal expansion mismatch and formation of APDs. The thermal mismatch results in micro cracks in the deposited thin film during the cool-down of the substrate from growth temperature to RT. AlSb has a smaller thermal expansion coefficient than Si, resulting in a small compressive rather than tensile strain at RT. Hence, in spite of ~13% lattice mismatch between AlSb and Si, the high bond strength of AlSb and the miniscule thermal mismatch make it highly suitable III–V material on Si.

Formation of APDs is an inherent issue with the growth of polar III-Vs on non-polar Si (100) due to the unavailability of step-free substrates [64]. The diamond cubic (Si) and zincblende (AlSb) structures are composed of two interpenetrating-face-centered cubic sub-lattices, one of which is occupied either by Al or Sb atom in the latter case. The single-atomic-layer steps on the Si (100) plane disrupt a proper registry of these atoms, causing domains to form with opposite sub-lattice arrangements. The interface between such domains results in a two-dimensional structural defect called anti-phase boundary (APB). Miscut substrates (2.5° to 5°) can be used to suppress these APDs [61, 65] as they are typically characterized by double atomic layer-steps [66]. High density of shortened (100) terraces with double atomic steps are formed by the miscut toward the [011] with the surface layers on all terraces belong to the same Si sub-lattice. Therefore, the two sub-lattices on the AlSb side are in registry again, leading to the suppression of the APDs. A 5° miscut Si substrate is found to be the optimal orientation to form the IMF array and grow high-quality GaSb layers without APDs, which will be discussed in the TEM analysis section.

The growth of AlSb on Si was first explored by van der Ziel *et al.* [15] where an optically pumped laser was demonstrated with AlSb/GaSb superlattice buffer layer. In 2004, Akahane *et al.* reported the XRD characterizations of GaSb grown on Si substrates [67]. Direct growth of thick GaSb layers on Si (100) resulted in very hazy surfaces, and a 5 nm GaSb deposition resulted in large islands. This is attributed to the excessive Ga atom surface diffusion on the Si (100) surface. A dramatic improvement in the crystal quality of the GaSb layer was achieved when an AlSb nucleation layer was used. In fact, the AlSb deposition of just 1 ML turned out to be in the form of islands, and additional material resulted in larger islands rather than a 2D layer. Nevertheless, these islands promoted the 2D growth of GaSb. The growth temperature and the thickness of the

AlSb layer was also found to be crucial in getting a narrow full width half maximum (FWHM) in the XRD analysis. However, the growth mechanisms of this highly mismatched epitaxy have not been identified. Balakrishnan *et al.* demonstrated high-quality GaSb/AlSb layers on Si by the IMF-growth technique [8] and investigated the strain relief mechanism. Similar 90° misfit dislocations were also observed by Kim *et al.* when a low temperature AlSb nucleation layer is used instead of growing GaSb directly on Si (100) substrates [52]. In the following sections, we will discuss the extensive work done by Balakrishnan *et al.* on the IMF arrays [8].

1.3.4.1 Growth of AlSb/Si IMF

The Si substrates are treated with diluted HF to remove the native oxide and terminate the surface with hydrogen until the sample is under high vacuum. The HF reacts with the SiO_2 and leaves behind a clean Si surface with the dangling bonds passivated by hydrogen atoms. The loosely bonded hydrogen is removed by heating the substrate to ~500°C in vacuum followed by a thermal cycle at 800°C to get rid of any oxide remnants. This is verified by the RHEED, which shows a (2 × 2) surface reconstruction on oxide desorption. The substrate is then cooled to 500°C followed by few seconds of Al and Sb soak. The initial Al soak of the Si surface enables Al-Si bond formation, while the Sb soak results in a 2D reconstruction on the surface through an interfacial array of misfit dislocations. This reconstruction involves the packing of the Sb atoms on the Si surface such that the Sb atoms skip every 9th Si atom in the substrate. The subsequent growth of AlSb on Si is initiated without any changes in the growth temperature, resulting in a strain-relaxed growth mode with extremely low defect density. The RHEED pattern proceeds through two distinct phases during the initial growth. The initial deposition (~3 MLs) of AlSb on Si results in interconnected spotty features and a superimposed (1 × 3) pattern as shown in Fig. 1.12a implying that AlSb islands with {111} facets truncated with (100) plane on top are formed. Continued deposition of AlSb causes the individual islands to coalesce but remain crystallographic. Figure 1.12b shows the RHEED pattern of ~54 MLs of AlSb, a pure (1 × 3) pattern indicating a planar growth mode.

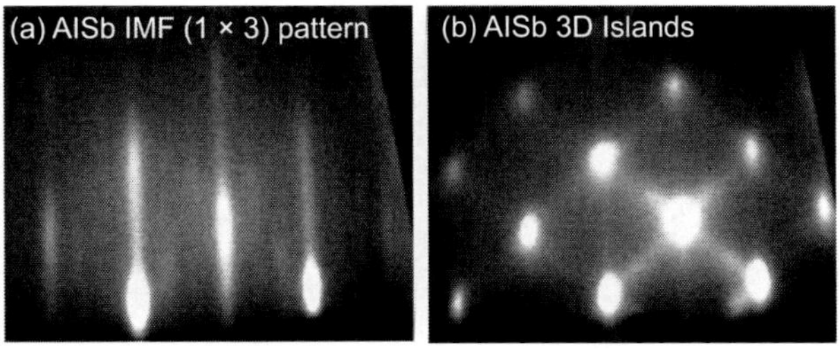

Figure 1.12. RHEED patterns of AlSb on Si with (a) Chevrons and superimposed (1×3) surface for 3 MLs thick islands and (b) a clear (1×3) pattern at 54 MLs indicating planarization of the islands [8].

1.3.4.2 Structural analysis of AlSb/Si IMF

Cross-sectional TEM image of a 0.5 μm thick AlSb layer grown under optimized conditions on Si (100) substrate using the IMF array resulting in a very low-defect density is shown in Fig. 1.13a. Undulations in semiconductors due to misfit dislocations have been noted and modeled by Jonsdottir [68]. Surface undulations on the AlSb surface after ~25 nm of AlSb growth on Si are observed in these samples. With continued growth, the surface undulations merge, become shallower and considerably broader until they can no longer be detected by TEM after a thickness of about 1 μm. However, evidence of the undulation is still visible in the RHEED pattern for AlSb layer thickness ~10 μm. These undulations can, however, be suppressed very effectively by keeping the AlSb nucleation layer to less than ~100 Å and immediately following it with a GaSb layer. The best quality GaSb is obtained when the AlSb growth is terminated at ~50Å. This effect has also been demonstrated by Akahane *et al.* using PL and XRD studies [69].

A high-resolution TEM image of a 10 nm thick, strain relaxed GaSb layer grown on a thinner 5 nm AlSb IMF nucleation/buffer layer is shown in Fig. 1.13. The misfit dislocations arranged in a highly periodic array and localized at the AlSb/Si interface can be seen. Burgers vector analysis of the interface identifies the misfit to be pure-edge type with a separation of ~3.46 nm, corresponding to exactly 8 AlSb lattice sites and 9 Si lattice sites. Similar misfit spacing of 3.46 nm was observed by Kim *et al.* compared to the theoretical spacing of 3.34 nm [52].

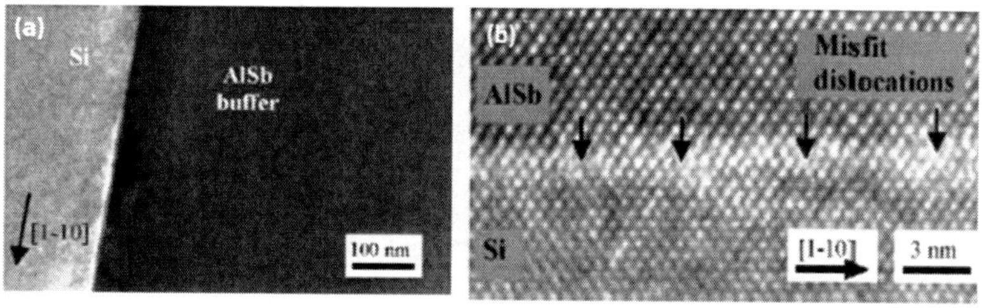

Figure 1.13. Cross-sectional TEM images of AlSb on Si, (a) defect-free buffer layer and (b) HR-TEM image of high-quality GaSb grown on AlSb/Si, with periodic misfit dislocations along the AlSb/Si interface [66].

A thicker bulk 100 nm GaSb layer grown on AlSb (10 nm) on Si (100) substrate is examined by XTEM in Fig. 1.14, where a high density of APDs are detected. The APDs appear to be propagating vertically indicating that the AlSb/Si interface is comprised of isolated 60° dislocations, pairs of 60° dislocations, and 90° dislocations. The APDs formation on the (100) surface is hence a hindrance to realize high-quality GaSb layers as the pure-edge-dislocation misfits are not generated over large areas. To improve the quality of the AlSb buffer on Si substrate, the APDs have to be suppressed and/or annihilated As mentioned

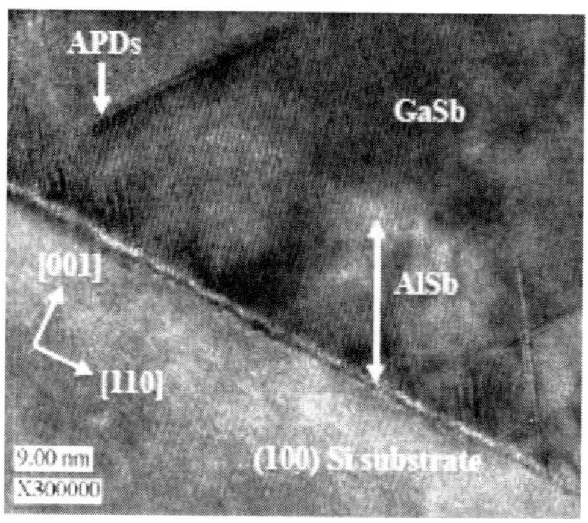

Figure 1.14. High-resolution TEM image of the structure with 100 nm GaSb/10 nm AlSb/ on (100) Si substrate indicating the formation of lots of APDs and defects [66].

earlier, miscut substrates are essential for eliminating the ADPs and stacking faults at the AlSb/Si interface. The major reason of material quality improvement is due to the high density of shortened (100) terraces with double atomic steps, formed by the miscut toward [011]. Additionally, an Al pre-layer and strong Al-Sb bond prior to AlSb growth contributes to the reduction in APDs.

AlSb bulk layers on Si substrates with 3° and 5° miscut toward [110] are grown using the IMF-growth mode to simultaneously demonstrate low APD density and achieve a high-quality IMF layer at the AlSb/Si interface [64]. The growth conditions and sample preparation are kept similar to those used for the Si (100) substrate. The interface on the 3° miscut Si substrate still has APDs and other defects indicating that the IMF does not form perfectly. It has to be pointed out that the defect density is lower than on Si (100) substrate. Figure 1.15 shows the XTEM images of the AlSb/Si interface on the 5° miscut Si substrate. The specific geometry of the 5° surface step seems to be well suited for IMF formation and highlights the sensitivity of the IMF formation conditions.

The bulk AlSb layer and the interface in Fig. 1.15a do not reveal APDs or any other defects. The IMF appears as periodic dark and light bands localized at the interface. The darker regions in this multi-beam image indicate a compressively strained atomic lattice relative to the surrounding material. The strained, dark areas are slightly elongated suggesting an asymmetric distribution of the strain-field amongst several atoms. This is different from the previous cases, AlSb/Si (100) and GaSb/GaAs, in which the dark strain-field images appear more circular and symmetric along the interface [4]. In the higher resolution image of Fig. 1.15b, the periodic IMF and associated strain field have a wave-like distribution of contrasting dark and light regions. At even higher magnification, in Fig. 1.15c, the atomic structure becomes clearer. Though the Si double steps are not discernable, the exact positions of the misfit dislocations can be detected and are marked with boxes. Two dislocations labeled “1” and “2” are identified by the contrast as the misfit dislocations adjacent to a high- and low-strain regions, respectively. The dark region is perhaps associated with the step–edge and results from the strained Al-Al or Al-Si bonds along the [100] direction.

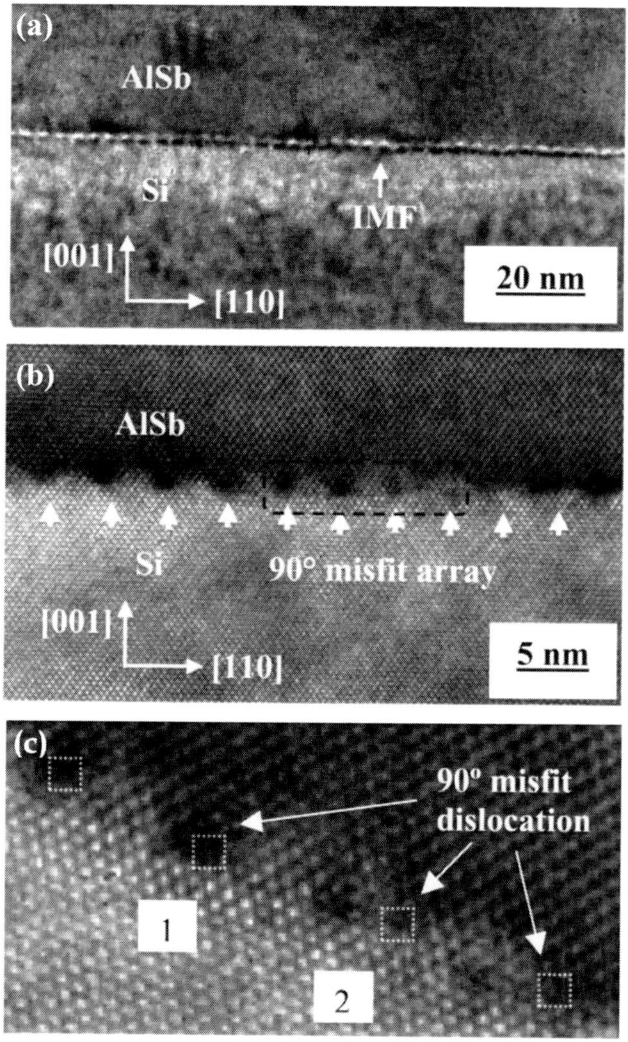

Figure 1.15. Cross-sectional TEM images of AlSb grown on 5° miscut Si (001) substrate under (a) low-resolution, (b) high-resolution conditions, and (c) highlighted section of figure (b), featuring the periodic IMF array. The squares represent the misfits and areas "1" and "2" correspond to misfits around high and low strain regions [64].

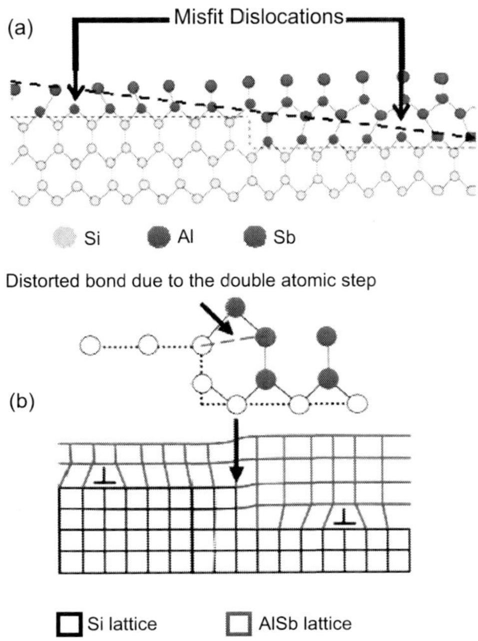

Figure 1.16. Schematics to elucidate atomic arrangement and resulting strain involved in AlSb on miscut Si. Part (a) shows a "ball and stick" schematic of Al, Sb, and Si sub-lattices. Part (b) indicates sources of strain at the step edge due to the step height mismatch [64]. See also Color Insert.

A schematic stick and ball illustration of a possible arrangement of the step geometry and IMF location along with atomic alignment of the Al and Sb sub-lattices at the step edges forming the strain field is shown in Fig. 1.16. Simple geometry of the atomic steps on the 5° miscut surface results in a (100) terrace step length of ~3.109 nm, which is fairly close to the misfit spacing of 3.46 nm that is assigned to AlSb/Si system. This implies that there could be potentially a single misfit dislocation per terrace step. The height of the Si double atomic step and the two atom layer height of AlSb in the [100] direction are different, causing the AlSb growing on the lower step to be compressed along [110]. Hence the strain field indicated in the TEM of Fig. 1.15 can be said to have generated at the step edge and is schematically shown in Fig. 1.16b. Since the miscut is 5° toward the [110] direction, this effect is not repeated along the

[1–10] direction. The double step height and the terrace lengths hence seem to play an important role in suppressing the APDs. This also supports lower-density APDs observed in the 3° miscut substrate as the step geometry is not suitable. Single and double atomic step lengths of 2.57 and 5.19 nm can be present, which are significantly different from the misfit spacing. The number of interaction points of the step edges (strain fields) and the 90° misfit dislocations increase, which may initiate generation of 60° misfits.

The growth mechanisms of highly mismatched AlSb on Si substrate have been examined in this section. A thin AlSb/Si IMF nucleation layer of ~5 nm produced the best quality GaSb growth. APDs are observed in the IMF-based epilayers on Si (100) substrates. To suppress or potentially eliminate the APDs and improve the IMF/epilayer quality, Si substrates misoriented toward [110] have been employed. The 5° miscut substrate is found to be the best option as the step dimensions allow the formation of a long array of pure 90° misfit dislocations without any APDs.

1.4 Strain Distribution-Atomistic Modeling

The lattice mismatch epitaxy approach involving IMF formation appears fundamentally different from the metamorphic approach as strain energy is immediately relieved at the interface by laterally propagating misfit dislocations confined to the epi-substrate interface. After IMF-array formation, subsequent material deposition proceeds in a strain-free layer-by-layer growth mode. The TEM observations suggest that the IMF arrays exist along [110] and [1–10] directions in GaSb/GaAs system. Figure 1.17 shows the cross-sectional TEM image of a typical GaSb/GaAs system IMF array with the misfits as two bright spots in contrast to surrounding material to indicate strain. The high-resolution image clearly shows individual lattice sites and the undistorted bonds between misfits, which appear every 14 Ga-sites (equivalent to 13 Sb-sites). The ratio of lattice constants, a_f: a_s is found to be equal to the ratio of lattice sites, $x_s : x_f$, where the subscripts f and s denote film and substrate, respectively. The ratio $x_s : x_f$ determines the misfit periodicity and equals 13:14 in the case of GaSb/GaAs, i.e., $x_f = 13$ lattice sites

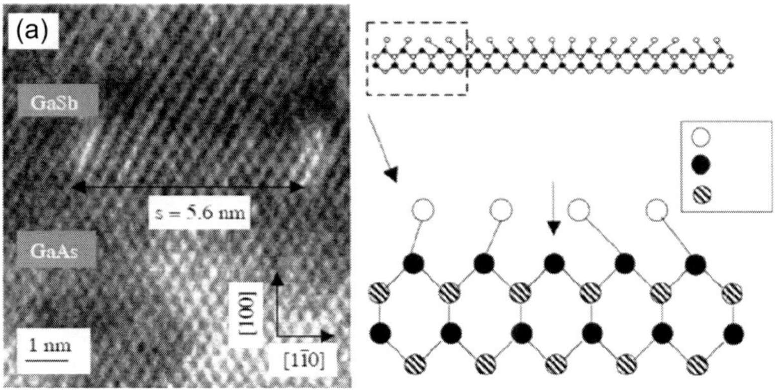

Figure 1.17. (a) Cross-sectional TEM images of 120 nm GaSb bulk material on a GaAs substrate that identifies the highly periodic array of misfit dislocations at the interface of the GaSb layer and the GaAs substrate and (b) a schematic of the atomic arrangement [71].

of GaSb and is equivalent in distance to 14 lattice sites in GaAs. For compressively strained material, $a_f > a_s$ and $x_s = x_f + 1$. In this section, we will describe these observations mathematically and provide a detailed analysis of the periodic strain distribution in the GaSb/GaAs IMF layer. The energy minimization calculations were performed on three systems, including GaSb/GaAs, AlSb/Si, and InAs/GaAs. Experimental data from Trampert *et al.* [70] is used as a reference for modeling InAs/GaAs system.

Models by van der Merwe and Matthews describe the critical thickness of lattice-mismatched systems. The former compares the energy of the composite system before and after the generation of misfit dislocations, while the latter considers forces acting on a dislocation to determine critical thickness. Matthews's theory can be used to calculate the misfit spacing of IMF arrays. Frank and van der Merwe proposed a model that describes the atomic position and uniformly distributed strain energy. Both these models, however, assume pseudomorphic growth up to critical thickness and hence are not useful for understanding periodic strain distribution at the IMF interface.

Kuronen *et al.* [72] introduced molecular mechanics, wherein the lattice is treated as a collection of weights (nuclei) connected with springs (bonds). This approach enabled the calculation of strain energy based on individual atomic bonds. According to the Kuronen's model, the 30° and 60° misfits are formed due to strain, and the interaction of the 60° misfits result in 90°. Since the 90° misfits form spontaneously at the heterointerface, it is not possible to use Kuronen's model to estimate strain in the IMF arrays. To describe the HR-TEM image results mathematically and provide a detailed analysis of the periodic strain distribution in the IMF layer, an atomistic model, which considers atom–atom interaction and bond-energetics, is necessary. Jallipalli *et al.* [71] presented a model based on molecular mechanics to suggest that the IMF formation is driven by strain energy minimization.

The strain energy and atomic geometry at the lattice-mismatched heterointerface are derived from isolated atomic parameters such as force constants and lattice spacing. The total strain energy for each atomic bond is expressed as the sum of several force potentials. These are:

E_s: Energy required for stretching the atoms;
E_θ: Energy required for bending the atoms;
E_{s_θ}: Energy required for stretch–bend interactions;
E_{tor}: Torsional strain;
E_{vdw}: van der Waals interactions and
E_{Dp_Dp}: Dipole–dipole interactions.

The physical system considered here is sensitive primarily to two components E_s and E_θ, i.e., $E_{strain} = E_s + E_\theta$. The energy required to stretch or compress an atomic bond is described using an equation similar to Hook's Law:

$$E_s = 143.82\left(\frac{K_s}{2}\right)(l - l_0)^2 \tag{1.8}$$

where K_s is the stretching force constant, l is the actual bond length and l_0 is the equilibrium bond length, and 143.82 is a constant used to convert the stretching energy to kcal/mol. Similarly the bond bending is given by

$$E_\theta = 0.21914 K_\theta (\theta - \theta_0)^2 \tag{1.9}$$

K_θ is the bending force constant, 0.21914 is a constant used to convert the bending energy to kcal/mol, θ is the distorted bond angle and $\theta_0 (= 90°)$ is the equilibrium bond angle. From the

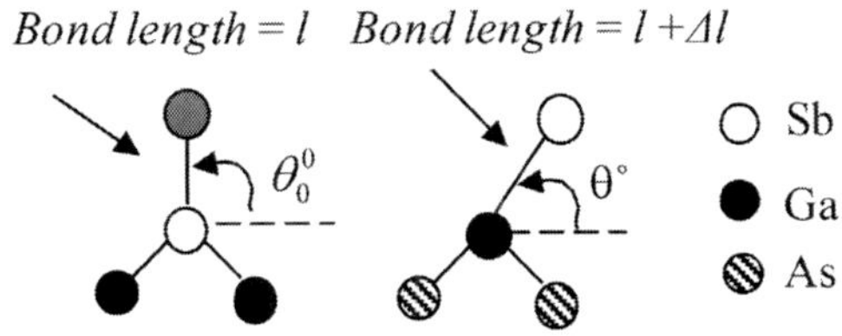

Figure 1.18. Schematic drawing of undistorted and distorted lattices, respectively, showing l, Δl, θ, and $\Delta\theta$ [71].

above equations, the distortion energy for bond stretching is directly proportional to the square of the deviation in bond length, $(l - l_0)$, where as distortion energy for bond bending is directly proportional to the square of the deviation in bond angle, $(\theta - \theta_0)$. Figure 1.18 illustrates the zincblende atomic lattice in equilibrium and distorted conditions with bond length, l; bond stretching, Δl; bond angle, θ; and bond bending, $\Delta\theta$. The covalent radius of Sb atom is larger (1.38 Å) compared to that of the Ga atom (1.26 Å). Therefore, in GaSb/GaAs system, when Sb atoms are deposited on a Ga-terminated GaAs (001) surface, each Sb atom must deviate slightly from the ideal zincblende structure resulting in a new bond length $l + \Delta$l and the new bond angle $\theta + \Delta\theta$ in order to accommodate the lattice and atomic size mismatch. The stretching and bending of the Ga–Sb bonds continue till a physical limit is surpassed and necessitates a skipped bond (misfit). This scenario results in a well-ordered array of periodic array of 90° misfit dislocations.

The strain energy minimization calculations were performed on three systems, including GaSb/GaAs, AlSb/Si, and InAs/GaAs, by allowing the epilayer atoms to relax using all possible periodicities. The least peak strain energy, E_{max}, in GaSb/GaAs system for a grid of 17×17 atoms along [110] and [1–10] is obtained for a 13:14 periodicity. Hence, we can say that the optimal strain energy distribution is always the same, 13:14 (105.36 kcal/mol) for GaSb/GaAs. The other two systems behave in similar fashion with optimal misfit spacing of 8:9 (AlSb/Si) and 14:15 (InAs/GaAs). The calculated stretching and bending energies are plotted as a function of Ga atom position at the GaSb/GaAs interface along [110] as shown

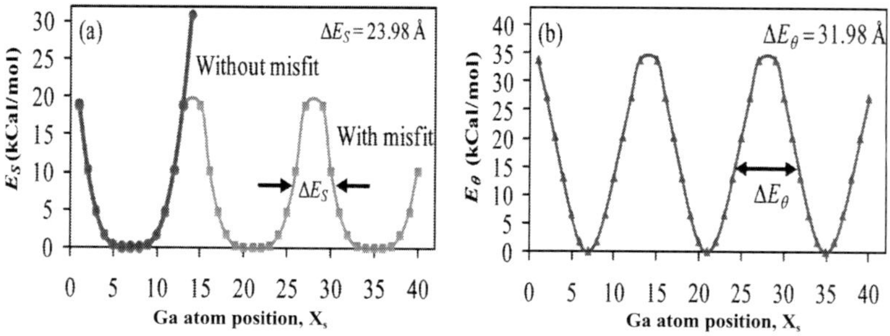

Figure 1.19. A 2D view of (a) stretching, and (b) bending energies plotted as a function of Ga atom position at the GaSb/GaAs interface along [110] [71]. See also Color Insert.

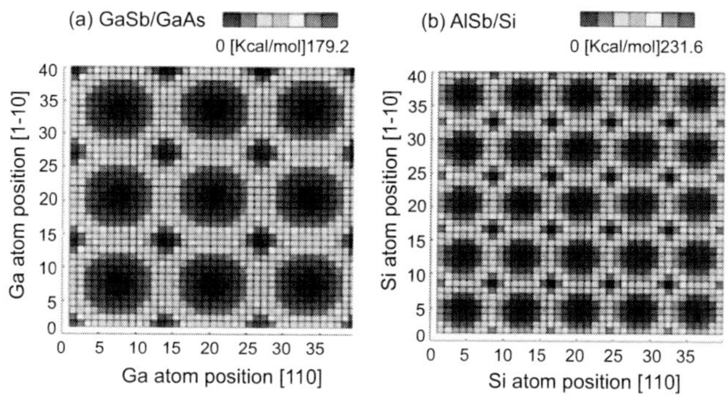

Figure 1.20. A 3D plot of total energy calculated at the GaSb/GaAs interface using molecular modeling for a grid of 40 $\times$ 40 atoms [71]. See also Color Insert.

in Fig. 1.19. The stretching energy increases continuously without the misfit, which can be reduced by skipping a Ga atom.

A 3D grid plots of total strain energy as a function of position is generated by adding energies in both the [110] and [1–10] directions. The GaSb/GaAs and AlSb/Si interface 3D strain plot is shown in Fig. 1.20a,b, respectively, indicating similar energies in either direction. The color scale from blue to red represents strain energy variation from low to high. The misfits are surrounded by highly strained areas indicated in red and mostly unstrained with regions of yellow and green in between them indicating residual strain.

The lattice mismatch of 7.2% and 7.8% for the InAs/GaAs and GaSb/GaAs systems and 13% for the AlSb/Si correspond to lattice ratios of 14:15, 13:14, and 8:9, respectively. The values for K_s and K_θ, which represent the amount of energy stored in the distorted bond, are rather similar for these materials: K_s range from 32 to 36 N/m and $K_\theta \sim 6$ N/m. Edge dislocation spacing or misfit spacing, s, of the IMF arrays for a totally relaxed epilayer can also be calculated theoretically using the formula [34, 52], $s = b/f$, where $b = a_f/\sqrt{2}$ along [110] or [1–10], f is the % of lattice mismatch and is given by $f = (a_f - a_s)/a_s$. The misfit periodicity calculated using MM model for AlSb/Si is 8:9 and hence is equal to the misfit spacing, $s = 8a_f = 9a_s = 3.47$ nm, where a_f and a_s are the lattice constants of the epi-layer (AlSb) and substrate (Si), respectively. Similarly, for InAs/GaAs, the periodicity 14:15 is 6 nm, and for GaSb/GaAs, the misfit periodicity 13:14 is 5.6 nm. The misfit spacing or misfit periodicity values calculated using MM method are in good agreement with experimental HR-TEM images for InAs/GaAs [50], GaSb/GaAs [7], and AlSb/Si [52] as well as with Matthew's theory.

These calculations result in periodic distribution of strain energy and indicate that total strain energy is strongly dependent upon material parameters, such as stretching and bending force constant as well as lattice mismatch. The simulations and modeling suggest that the misfit array formation is driven by strain energy minimization.

1.5 Defect Analysis of Bulk IMF: GaSb/GaAs

The quality of the GaSb epilayers grown using the IMF-array technique on GaAs and Si substrates is verified by defect density measurements and high-resolution XRD. In highly lattice mismatched systems, 60° and 90° dislocations form to relieve strain at the critical thickness. The 90° dislocations can also be formed by the interaction between two 60° dislocations, where the dislocation line is perpendicular to the interface. Threading dislocations that are formed by the 60° dislocations propagating through the bulk material all the way to the surface can be measured by two methods: The first method is EPD. In this method, the layer surface is etched by chemicals like potassium hydroxide (KOH) to reveal defects that can be observed under Nomarski phase contrast microscope. The second method is to use plan-view TEM to observe the threading defects directly on the surface. The EPD technique allows analyzing large areas and is reliable for defect densities in the range of 10^5–10^6 cm^{-2}. Plan-view TEM is limited by the scan area but very accurately determines defect densities. These characterizations are also helpful to understand the effect of the epilayer thickness on the strain relieving mechanism and defect suppression and thus are useful to optimize the IMF-based GaSb layer growth conditions.

1.5.1 *Etch Pit Density*

GaSb epilayers with thickness of ~100, 1000, and 3000 nm grown on GaAs at different growth temperatures of 480°C, 510°C, and 540 °C are used for EPD analysis. The wafers are etched in 20% KOH solution for 10 minutes with an etch rate of ~10 nm/minute. The defect density results are plotted in Fig. 1.21 and shows that the optimal growth temperature is ~510°C, where the EPD is the lowest on the order of 10^5 cm^{-2}. Additionally, the defect density tends to decrease for thicker epilayers.

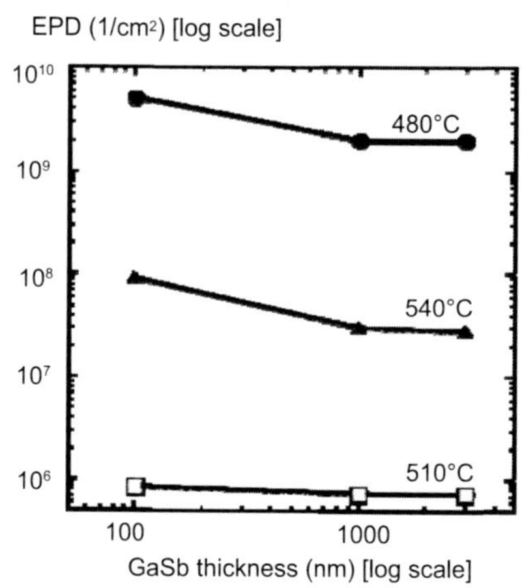

Figure 1.21. Defect densities at different thicknesses of GaSb on GaAs using IMF for various GaSb growth temperatures [73].

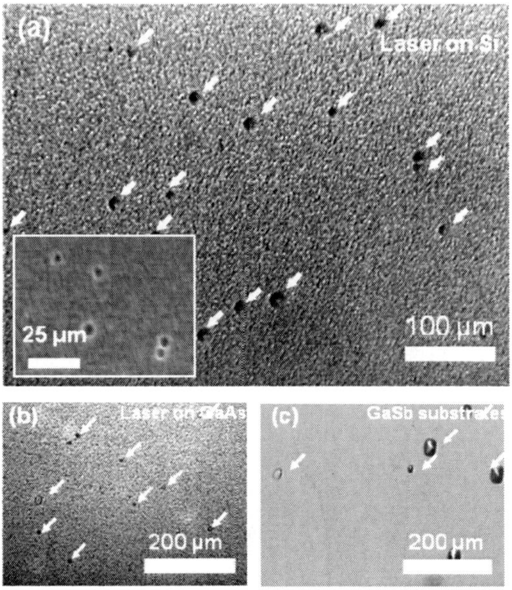

Figure 1.22. Surface morphology of etched (a) GaSb/AlSb on Si substrate, (b) GaSb on GaAs substrate, and (c) commercial GaSb substrate. Average EPDs are (a) 3.5×10^4 pits/cm^2, (b) 9.0×10^3 pits/cm^2, and (c) 2.7×10^3 pits/cm^2. Inset of (a) is an SEM image of the surface morphology of GaSb/AlSb on Si substrates [74].

EPD characterization of the GaSb buffer layer is carried out on fabricated laser structures on 5° miscut Si [74]. EPD values are determined by Nomarski optical microscope on the laser samples immersed into a solution of hydrochloric acid (HCl) and hydrogen peroxide (H_2O_2) ($HCl:H_2O_2:H_2O = 1:1:2$) for 1 minute at RT. The EPD for two kinds of laser structures grown on Si and GaAs substrates along with commercial GaSb substrates are estimated by averaging the EPDs observed on several randomly chosen 500 μm × 500 μm areas as shown in Fig. 1.22. Both of the laser structures are etched through the *n*-GaSb contact layer (~500 nm above from AlSb/Si interface) to evaluate the EPD at GaSb buffer layers and compare with commercial GaSb substrates. Figure 1.22a shows the EPD of $\sim 3.5 \times 10^4$ cm^{-2} from the laser structures grown on 5° miscut Si, which is approximately four times higher than that of the lasers on GaAs substrates (Fig. 1.22b) and 13 times higher than that of commercial GaSb substrates (Fig. 1.22c).

1.5.2 *Plan-View TEM*

As pointed out in Section 1.3.1, the growth parameters are very critical to the formation of IMF arrays comprised only of pure-edge-type-misfit dislocations. To highlight the difference between the formation of 60° and 90° misfits, a GaSb layer was grown without the IMF formation by directly depositing GaSb on GaAs. The excess strain is relieved in this case by 60° misfits and threading dislocations. This non-IMF GaSb epilayer is included in the defect analysis by plan-view TEM and XRD analysis for comparison.

Figure 1.23 shows the bright field plan-view TEM images of 5 μm GaSb grown on GaAs by IMF and non-IMF-growth modes. Figure 1.23a shows the surface area with no defects from the wafer center. Some defects are visible around the edge region of the wafer, which can be expected as unoptimized growth conditions exist around the wafer edges. Scanning different areas of the wafer a defect density of $\sim 5 \times 10^5$ cm^{-2} is estimated, which agrees fairly with the EPD analysis. However, the dislocation density for the non-IMF-growth mode shown in Fig. 1.23b goes up to $\sim 5 \times 10^9$ cm^{-2}. Figure 1.23 demonstrates that the IMF array can greatly reduce the propagation

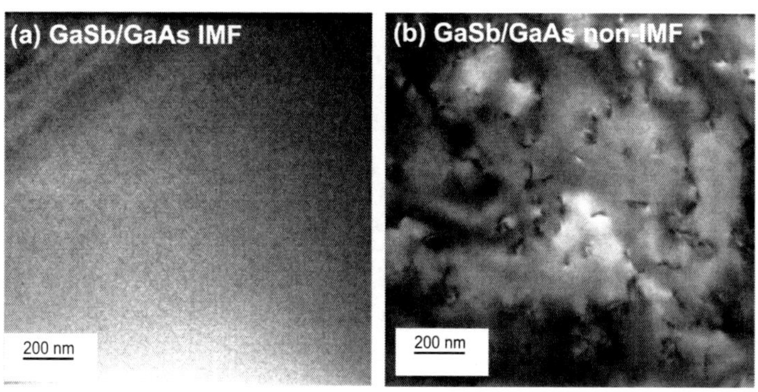

Figure 1.23. Plan-view TEM showing (a) threading dislocation free area of IMF-based 5 μm GaSb layer surface and (b) non-IMF GaSb layer with a defect density of $\sim 5 \times 10^9$ cm^{-2} [61].

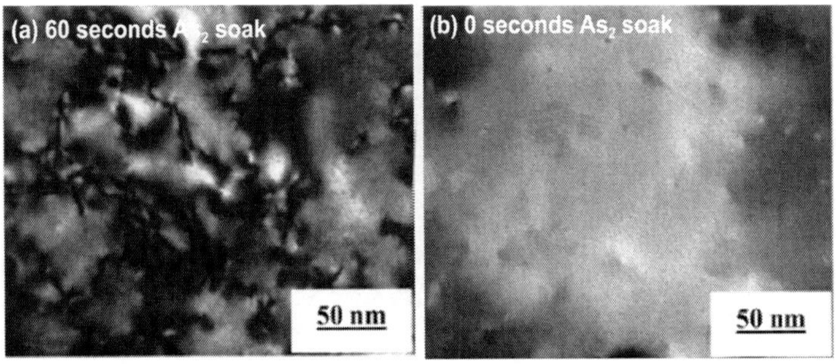

Figure 1.24. Plan-view TEM images of the GaAs surface after GaSb surface (a) 60 sec, (b) 0 sec As_2 soak, respectively, prior to the growth of GaAs epilayer [55].

of threading dislocation to the epilayer, improving the quality of bulk materials.

Figure 1.24 shows plan-view TEM images of the GaAs surface after GaSb surface (a) 60 sec, (b) 0 sec As_2 soak, respectively, prior to the growth of GaAs epilayer. These images indicate threading dislocations that appear as dark, squiggly lines on the shiny GaAs surface and enable the density of threading dislocations to be calculated. The 60 sec As_2 soak condition yields a threading dislocation density $\sim 1 \times 10^9$ cm^{-2}, while without As_2 soak, the density of threading dislocation is reduced to $\sim 3 \times 10^6$ cm^{-2}. This result is consistent with that observed in Fig. 1.8 with cross-sectional TEM analysis, where nanovoids were observed at the interface with As_2 soak.

Compared to the formation of compressive IMF array, the tensile IMF array produces more misfit dislocations. The Sb atoms have good surface mobility and the single Sb atomic layer can easily form on the Ga atomic layer. While the single As atomic layer seems to form on the Ga atomic layer by ambient As overpressure, it appears to lack long-range uniformity. It is the reason why the tensile IMF array has higher density of misfit dislocation than that of compressive IMF array as indicated in plan-view TEM analysis. However, to realize high-performance electronic devices using the tensile IMF, the defect density of the GaAs on GaSb needs to be reduced $<10^5$/cm^2. To achieve the higher quality GaAs on GaSb, a

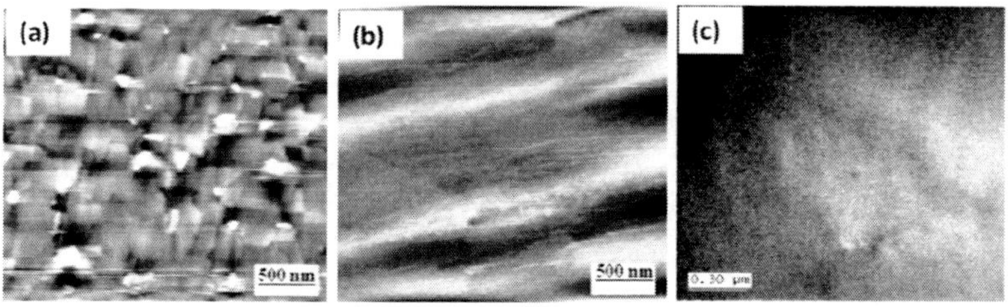

Figure 1.25. AFM images of the 1 μm GaSb grown on AlSb on (a) Si (001) surface, (b) 5° miscut Si surface, indicating the use of misoriented substrates is essential for the suppression of antiphase domains. (c) Plan-view TEM image of the GaSb surface grown on AlSb on 5° miscut Si surface indicating a very low-defect density [66].

non-reactive atom can be used either as a surfactant or catalyst. A thin buffer of AlSb, with greater bond strength compared to GaSb, may also reduce the As-surface reactivity enabling long-range uniform IMF formation.

To verify the quality of the AlSb epilayer on 5° miscut Si substrate, AFM and plan-view TEM images are used to measure the dislocation density. The experimental results, shown in Fig. 1.25a,b, indicate that the AlSb bulk layer grown on 5° miscut Si (100) substrate has very low density of APDs, compared to that grown on Si (100) substrate. The average surface roughness of AlSb bulk layer grown on 5° miscut Si (100) substrate is only ~0.6 nm, indicating a good surface quality. Figure 1.25c shows plan-view TEM image of GaSb bulk material (1 μm) on AlSb (50 nm)/5° miscut Si at the center wafer area with no defects. Based on scanning different areas of the center of the wafer a defect density of $\sim 7 \times 10^5$ defects/cm^2 is estimated. The images show that the AlSb bulk layer grown on 5° miscut Si planes is free of APDs.

1.5.3 *X-ray Diffraction*

Further evaluation of the IMF-based GaSb layers on GaAs substrate is obtained from the ω–2θ scans of the symmetric (004) XRD spectra. Figure 1.26 shows the ω-2θ (004) curve of 700 nm GaSb/GaAs layer grown with IMF-array technique. The GaSb peak is at ~30.36° (~9614 arcseconds) from the GaAs substrate peak indicating almost

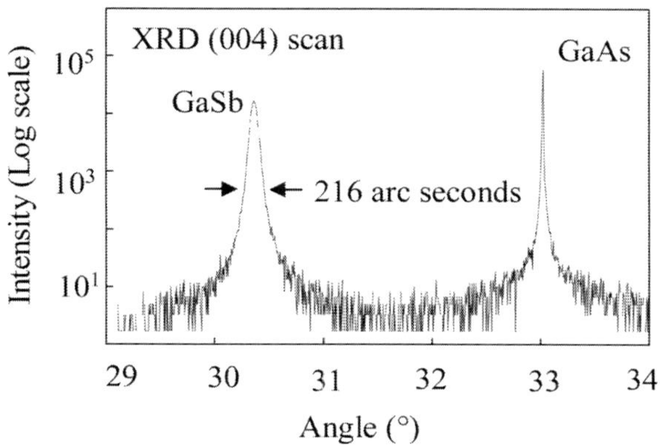

Figure 1.26. Symmetric (004) $\omega - 2\theta$ scan of XRD 700 nm GaSb grown on GaAs substrate using IMF illustrating highly relaxed GaSb [76].

full relaxation. A relatively broader full width at half maximum (FWHM) of ~216 arcseconds is exhibited by IMF GaSb layers for thicknesses less than 1 μm. However, a 5 μm thick GaSb layer showed an FWHM of ~20 arcseconds in a triple axis ω-2θ scan, signifying that the residual strain is relieved with increasing thickness. Calculations from the symmetric (004) and asymmetric (115) XRD measurements show 98.5% relaxation for the IMF-grown GaSb. Similar type of relaxation is also observed in GaSb/GaAs system with AlSb nucleation layer [75].

All the strain energy generated by the GaSb/GaAs lattice mismatch should be dissipated by the misfit array immediately at the interface in order to achieve relaxed GaSb growth. This strain energy is calculated to be 0.18983 J/m^2 for a 1 ML thick GaSb on GaAs. Since the IMF arrays form within the first ML, this is a good estimate as to what percentage of the strain generated by the mismatch is relieved at the interface. The dislocation energy dissipated by a 2D misfit array is found to be 0.186852 J/m^2, implying that ~0.03 J/m^2 is not relieved through misfit dislocations. Hence, the 2D GaSb/GaAs IMF array relieves >98% of the strain mismatch at the interface, and the residual strain is dissipated gradually with increasing thickness. Thus, the relaxation of the GaSb layer can be said to improve with thickness, which supports the observed trends of smaller EPD and decreasing FWHM in XRD with thickness.

1.6 IMF-Based Devices

Various optoelectronic devices, e.g., edge emitting lasers, vertical cavity lasers, LEDs, solar cells, and detectors based on III-Sb active regions, have been demonstrated using the IMF-array technique on Si and GaAs. In this section, we will briefly highlight the performance of these devices, and more detailed information about the device growth/fabrication/testing can be found in the corresponding references.

1.6.1 *Near-IR Edge-Emitters*

An optically pumped RT edge-emitting laser with Sb-based active region on Si (100) substrate using the AlSb/Si IMF was demonstrated by Balakrishnan *et al.*, in 2005 [76]. The 2 μm laser devices consisted of three $In_{0.2}Ga_{0.8}Sb$ QWs separated by $Al_{0.3}Ga_{0.7}Sb$ barriers in the active region. Later, an electrically injected laser at 1.54 μm with a threshold current density (J_{th}) of 2 kA/cm^2 under pulse condition was demonstrated at 77 K on 5° miscut Si substrate. The maximum peak output power ~20 mW was obtained from the fabricated device [74].

The GaSb/GaAs IMF was also successfully used to grow laser devices. A 1.65 μm electrically injected laser was demonstrated by Mehta *et al.* [77] at RT on GaAs substrate. The active region consisting of six $GaSb/Al_{0.35}Ga_{0.65}Sb$ QWs generated an output power of 15 mW at −10°C in pulsed regime with a J_{th} of ~2 kA/cm^2. Indium was then incorporated into the QW layer to increase the compressive strain in the well and suppresses the Auger recombination process. The schematic of the device structure and the laser characteristics of a 1.25 mm × 25 μm device are shown in Fig. 1.27. The active region consists of six $Ga_{0.9}$ $In_{0.1}Sb$ QWs separated by $Al_{0.35}Ga_{0.65}Sb$ barriers. RT lasing with peak power of ~1 mW and a J_{th} of 1.27 kA/cm^2 are obtained under pulsed condition as shown in Fig. 1.27b. The inset of Fig. 1.27b shows the electroluminescence (EL) spectra at various current densities ranging from 0.5 J_{th} to 1.05 J_{th} with a peak wavelength of 1.82 μm and a lasing peak at 1.816 μm [78].

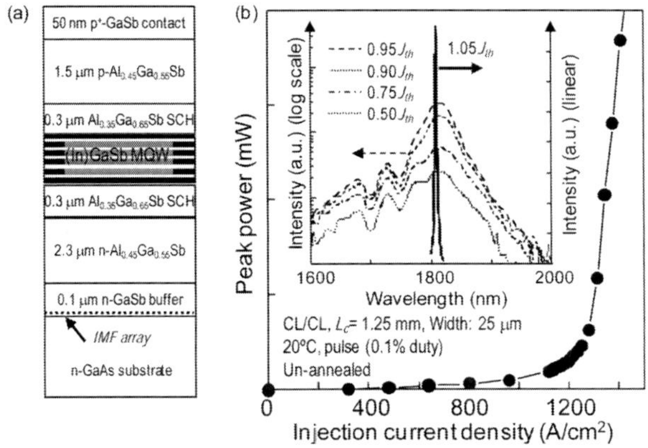

Figure 1.27. (a) Schematic illustration, and (b) L-I characteristics of the fabricated 6-layer $Ga_{0.9}In_{0.1}Sb/Al_{0.35}Ga_{0.65}Sb$ QW laser structure. The $L-I$ characteristics are measured at 20°C under a pulsed condition and 0.1% duty cycle from a diode with a cavity length of 1.25 mm. Inset shows EL spectra with different current densities ranging from 0.5 to 1.05 J_{th} [73, 78].

Using the IMF-growth technique, Rodriguez *et al.* demonstrated edge-emitting lasers comprised of GaInAsSb/AlGaAsSb QW active regions on GaAs as well as Si (4° miscut) substrates [79, 80]. The laser device on GaAs exhibited continuous wave lasing operation at RT with a J_{th} in the range of 1.5–2.2 kA/cm^2, a characteristic temperature of 86 K, and maximum output power of 3.7 mW at ~2.2 μm. The laser diodes on Si operated in pulsed regime at a wavelength of 2.3 μm. The J_{th} in the range of 4.5–8.2 kA/cm^2 are obtained, which are within a factor 50 and 10 of that obtained in the case of similar lasers grown directly on GaSb and GaAs substrates, respectively.

The relatively higher J_{th} values are usually observed due to the difficulty in cleaving good facets in Si substrates, while in the GaAs case due to the voltage drop at the heterointerface (discussed later). However, these edge-emitting device demonstrations show the true potential of the IMF-growth mode to realize III-Sb-based lasers on GaAs and Si substrates.

1.6.2 *Surface Emitters at Near-IR Wavelengths*

A monolithic, electrically injected vertical cavity LED emitting at 1.6 μm comprised of hybrid GaAs/GaSb-based structure has been

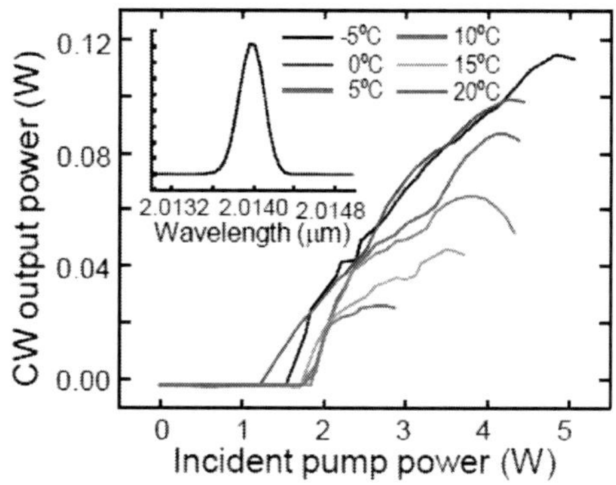

Figure 1.28. CW output power versus incident pump power of the VECSEL measured at different heat sink temperatures. The inset shows the lasing spectrum of the fabricated VECSEL just above the threshold pump power density [81]. See also Color Insert.

demonstrated by Mehta *et al.* [56]. The GaSb/$Al_{0.3}Ga_{0.7}Sb$ active region is sandwiched on the top and bottom by both compressive and tensile IMFs, respectively. At a current density of 100 mA/cm^2, the EL at 1.6 μm showed a FWHM of 78 meV. The forward bias characteristics show a turn-on voltage of ~3 V and a differential resistance of ~30 Ω at 8 V. The rather high voltage is attributed partially due to ~1 V drop per IMF.

Optically pumped vertical external cavity surface emitting lasers (VECSELs) with high gain III–V QW regions and high beam-quality are used for high power applications. Recently, an Sb-based VECSEL deposited on a GaAs substrate using IMF-growth mode on a GaAs/AlGaAs distributed Bragg reflector (DBR) has been demonstrated to emit at 2 μm under pulsed conditions with a record peak power of 70 W [32]. The CW operation is achieved by the deposition of processed heat spreaders on the top surface of the VECSEL with a maximum output power exceeding 100 mW at −5°C [81]. A GaSb active region is deposited on a commercially acquired bottom DBR. The active region consists of nine $In_{0.2}Ga_{0.8}Sb/Al_{0.25}Ga_{0.75}Sb$ QWs. The continuous output power of the VECSEL as a function of incident pump power for wafer temperatures ranging from −5°C to 20°C is shown in Fig. 1.28. The lasing spectrum is shown in the inset of Fig. 1.28 and is measured above the threshold pump power density of 13.9 kW/cm^2 at 20°C. Lasing is observed at 2.014 μm with a linewidth of 0.3 nm and a maximum output power of ~120 mW at −5°C.

1.6.3 *Detectors*

Strained layer superlattices of type-II InAs/GaSb-based detectors are currently researched as an alternative to HgCdTe and quantum well IR photo detector technologies for IR imagers. Several advantages such as simple band structure tuning, reduced Auger recombination rates, and improved quantum efficiency enable the fabrication of high-performance focal plane arrays (FPAs) operating in the mid-IR and long-IR wavelengths. In order to realize high performance and commercialize these SLS structures, GaAs substrates offer critical features. Apart from the lower cost and better quality of GaAs substrates, the absorption coefficients (>5 μm) are nearly two orders of magnitude lower than GaSb substrates. Mass production of FPAs is made extremely attractive due to the commercially available larger diameter wafers, reducing the manufacturing costs.

Bao *et al.* demonstrated short period InAs/GaSb SLS for short-wave IR wavelengths with IMF mode AlSb QD layer and GaSb/AlSb buffer layer on GaAs substrates [10]. Peak detectivity of the photoconduction detectors at 50% cutoff wavelength of 2.25 μm is 2×10^8 cmHz$^{1/2}$/W at 300K and 4×10^9 cmHz$^{1/2}$/W at 77 K. Plis *et al.* used the GaSb/GaAs IMF growth to develop the InAs/GaSb SLS detectors with 18% quantum efficiency at zero bias and a specific detectivity (D^*) of ~10^{11} Jones at 77 K [82]. The D^* at 293 K was within a factor of 2 lower than that measured for the mid-IR $p-i-n$ detector grown on the GaSb substrate.

1.6.4 *Solar Cells*

The type II nature of GaSb/GaAs QDs provides a long carrier lifetime that is suitable for photo-generated carrier collection. The QD ensembles are incorporated as IR absorption material into solar cells to increase their efficiency, and the absorption volume can be increased by stacking the QD layers. In traditional SK mode

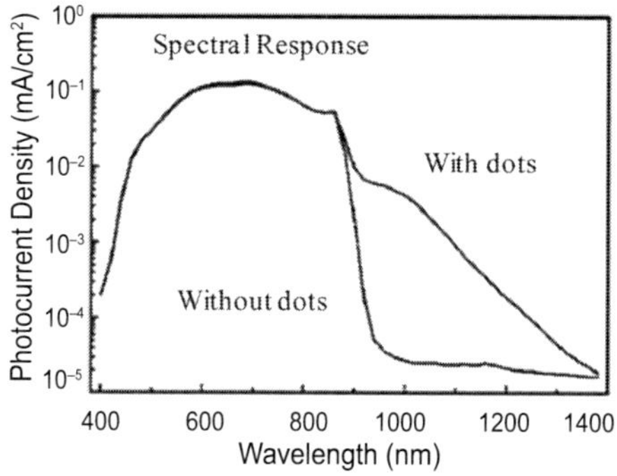

Figure 1.29. Spectral response of the solar cells with and without the inclusion of type-II GaSb/GaAs IMF QDs [56].

QDs, the strain builds up as the number of stacks is increased, causing defects and hence lowering the overall efficiency. On the other hand, the IMF QDs are strain relaxed and therefore are highly suitable to embed as many QD layer stacks as possible without degrading the material quality. Laghumavarapu *et al.* demonstrated 10 stacks of IMF QDs inserted into the *i*-region of a *p*–*i*–*n* GaAs solar cell heterostructure [56]. The solar cells with IMF QDs exhibited enhanced photo conversion at longer wavelengths over the cells without dots as shown in Fig. 1.29. Short-circuit current density and open circuit voltage of the solar cells with and without dots of 1.29 mA/cm^2, 0.37 V and 1.17 mA/cm^2, 0.6 V, respectively, are achieved. The increased response at 1.3 μm is attributed to the ground state transition absorption while the response at 1.1 μm is ascribed to the higher energy level transitions.

1.7 Electrical Performance of IMF

The several advantages of transferring III-Sb devices on to GaAs substrates can be utilized by not only understanding the IMF array formation but also identifying the effects of the IMF array on the performance of these optoelectronic devices. The electronic nature of the IMF array is especially important in electrically injected devices, where the highly ohmic and mature *n*-GaAs metal contacts can be substituted for the poor n-GaSb contacts. In this section, we will analyze the GaSb/GaAs IMF array from a device point of view and characterize the electrical properties of the heterointerface. Simple GaSb-based p–n junctions grown on the IMF are characterized by electronic band structure modeling and current-voltage (I–V) measurements to study the conduction mechanisms.

1.7.1 *Potential Barrier at the GaSb/GaSb IMF*

To differentiate the characteristics of the laser with and without the inclusion of the IMF array, the 1.8 μm laser diode shown in Fig. 1.27a is processed using both top-top and top-bottom approaches. The former scheme has metal contact above the IMF array such that no current flow through the heterointerface. The lasing threshold current density is found to be lower: 0.86 kA/cm^2 (as opposed to 1.27 kA/cm^2) when the laser diode is contacted above the IMF. Additionally, a smaller redshift in the lasing wavelength as a function of the current injection duty cycle is observed. Gallium dangling bonds are present at the IMF array due to skipped Sb-Ga bond, which can vact as non-radiative recombination centers resulting in device heating. Moreover, any threading dislocations present at the interface can also contribute to the leakage currents. Thus, the higher J_{th} and wavelength redshift can be attributed to the IMF array [78].

To investigate these IMF features, a GaSb/GaAs, p–n heterojunction is grown by IMF-array technique such that the interface is located in the depletion region [83]. The diode structure schematic and the simulated band diagram are shown in Fig. 1.30a,b, respectively. At the heterojunction, a barrier potential of 0.7 eV is estimated, which includes an additional 0.2 eV that is contributed by the IMF array. This is also observed in the VLEDs as mentioned in Section 1.6.2, where a voltage drop of ~1 V is associated with the dual IMF arrays [56]. Simulated band diagrams for an equally doped n-GaSb/n-GaAs interface illustrated that IMF possess a localized hole density of 2 × 10^{20} cm^{-3} resulting from the Ga dangling bonds that pin the Fermi level ~0.1 eV below the valence band. As a result,

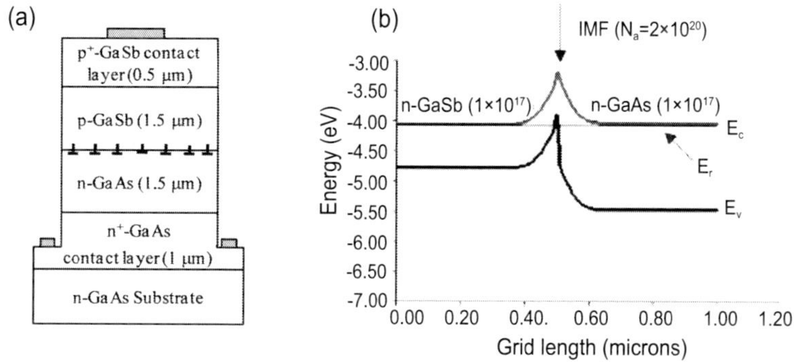

Figure 1.30. (a) shows a schematic illustration of the GaSb/GaAs *p–n* diode structure [83] and (b) shows the simulated band structure depicting the barrier spike at the interface and Fermi level pinning below the valence band [56].

a voltage barrier (seen as a spike in Fig. 1.30b) for carrier flow across the IMF is created resulting in high forward resistances in the diodes.

1.7.2 *Interfacial States*

The IMF array, comprised of Ga-dangling bonds with a sheet density of 3×10^{12} cm^{-2} creates interfacial states that alter the electronic properties of the devices. To estimate these energy levels of these misfit dislocations, surface photo voltage (SPV) technique is used [67]. SPV defined as the illumination-induced change in the surface potential has been used to identify interface energy states in the bandgap of semiconductors. This change is measured as the contact potential difference (CPD) between a reference (Kelvin Probe) electrode and the sample under consideration as a function of illumination wavelength. Undoped GaSb is *p*-type in nature with a residual acceptor concentration of $\sim$1–5 $\times$ 10^{16} cm^{-3}, which are related to gallium vacancies (V_{Ga}) and Ga at Sb sites (GaSb) [9, 69]. The energy levels of GaSb have been measured to be at 0.25, 0.3–0.35, 0.63 eV [9].

Figure 1.31 shows the slope of CPD as a function of the energy of the incident light for an undoped GaSb grown on semi-insulating GaAs substrate using IMF array growth mode. The energy transitions between energy bands to and from interface states are

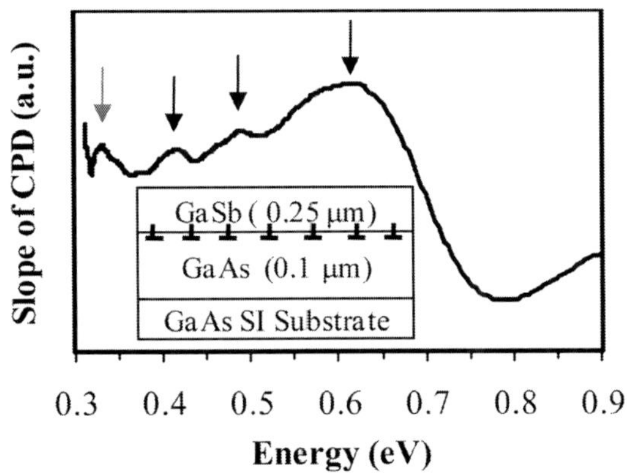

Figure 1.31. SPV measurements of the interfacial states of the IMF array located at 0.41, 0.49, and 0.61 eV. The structure of the sample used for these measurements is shown in the inset and consists of unintentionally doped GaSb grown SI-GaAs substrate using IMF growth [84].

seen as points of inflection in the CPD plots and therefore as peaks in the plot of the slope of CPD. The point at which CPD plot crosses zero is typically the bulk bandgap and in this measurement that occurs near 0.69 eV, which is near the anticipated bandgap for GaSb of 0.726 eV. Energy transitions were observed at 0.33, 0.41, 0.49, 0.61 eV and represent the energy levels of the IMF array states.

1.7.3 *Compensation of the IMF*

Theoretically, Fermi level can be unpinned by the compensation of the acceptor levels of the IMF via the addition of donors at the GaSb/GaAs interface, where the density of the donors (3×10^{12} cm^{-2}) is analogous to that of the IMF. Hence, by compensating the Ga-dangling bonds, it is possible to achieve better device characteristics.

The interfacial states are shown at the *p*-GaSb/*n*-GaAs interface in Fig. 1.32a, where a tunneling path can be created by electrons hopping through the energy levels. Sufficient energy is provided by either forward-biasing or annealing the diodes, due to which the electrons from *n*-GaAs can easily be transferred into *p*-GaSb. In addition to these two methods, the interfacial energy levels

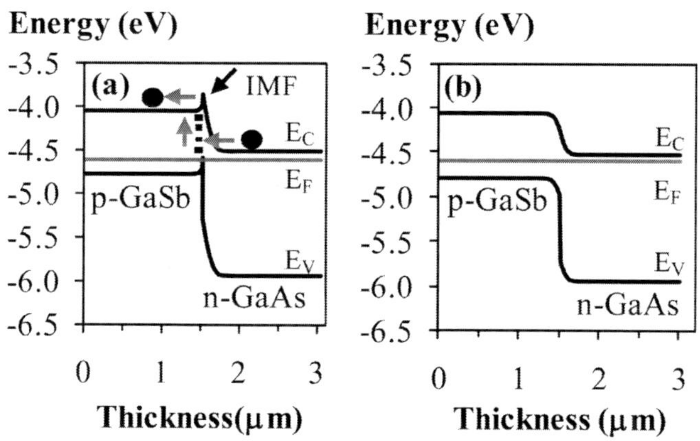

Figure 1.32. Band diagrams of (a) uncompensated and (b) compensated p-GaSb/n-GaAs diodes [83].

can also be compensated by δ-doping the heterointerface with Te atoms (sheet density $\sim 3 \times 10^{12}$ cm^{-2}). Band diagram of the compensated p-GaSb/n-GaAs diode is shown in Fig. 1.32b. The Fermi-level pinning at 0.1 eV below the valance band and the extra barrier (0.2 V) provided by the interfacial states are compensated, making the effective barrier 0.5 V.

The I–V characteristics of the compensation mechanisms of *p-GaSb/n-GaAs* diodes by annealing, electrical injection, and delta doping are illustrated in Fig. 1.33a,b, respectively. The turn-on voltage and reverse bias (RB) leakage of the respective diodes are shown in the corresponding insets of Fig. 1.33. The curves PN_{IMF}-BC and $P\,N_{IMF}$-AC represent the I–V characteristics of the diode before and after compensation, respectively. The turn-on voltage drops to 0.6 V after the first sweep indicating that the barrier to block the carriers is absent.

Similar situation is observed after annealing the diodes, where the turn-on voltage is reduced to 0.35 V. The metal contact alloying is also accomplished during the annealing step along with the barrier compensation. However, the reverse leakage currents increase. In the δ-doped diodes, initially the turn-on voltage is similar to that of the unannealed diode (4.45 V) as the Te atoms are believed to bond with Ga atoms only after annealing the sample. Hence, delta doping the diodes appears to the best mechanism to compensate the

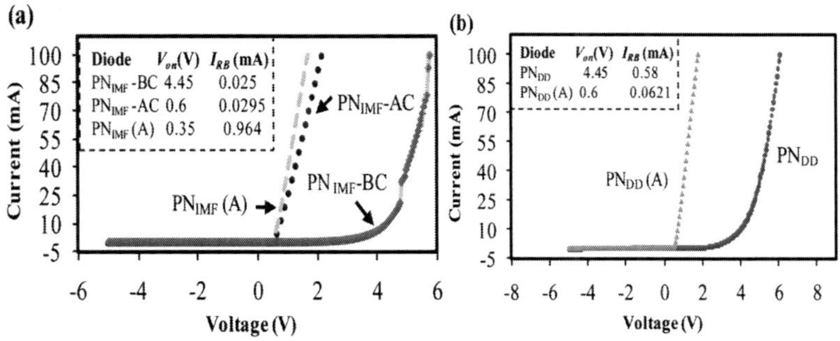

Figure 1.33. I–V characteristics of p-GaSb/n-GaAs IMF 400 × 400 μm^2 square diodes. (a) Shows barrier compensation by annealing and electrical injection, and (b) shows compensation by δ-doping. The turn-on voltages (V_{on}) and RB leakage currents (I_{RB}) of the corresponding diodes are shown in the insets [83].

potential barrier created by the IMF and achieve low reverse leakage currents.

1.8 Summary of IMF-Array Technique

The IMF-dislocation arrays generated in highly lattice mismatched GaSb/GaAs and AlSb/Si systems demonstrate that the excess strain in the epilayers is almost completely relived instantaneously at the heterointerface. Device quality GaSb epitaxial layers with a very low threading dislocation density of $<10^5$ cm^{-2} can be achieved on GaAs substrates via compressive IMF growth despite the high lattice-mismatch (7.8%). Similar quality GaSb epilayers are possible on Si substrates by using the IMF based on AlSb nucleation layer. Additionally a 5° miscut Si substrate annihilates the detrimental antiphase domains, a problem encountered in III–V material growth on Si (100) substrates. MBE is a highly suitable growth technique as it provides critical real-time information about the surface reconstructions based on which the growth conditions for highly periodic IMF-array formation can be monitored and optimized. Structural characterizations by high-resolution TEM confirm that the IMF array comprises of pure-edge 90° misfit dislocations and its 2D nature along both [110] and [1–10] directions. XRD and EPD studies also confirm the high crystalline nature of the GaSb epilayers grown using the IMF-array method.

Various optoelectronic devices like solar cells, detectors, lasers, and transistors with enhanced device properties such as reduced forward resistances, low leakage currents have been successfully demonstrated using the IMF-growth technique. The electronic properties of the IMF have been investigated to understand the effects of the IMF array on device performances. A barrier potential introduced due to the IMF has been identified as a problem, especially in electrically injected devices. Delta doping the IMF array with Te atoms has been successfully shown as the best solution to compensate this barrier.

In conclusion, this chapter highlights the importance of lattice engineering in realizing extreme device integration and drastic reduction in manufacturing costs. The *IMF-array* technique is a very efficient method to achieve device quality lattice mismatch material growth. The IMF mode is readily available to realize III-Sb active region based optoelectronic devices on GaAs and Si platforms to utilize the exceptional device and commercial advantages as well as realize next generation monolithically integrated (photonic) circuits.

Acknowledgments

The authors gratefully acknowledge the financial support of this research by the AFOSR through Grant No. FA9550-08-1-0198. The authors would like to acknowledge and thank *Dr. Ganesh Balakrishnan* — pioneering IMF technology research, *Dr. S. H. Huang* for all the excellent transmission electron microscopy, *Dr. A. Jallipalli, Dr. R. B. Laghumavarapu, Dr. M. Mehta, Maya. N. Kutty* for the design, fabrication, processing, and characterization of various devices, *Dr. Jun Tatabayashi* for IMF device research on GaAs/Si substrates, *Dr. A. Khoshaklagh, Dr. Thomas J. Rotter,* for epitaxial growth, *Dr. B. L. Liang* and *Charles J. Reyner* for many fruitful discussions and valuable suggestions. The authors gratefully thank *Dr. L. R. Dawson*, Center for High Technology Materials, University of New Mexico, Albuquerque, for his continuous support and extraordinary guidance, *Dr. G. B. Lush* from University of Texas at El-Paso for his assistance with SPV measurements.

References

1. T. P. Chin, J. C. P. Chang, K. L. Kavanagh, C. W. Tu, P. D. Kirchner, and J. M. Woodall, Gas-source molecular beam epitaxial growth, characterization, and light-emitting diode application of $In_xGa_{1-x}P$ on GaP(100). *Appl. Phys. Lett.*, 1993. **62**(19): pp. 2369–2371.
2. E. A. Fitzgerald, Y.-H. Xie, D. Monroe, P. J. Silverman, J. M. Kuo, A. R. Kortan, *et al.*, Relaxed Ge_xSi_{1-x} structures for III–V integration with Si and high mobility two-dimensional electron gases in Si. *J. Vac. Sci. Technol. B*, 1992. **10**, p. 1807.
3. J. C. P. Chang, Jianhui Chen, J. M. Fernandez, H. H. Wieder, and K. L. Kavanagh, Strain relaxation of compositionally graded In_xGa_{1-x} As buffer layers for modulation-doped $In_{0.3}Ga_{0.7}As/In_{0.29}Al_{0.71}As$ heterostructures. *Appl. Phys. Lett.*, 1992. **60**(9): pp. 1129–1131.
4. M. E. Groenert, C. W. Leitz, A. J. Pitera, V. Yang, H. Lee, R. J. Ram, *et al.*, Monolithic integration of room-temperature cw GaAs/AlGaAs lasers on Si substrates via relaxed graded GeSi buffer layers. *J. Appl. Phys.*, 2003. **93**(1): pp. 362–367.
5. T. Kimura, T. Kimura, E. Ishimura, F. Uesugi, M. Tsugami, K. Mizuguchi, *et al.*, Improvement of InP crystal quality grown on GaAs substrates and device applications. *J. Cryst. Growth*, 1991. **107**(1–4): pp. 827–831.
6. N. Y. Jin-Phillipp, W. Sigle, A. Black, D. Babic, J. E. Bowers, E. L. Hu, *et al.*, Interface of directly bonded GaAs and InP. *J. Appl. Phys.*, 2001. **89**(2): pp. 1017–1024.
7. S. H. Huang, G. Balakrishnan, A. Khoshakhlagh, A. Jallipalli, L. R. Dawson, and D. L. Huffaker, Strain relief by periodic misfit arrays for low defect density GaSb on GaAs. *Appl. Phys. Lett.*, 2006. **88**(13): p. 131911.
8. G. Balakrishnan, S. Huang, L. R. Dawson, Y.-C. Xin, P. Conlin, and D. L. Huffaker, Growth mechanisms of highly mismatched AlSb on a Si substrate. *Appl. Phys. Lett.*, 2005. **86**(3): p. 034105.
9. P. S. Dutta, H. L. Bhat, and V. Kumar, The physics and technology of gallium antimonide: An emerging optoelectronic material. *J. Appl. Phys.*, 1997. **81**(9): p. 5821.
10. T. Bao, Y.-Q. Xu, Z.-Q. Zhou, R.-T. Hao, G.-W. Wang, Z.-W. Ren, *et al.*, GaAs Based InAs/GaSb Superlattice Short Wavelength Infrared Detectors

Grown by Molecular Beam Epitaxy. *Chin. Phys. Lett.*, 2009. **26**(2): p. 028102.

11. H. Mohseni, A. Tahraoui, J. Wojkowski, M. Razeghi, G. J. Brown, W. C. Mitchel, *et al.*, Very long wavelength infrared type-II detectors operating at 80 K. *Appl. Phys. Lett.*, 2000. **77**(11): p. 1572.
12. E. Plis, J. B. Rodriguez, H. S. Kim, G. Bishop, Y. D. Sharma, L. R. Dawson, *et al.*, Type II InAs/GaSb strain layer superlattice detectors with p-on-n polarity. *Appl. Phys. Lett.*, 2007. **91**(13): p. 133512.
13. J. P. Prineas, J. Yager, S. Seyedmohamadi, and J. T. Olesberg, Leakage mechanisms and potential performance of molecular-beam epitaxially grown GaInAsSb 2.4 μm photodiode detectors. *J. Appl. Phys.*, 2008. **103**(10): p. 104511.
14. M. Walther, J. Schmitz, R. Rehm, S. Kopta, F. Fuchs, J. FleiXner, *et al.*, Growth of InAs/GaSb short-period superlattices for high-resolution mid-wavelength infrared focal plane array detectors. *J. Cryst. Growth*, 2005. **278**(1–4): pp. 156–161.
15. J. P. van der Ziel, R. J. Malik, J. F. Walker, and R. M. Mikulyak, Optically pumped laser oscillation in the 1.6–1.8 μm region from $Al_{0.4}Ga_{0.6}Sb/GaSb/Al_{0.4}Ga_{0.6}Sb$ double heterostructures grown by molecular beam heteroepitaxy on Si. *Appl. Phys. Lett.*, 1986. **48**(7): pp. 454–456.
16. A. Salhi, Y. Rouillard, J. Angellier, and M. Garcia, Very-Low-Threshold 2.4-mm GaInAsSb–AlGaAsSb Laser Diodes Operating at Room Temperature in the Continuous-Wave Regime. *IEEE Photon. Technol. Lett.*, 2004. **16**(11): p. 2424.
17. T. F. Refaat, M. N. Abedin, O. V. Sulima, S. Ismail, and U. N. Singh, AlGaAsSb/InGaAsSb phototransistors for 2-μm remote sensing applications. *Opt. Eng.*, 2004. **43**(7): p. 1647.
18. E. A. Pease, L. R. Dawson, L. G. Vaughn, P. Rotella, and L. F. Lester, 2.5–3.5 μm optically pumped GaInSb/AlGaInSb multiple quantum well lasers grown on AlInSb metamorphic buffer layers. *J. Appl. Phys.*, 2003. **93**(6): pp. 3177–3181.
19. A. Härkönen, M. Guina, O. Okhotnikov, K. Rößner, M. Hümmer, T. Lehnhardt, *et al.*, 1-W antimonide-based vertical external cavity surface emitting laser operating at 2-μm. *Opt. Express*, 2006. **14**(14): pp. 6479–6484.
20. H. K. Choi, G. W. Turner, and S. J. Eglash, High-power GaInAsSb-AlGaAsSb multiple-quantum-well diode lasers emitting at 1.9 μm. *IEEE Photon. Technol. Lett.*, 1994. **6**(1): pp. 7–9.

21. J. E. Bowers, A. K. Srivastava, C. A. Burrus, J. C. Dewinter, M. A. Pollak, and J. L. Zyskind, High-Speed GalnAsSb/GaSb PIN Photodetectors for Wavelengths to 2-3 mm. *Electron. Lett.*, 1985. **22**(3): p. 137.
22. B. Rösener, N. Schulz, M. Rattunde, C. Manz, K. Köhler, and J. Wagner, High-Power High-Brightness Operation of a 2.25-mm (AlGaln)(AsSb)-Based Barrier-Pumped Vertical-External-Cavity Surface-Emitting Laser. *IEEE Photon. Technol. Lett.*, 2008. **20**(7): p. 502.
23. C. A. Wang, H. K. Choi, S. L. Ransom, G. W. Charache, L. R. Danielson, and D. M. DePoy, High-quantum-efficiency 0.5 eV GalnAsSb/GaSb thermophotovoltaic devices. *Appl. Phys. Lett.*, 1999. **75**(9): p. 1305.
24. B.-M. Nguyen, D. Hoffman, E. K.-w. Huang, S. Bogdanov, P.-Y. Delaunay, M. Razeghi, *et al.*, Demonstration of midinfrared type-II InAs/GaSb superlattice photodiodes grown on GaAs substrate. *Appl. Phys. Lett.*, 2009. **94**(22): p. 223506.
25. B. R. Bennett, R. Magno, J. B. Boos, W. Kruppa, and M. G. Ancona, Antimonide-based compound semiconductors for electronic devices: A review. *Solid-State Electron.*, 2005. **49**(12): pp. 1875–1895.
26. G. Delhaye, L. Desplanque, and X. Wallart, Metamorphic high electron mobility Te-doped AlInSb/GalnSb heterostructures on InP (001). *J. Appl. Phys.*, 2008. **104**(6): p. 066105.
27. B. P. Tinkham, B. R. Bennett, R. Magno, B. V. Shanabrook, and J. B. Boos, Growth of InAsSb-channel high electron mobility transistor structures. *J. Vac. Sci. Technol. B Microelectron. Nanometer Struct.*, 2005. **23**(4): p. 1441.
28. L. M. Fraas, G. R. Girard, J. E. Avery, B. A. Arau, V. S. Sundaram, A. G. Thompson, *et al.*, GaSb booster cells for over 30% efficient solar-cell stacks. *J. Appl. Phys.*, 1989. **66**(8): pp. 3866–3870.
29. C. A. Wang, D. A. Shiau, and D. R. Calawa, Growth and characterization of InAsSb/GalnAsAb/AlGaAsAb/GaSb heterostructures for wafer-bonded thermophotovoltaic devices. *J. Cryst. Growth*, 2004. **261**(2–3): pp. 372–378.
30. V. P. Kunets, S. Easwaran, W. T. Black, D. Guzun, Y. I. Mazur, N. Goel, *et al.*, InSb Quantum-Well-Based Micro-Hall Devices: Potential for pT Detectivity. *IEEE Trans. Electron Devices*, 2009. **56**(4): p. 683.
31. E. Plis, J. B. Rodriguez, G. Balakrishnan, Y. D. Sharma, H. S. Kim, T. Rotter, *et al.*, Mid-infrared InAs/GaSb strained layer superlattice detectors with nBn design grown on a GaAs substrate. *Semicond. Sci. Technol.*, 2010. **25**(8): p. 085010.

32. J. M. Yarborough, Y.-Y. Lai, Y. Kaneda, J. Hader, J. V. Moloney, T. J. Rotter, *et al.*, Record pulsed power demonstration of a 2 μm GaSb-based optically pumped semiconductor laser grown lattice-mismatched on an AlAs/GaAs Bragg mirror and substrate. *Appl. Phys. Lett.*, 2009. **95**(8): p. 081112.
33. R. J. Malik, J. P. van der Ziel, B. F. Levine, C. G. Bethea, and J. Walker, Molecular-beam epitaxy of GaSb/AlSb optical device layers on Si(100). *J. Appl. Phys.*, 1986. **59**(11): p. 3909.
34. J. W. Matthews and A. E. Blakeslee, Defects in epitaxial multilayers: I. Misfit dislocations. *J. Cryst. Growth*, 1974. **27**: pp. 118–125.
35. J. C. P. Chang, T. P. Chin, C. W. Tu, and K. L. Kavanagh, Multiple dislocation loops in linearly graded $In_xGa_{1-x}As$ ($0 \leq x \leq 0.53$) on GaAs and $In_xGa_{1-x}P$ ($0 \leq x \leq 0.32$) on GaP. *Appl. Phys. Lett.*, 1993. **63**(4): pp. 500–502.
36. T. P. Chin, H. Q. Hou, C. W. Tu, J. C. P. Chang, and N. Otsuka, InGaAs/InP and InAsP/InP quantum well structures on GaAs (100) with a linearly graded InGaP buffer layer grown by gas-source molecular beam epitaxy. *Appl. Phys. Lett.*, 1994. **64**(15): pp. 2001–2003.
37. W. E. Hoke, P. J. Lemonias, T. D. Kennedy, A. Torabi, E. K. Tong, R. J. Bourque, *et al.*, Metamorphic heterojunction bipolar transistors and P–I–N photodiodes on GaAs substrates prepared by molecular beam epitaxy. 2001. Tempe, Arizona (USA): AVS.
38. Y.-C. Xin, L. G. Vaughn, L. R. Dawson, A. Stintz, Y. Lin, L. F. Lester, *et al.*, InAs quantum-dot GaAs-based lasers grown on AlGaAsSb metamorphic buffers. *J. Appl. Phys.*, 2003. **94**(3): pp. 2133–2135.
39. G. Balakrishnan, S. Huang, T. J. Rotter, A. Stintz, L. R. Dawson, K. J. Malloy, *et al.*, 2.0 μm wavelength InAs quantum dashes grown on a GaAs substrate using a metamorphic buffer layer. *Appl. Phys. Lett.*, 2004. **84**(12): pp. 2058–2060.
40. W. Qian, M. Skowronski, and R. Kaspi, Dislocation Density Reduction in GaSb Films Grown on GaAsSubstrates by Molecular Beam Epitaxy. *J. Electrochem. Soc.*, 1994. **144**(4): pp. 1430–1434.
41. J. M. Kang, S.-K. Min, and A. Rocher, Asymmetric tilt interface induced by 60° misfit dislocation arrays in GaSb/GaAs(001). *Appl. Phys. Lett.*, 1994. **65**(23): pp. 2954–2956.
42. A. Rocher, J. M. Kang, and A. Ponchet, *Mater. Res. Soc. Symp. Proc.*, 1992. **238**: p. 91.
43. A. Bourret and P. H. Fuoss, Structure of GaAs/GaSb incoherent interface after epltaxial growth. *Appl. Phys. Lett.*, 1992. **61**(9): p. 1034.

44. J. M. Kang, M. Nouaoura, L. Lassabatère, and A. Rocher, Accommodation of lattice mismatch and threading of dislocations in GaSb films grown at different temperatures on GaAs (001). *J. Cryst. Growth*, 1994. **143**(3–4): pp. 115–123.
45. W. Qian, M. Skowronski, R. Kaspi, M. De Graef, and V. P. Dravid, Nucleation of misfit and threading dislocations during epitaxial growth of GaSb on GaAs(001) substrates. *J. Appl. Phys.*, 1997. **81**(11): p. 7268.
46. R. E. Mallard, P. R. Wilshaw, N. J. Mason, P. J. Walker, and G. R. Booker, *Microsc. Semicond. Mater.*, 1989. **100**, p. 331.
47. J. H. Kim, T. Y. Seong, N. J. Mason, and P. J. Walker, *J. Electron. Mater.*, 1998. **27**: p. 466.
48. A. E. Blakslee, M. M. Al-Jassim, J. M. Olson, K. M. Jones, and S. M. Vernon, *Mater. Res. Soc. Symp. Proc.*, 1988. **116**: p. 313
49. M. M. Al-Jassim, T. Nishioka, Y. Itoh, A. Yamamoto, Masafumi, and Yamaguchi, *Mater. Res. Soc. Symp. Proc.*, 1988. **116**: p. 141.
50. A. Trampert, E. Tournie, and K. H. Ploog, Novel plastic strain-relaxation mode in highly mismatched III-V layers induced by two-dimensional epitaxial growth. *Appl. Phys. Lett.*, 1995. **66**(17): p. 2265.
51. J. C. P. Chang, T. P. Chin, and J. M. Woodall, Incoherent interface of InAs grown directly on GaP(001). *Appl. Phys. Lett.*, 1996. **69**(7): pp. 981–983.
52. Y. H. Kim, J. Y. Lee, Y. G. Noh, M. D. Kim, S. M. Cho, Y. J. Kwon, *et al.*, Growth mode and structural characterization of GaSb on Si (001) substrate: A transmission electron microscopy study. *Appl. Phys. Lett.*, 2006. **88**(24): p. 241907.
53. A. M. Rocher, *Solid State Phenomena*, 1991. **19/20**: p. 563.
54. M. Aindow, T. T. Cheng, N. J. Mason, T. Y. Seong, and P. J. Walker, Geometry and interface structure of island nuclei for GaSb buffer layers grown on (001) GaAs by metalorganic vapour phase epitaxy. *J. Cryst. Growth*, 1993. **133**(1–2): pp. 168–174.
55. S. H. Huang, G. Balakrishnan, M. Mehta, A. Khoshakhlagh, L. R. Dawson, D. L. Huffaker, *et al.*, Epitaxial growth and formation of interfacial misfit array for tensile GaAs on GaSb. *Appl. Phys. Lett.*, 2007. **90**(16): p. 161902.
56. M. Mehta, G. Balakrishnan, S. Huang, A. Khoshakhlagh, A. Jallipalli, P. Patel, *et al.*, GaSb quantum-well-based "buffer-free" vertical light emitting diode monolithically embedded within a GaAs cavity incorporating interfacial misfit arrays. *Appl. Phys. Lett.*, 2006. **89**(21): p. 211110.
57. J. Tatebayashi, A. Khoshakhlagh, S. H. Huang, L. R. Dawson, G. Balakrishnan, and D. L. Huffaker, Formation and optical characteristics

of strain-relieved and densely stacked GaSb/GaAs quantum dots. *Appl. Phys. Lett.*, 2006. **89**(20): p. 203116.

58. G. Balakrishnan, J. Tatebayashi, A. Khoshakhlagh, S. H. Huang, A. Jallipalli, L. R. Dawson, *et al.*, III/V ratio based selectivity between strained Stranski–Krastanov and strain-free GaSb quantum dots on GaAs. *Appl. Phys. Lett.*, 2006. **89**(16): p. 161104.
59. M. Losurdo, P. Capezzuto, G. Bruno, A. S. Brown, T. Brown, and G. May, Fundamental reactions controlling anion exchange during mixed anion heterojunction formation: Chemistry of As-for-Sb and Sb-for-As exchange reactions. *J. Appl. Phys.*, 2006. **100**(1): p. 013531.
60. F. Maeda and Y. Watanabe, Sb-induced reconstruction on Sb-terminated GaAs(001). *Phys. Rev. B*, 1999. **60**(15): p. 10652.
61. A. Jallipalli, G. Balakrishnan, Sh Huang, T. Rotter, K. Nunna, B. Liang, *et al.*, Structural Analysis of Highly Relaxed GaSb Grown on GaAs Substrates with Periodic Interfacial Array of 90 degrees Misfit Dislocations. *Nanoscale Res. Lett.*, 2009. **4**(12): pp. 1458–1462.
62. S. Huang, G. Balakrishnan, and D. L. Huffaker, Interfacial misfit array formation for GaSb growth on GaAs. *J. Appl. Phys.*, 2009. **105**(10): p. 103104.
63. S. H. Huang, G. Balakrishnan, M. Mehta, L. R. Dawson, D. L. Huffaker, and P. Li, Arsenic-induced etched nanovoids on GaSb(100). *J. Appl. Phys.*, 2007. **102**(4): p. 044312.
64. S. H. Huang, G. Balakrishnan, A. Khoshakhlagh, L. R. Dawson, and D. L. Huffaker, Simultaneous interfacial misfit array formation and antiphase domain suppression on miscut silicon substrate. *Appl. Phys. Lett.*, 2008. **93**(7): p. 071102.
65. R. M. Sieg, S. A. Ringel, S. M. Ting, S. B. Samavedam, M. Currie, T. Langdo, *et al.*, Toward device-quality GaAs growth by molecular beam epitaxy on offcut Ge/$Si_{1-x}Ge_x$/Si substrates. *J. Vac. Sci. Technol., B*, 1998. **16**: pp. 1471–1474.
66. S. Huang, Microscopic Study of Extreme Lattice Mismatched Heteroepitaxy Using Interfacial Misfit Arrays, in *Engineering* 2007, University of New Mexico: Albuquerque.
67. L. Kronik and Y. Shapira, *Surf. Sci. Rep.*, 1999. **37**: p. 1.
68. F. Jonsdottir, Computation of equilibrium surface fluctuations in strained epitaxial films due to interface misfit dislocations. *Modell. Simul. Mater. Sci. Eng.*, 1995. **3**: p. 503.
69. E. Kuramochi, N. Kondo, Y. Takanashi, and M. Fujimoto, *Appl. Phys. Lett.*, 1993. **63**: p. 2664.

70. A. Trampert, K. H. Ploog, and E. Tournié, Anisotropic misfit dislocation nucleation in two-dimensional grown InAs/GaAs (001) heterostructures. *Appl. Phys. Lett.*, 1998. **73**: p. 1074.

71. A. Jallipalli, G. Balakrishnan, S. H. Huang, A. Khoshakhlagh, L. R. Dawson, and D. L. Huffaker, Atomistic modeling of strain distribution in self-assembled interfacial misfit dislocation (IMF) arrays in highly mismatched III–V semiconductor materials. *J. Cryst. Growth*, 2007. **303**(2): pp. 449–455.

72. A. Kuronen, K. Kaski, L. F. Perondi, and J. Rintala, Atomistic modelling of interaction between dislocations and misfit interface. *Europhys. Lett.*, 2001. **55**(1): pp. 19–25.

73. D. Huffaker and J. Tatebayashi, (2012), "Silicon Photonics for Telecommunications and Biomedicine," Sasan Fathpour, Barham Jalali, Chapter 8, "Novel III–V on Si Growth Techniques." (CRC Press, Taylor and Francis Group) pp. 255–296.

74. J. Tatebayashi, A. Jallipalli, M. N. Kutty, S. Huang, K. Nunna, G. Balakrishnan, *et al.*, Monolithically Integrated III-Sb-Based Laser Diodes Grown on Miscut Si Substrates. *IEEE J. Sel. Topics Quantum Electron.*, 2009. **15**(3): pp. 716–723.

75. Z.-Q. Zhou, Y.-Q. Xu, R.-T. Hao, B. Tang, Z.-W. Ren, and Z.-C. Niu, Molecular Beam Epitaxy of GaSb on GaAs Substrates with AlSb Buffer Layers. *Chin. Phys. Lett.*, 2009. **26**: p. 018101.

76. A. Jallipalli, Periodic Misfit Dislocation Arrays Localized at the GaSb/GaAs Interface, in *Optical Sciences and Engineering* 2008, Univeristy of New Mexico: Albuquerque.

77. M. Mehta, A. Jallipalli, J. Tatebayashi, M. N. Kutty, A. Albrecht, G. Balakrishnan, *et al.*, Room-Temperature Operation of Buffer-Free GaSb–AlGaSb Quantum-Well Diode Lasers Grown on a GaAs Platform Emitting at 1.65 μm. *IEEE Photon. Technol. Lett.*, 2007. **19**(20): p. 1628.

78. J. Tatebayashi, A. Jallipalli, M. N. Kutty, S. H. Huang, T. J. Rotter, G. Balakrishnan, *et al.*, Device Characteristics of GaInSb/AlGaSb Quantum Well Lasers Monolithically Grown on GaAs Substrates by Using an Interfacial Misfit Array. *J. Electron. Mater.*, 2008. **37**(12): pp. 1758–1763.

79. J. B. Rodriguez, L. Cerutti, and E. Tournié, GaSb-based, 2.2 μm type-I laser fabricated on GaAs substrate operating continuous wave at room temperature. *Appl. Phys. Lett.*, 2009. **94**(2): p. 023506.

80. J. B. Rodriguez, L. Cerutti, P. Grech, and E. Tournié, Room-temperature operation of a 2.25 μm electrically pumped laser fabricated on a silicon substrate. *Appl. Phys. Lett.*, 2009. **94**(6): p. 061124.

81. T. J. Rotter, J. Tatebayashi, P. Senanayake, G. Balakrishnan, M. Rattunde, J. Wagner, *et al.*, Continuous-Wave, Room-Temperature Operation of 2-μm Sb-Based Optically-Pumped Vertical-External-Cavity Surface-Emitting Laser Monolithically Grown on GaAs Substrates. *Appl. Phys. Express*, 2009. **2**(11): p. 112102.

82. E. Plis, J. B. Rodriguez, G. Balakrishnan, Y. D. Sharma, H. S. Kim, T. Rotter, *et al.*, Mid-infrared InAs/GaSb strained layer superlattice detectors with nBn design grown on a GaAs substrate. *Semicond. Sci. Technol.*, 2010. **25**(8): p. 085010.

83. A. Jallipalli, K. Nunna, M. N. Kutty, G. Balakrishnan, L. R. Dawson, and D. L. Huffaker, Electronic characteristics of the interfacial states embedded in "buffer-free" GaSb/GaAs (001) heterojunctions. *Appl. Phys. Lett.*, 2009. **95**(20): p. 202107.

84. A. Jallipalli, K. Nunna, M. N. Kutty, G. Balakrishnan, G. B. Lush, L. R. Dawson, *et al.*, Compensation of interfacial states located inside the "buffer-free" GaSb/GaAs (001) heterojunction via δ-doping. *Appl. Phys. Lett.*, 2009. **95**(7): p. 072109.

Chapter 2

Compliant Substrates

Shumin Wang
Department of Microtechnology and Nanoscience, Chalmers University of Technology, 41296 Göteborg, Sweden
shumin@chalmers.se

This chapter provides an overview of compliant substrate, its theoretical concept, and practical implementation methods. The concept was originally proposed to offer a versatile technology beyond the limitation of lattice mismatch that has been encountered in most heterostructures. Different fabrication schemes to prepare compliant substrates in various materials are reviewed and assessed. Although there have been plenty of evidence showing reduction of threading dislocations (TDs) using a compliant substrate, its physical origin is still under debate.

2.1 Introduction

It was more than 60 years ago that Frank and van der Merve predicted that a thin film could be grown on a lattice mismatched substrate if the lattice mismatch was less than about 15% [1]. For semiconductor heterostructures, strain provides great freedom for tailor-making band gaps and improving device performance

Lattice Engineering: Technology and Applications
Edited by Shumin Wang

ISBN 978-981-4316-29-3 (Hardcover), 978-981-4364-25-6 (eBook)
www.panstanford.com

provided that no adverse strain relaxation occurs [2, 3]. Depending on the lattice mismatch, initial strain relaxation can be in the form of surface undulation, three-dimensional islanding or quantum dots (QDs) in the case when the lattice mismatch is larger than about 2%, or formation of a network of misfit dislocations (MDs) if the lattice mismatch is smaller [4]. For a substrate with a lateral size of sub-millimeter or larger, MDs are unlikely to glide across the whole wafer, thus leaving threading components of dislocations (TDs) penetrating through the epi-layer. While formation of self-assembled QDs via Stranski-Krastanov growth mode has received enormous attention due to their potentials for optoelectronic devices [5, 6], TDs can trap free carriers and act as extra scattering centers for carrier transport, and thus are deleterious for many device applications. There have been extensive research efforts both theoretically and experimentally on predicting and measuring the critical thickness at which initial MDs are generated and the subsequent strain relaxation. General experimental findings are that MDs are generated at the epi-layer/substrate interface at the critical thickness with TDs penetrating through the epi-layer and terminating at the surface, if the epi-layer thickness (in the range of μm or thinner) is much less than that of the substrate (often 300 μm or thicker).

It is wise to envisage the opposite situation where the roles of the epi-layer and the substrate discussed above are interchanged, i.e., to grow a thick film on a very thin substrate that is *free-standing*. If the substrate is thin enough, the critical thickness of the thick film will become infinity, and if there would be any strain relaxation caused by MDs, all the TDs will penetrate through the thin substrate rather than the thick film. This was the idea when Lo proposed the concept of compliant substrate (CS) in 1991 [7]. Below, we will first present theoretical understanding of the CS followed by a brief review of different techniques to prepare CSs. Although it has been nearly 20 years since the first proposal, the exact physical mechanisms (elastic or plastic deformation) are still under debate as will be shown below. Finally, we will discuss current trends on the CS research and some new ideas. There are several good review papers on this topic [8–10] including a recent one by Ayers in 2008 [10].

2.2 Theoretical Understanding of Compliant Substrates

Figure 2.1a shows a schematic of a thick film grown on a CS. When the thicknesses of the film and the substrate are comparable, strain will be accommodated by both layers, the so-called strain partitioning. However, if the CS is suspended without any mechanical support, elastic deformation occurs in all three dimensions resulting in buckling or warping. Figure 2.1b shows such an example of a 1 μm-thick $In_{0.5}Ga_{0.5}As$ grown on a 10 nm-thick GaAs CS with a size of 5 × 5 μm^2 simulated by finite element method. Such suspended CSs were first tested [11, 12] after the proposal by Lo [7] and were soon abandoned as it was difficult to handle such fragile structures practically. A more robust approach is to prepare a CS on a mechanical host substrate (MHS) with a buffer layer or interface in between, see Fig. 2.1c. This in principle eliminates deformation of the CS and the subsequent epi-layers along the growth directions. In the following, we present theoretical treatments based on biaxial deformation of epi-layers grown on a large CS without surface undulation, following Lo's original paper [7]. We use the subscript *e* for the film and *cs* for the CS below.

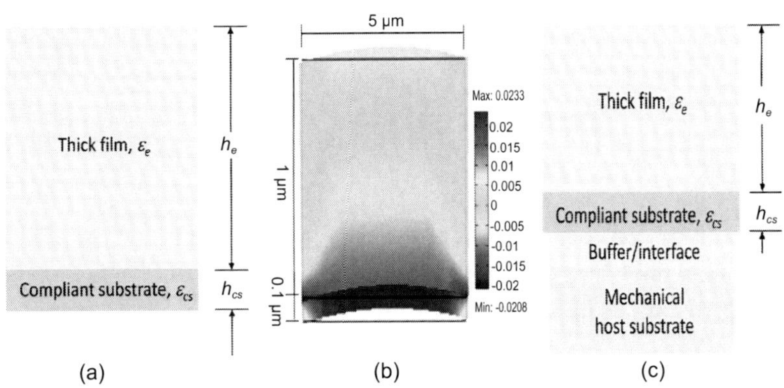

Figure 2.1. Schematic diagrams of a thick film grown on (a) a free-standing CS and (c) a CS with a MHS. In (b) elastic deformation is simulated by finite element method showing the in-plane strain in a 1 μm-thick $In_{0.5}Ga_{0.5}As$ grown on a 100 nm-thick GaAs CS with a lateral size of 5 × 5 μm^2. The lateral scale is contracted by 20 times. See also Color Insert.

As the lattice mismatch, f, is partitioned by the film and the CS, we have:

$$f = \varepsilon_e + \varepsilon_{cs} \tag{2.1}$$

where ε is the strain. At the equilibrium, the macroscopic force balance requires:

$$\sigma_e h_e + \sigma_{cs} h_{cs} = 0 \tag{2.2}$$

where h is the thickness and σ is the stress that is related to Young's module, Y, the Poisson's ratio, ν, and the strain by

$$\sigma = \frac{Y}{1-\nu}\varepsilon \tag{2.3}$$

The total strain energy, E_{strain}, is then:

$$E_{strain} = \frac{Y_e}{1-\nu_e} h_e \varepsilon_e^2 + \frac{Y_{cs}}{1-\nu_{cs}} h_{cs} \varepsilon_{cs}^2 \tag{2.4}$$

For most semiconductor heterostructures, the ratio in Eq. 2.3 is quite similar and can be assumed as a constant. Solving Eq. 2.1 to Eq. 2.3, the Eq. 2.4 reduces to

$$E_{strain} = \frac{Y}{1-\nu} f^2 \frac{h_e h_{cs}}{h_e + h_{cs}} \tag{2.5}$$

Equation 2.5 is the same as the strain energy of a pseudomorphic layer grown on a conventional substrate if the term $h_e h_{cs}/(h_e + h_{cs})$ is replaced by the epi-layer thickness. The critical thicknesses of the h_e and the h_{cs} can then be derived in a similar fashion for a lattice mismatched layer grown on an infinitively thick substrate. If we denote the critical thickness of the latter case as h_c, then we get

$$h_e^{-1} = h_c^{-1} - h_{cs}^{-1} \tag{2.6}$$

Figure 2.2 shows the normalized critical thickness h_e/h_c for a CS as a function of the normalized substrate thickness h_{cs}/h_c (in Lo's paper [7], he used h_a and h_b for the epi-layer and the CS thickness, respectively, $t_c \rightarrow h_c$ and $t_{eff} = h_c$). For a very thick substrate, it approaches to the critical thickness predicted by the Matthews' force balance model [13]. When the CS thickness is reduced, the effective critical thickness increases and eventually becomes infinity when it is equal to or below h_c.

Another important argument from Lo's paper is that the generated TDs will penetrate into the CS rather than into the thick

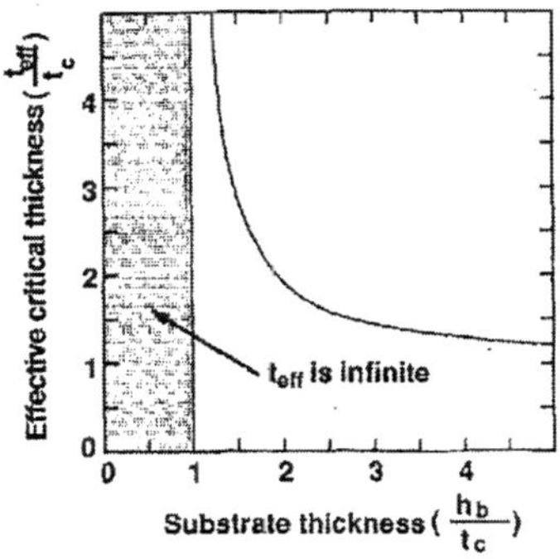

Figure 2.2. Dependence of the effective critical thickness as a function of the CS thickness normalized to the critical thickness for an infinitively thick substrate predicted by the Matthews' model. (Ref. [7], reprinted with permission from AIP).

film if the former is thin enough. For a free-standing CS, the bottom of the CS is regarded as a free surface similar to the surface of the epi-layer. Hence, an MD at the interface receives image forces from both interfaces. In dislocation theory [14], an image dislocation and an image force are introduced to ensure the boundary conditions for the strain field. The situation is similar to the case of a point electric charge placed in a dielectric medium, and therefore the image force becomes large when the MD is close to the surface. Lo calculated the net image force (per unit length) in a compliant CS [7] and the result is shown in Fig. 2.3. When $h_{cs} > h_e$, the net image force is positive and will therefore attract dislocations penetrating toward the epi-layer surface to form TDs. However, when $h_{cs} < h_e$, the net force is negative, and the TDs will then be threaded into the CS instead.

The above model is in an ideal case. As mentioned above, a practical CS is often rested on a MHS shown in Fig. 2.1c to ensure mechanically robust. The question arises then at the CS and the MHS interface: how does the bonding at the interface affect the pre-condition (free-standing) for a CS and what are physical mechanisms for reduction of TDs in epi-layers grown on a CS? Several authors [8, 15–17] have discussed the CSs made of (1) amorphous oxides and (2) twist bonding, which were experimentally successful at the time, and proposed new explanations. The free-standing assumption assumed by Lo implies that the CS can glide freely on the MHS by slip.

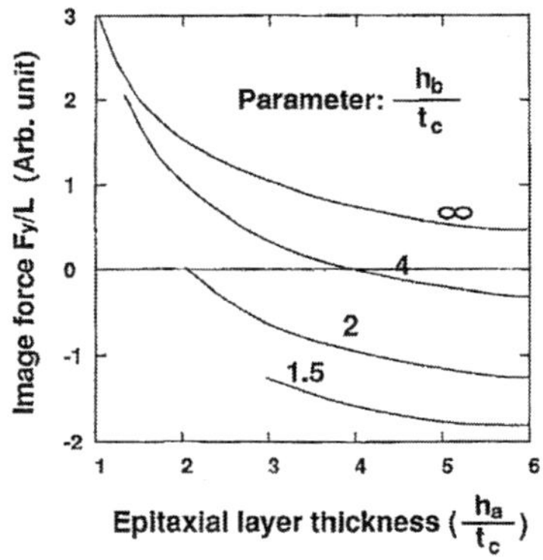

Figure 2.3. Dependence of the image force as a function of epitaxial layer thickness normalized to the critical thickness for an infinitively thick substrate predicted by the Matthews' model with the CS thickness as a parameter. (Ref. [7], reprinted with permission from AIP).

This is a macroscopic phenomenon and should be easily detected, in particular for a large CS. For example, for a film with a lattice mismatch of 1% grown on a 1 inch diameter CS, the required lateral slip displacement (also called absolute compliance) is 0.127 mm. However, none of the reported experiments to date have confirmed such an absolute compliance. Jesser *et al.* [15] have studied the twist-bonded CSs and suggested several criteria for design of such a CS (summarized below) among which a soft CS with the lattice mismatch closer to that of the epi-layer than the MHS and with a large twist angle ($>8^\circ$) is preferred. Kästner and Gösele proposed what they claimed more realistic mechanisms [16]. According to them, there exists a dynamic sliding resistance at the CS and the MHS interface and the shear stress required for the slide decays exponentially with the distance from the rim. Detailed calculations based on the bimetal model show that the maximum shear stress is only a few percents of the biaxial stress of the epi-layer film. This real condition is far from the idea macroscopic free slip of a CS. They pointed out [17] that the actual critical thickness of the film grown on a CS decreases rather than increases as shown in the other models based on the free-standing condition. A schematic illustration is depicted in Fig. 2.4. The image force due to the thin CS attracts the MD moving toward the interface *R* and allows the

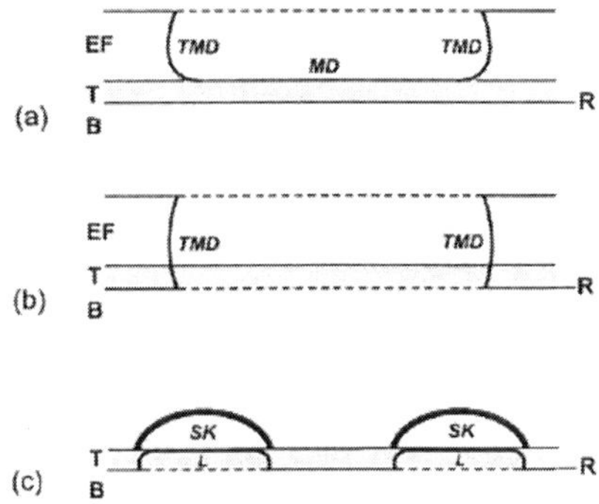

Figure 2.4. Schematics of MDs and their threading segments (thick solid lines). Broken lines indicate relaxed dislocation lines or slip steps: (a) usual plastic relaxation; (b) the MD escaped at the interface R on a CS; and (c) 3D islands formed and dislocation loops generated inside the CS under the islands. (Ref. [16], reprinted with permission from AIP).

TDs to be terminated at the *R* (Fig. 2.4a,b). Thus, there is no MD line tension that is the main force counteracting against the strain force in the Matthews' model on critical layer thickness [13]. The small Peierls friction force against dislocation gliding balances the strain force leading to a reduced critical thickness. At such an early plastic relaxation stage, dislocation multiplicative interaction will hardly occur, and the TDs can glide easily and fast to the rim or surface or annihilate when two TDs with opposite Burgers vectors meet. For large mismatch, 3D islands can be formed as shown in Fig. 2.4c. This will generate a dislocation loop inside the thin CS under the islands.

These understandings are essential for design of CSs. Ayers introduced relative compliance and absolute compliance [10]. From the Matthews' model [13], the critical thickness of a thin epi-layer on an infinitively thick substrate is:

$$h_c = \frac{A\,[\ln(h_c/b) + 1]}{f} \tag{2.7}$$

where *A* is a constant and *b* is the Burgers vector. The dimensionless relative compliance, *e*, is defined as:

$$e = \frac{A\,[\ln(h_{cs}/b) + 1]}{h_{cs}} \tag{2.8}$$

For a very thick substrate, the relative compliance approaches to zero while it is equal to the lattice mismatch, f, when $h_{cs} = h_c$.

The relative compliance is entirely determined by the thickness and the elastic properties of the CS and represents the ability of strain accommodation by the CS. The absolute compliance, s, is the lateral slip displacement and depends on both the lattice mismatch and the size of the CS by:

$$s = fD/2 \quad (2.9)$$

where D is the diameter of the CS. Experimentally, it is relatively easy to achieve the relative compliance, but absolute compliance is apparently much difficult to achieve, especially for a large wafer size. This implies that the concept of CS can be feasible for small-size substrates (lateral dimensional comparable to average gliding distance of TDs such that all the TDs can glide to the edge of the wafer). For a large wafer size, the absolute compliance must be allowed. This can be achieved, for example, by growth on patterned substrates.

2.3 Preparation of Compliant Substrates

There are several different approaches demonstrated so far to prepare CSs. The initial attempt was to fabricate free-standing CSs, but soon the focus was moved to preparing large-size CSs on MHSs. In the latter case, it is essential to create a buffer or an interfacial layer between the CS and the MHS as sketched in Fig. 2.1c such that the lattice structure of the CS is decoupled from that of the MHS. The buffer/interface layer can be of different bonding nature and prepared by different methods. These include use of an amorphous oxide buffer, a viscous material, porous surface structures prepared by ion beam treatment and twist bonding. We will review all these methods below.

2.3.1 *Free-Standing Thin Films*

The first experimental evidence showing the enhanced critical thickness when grown on a CS was reported from Lo's group in 1994 [12]. They first prepared a thin membrane of GaAs with a size of 10 $\times$ 5 μm^2 as illustrated in Fig. 2.5. The thickness of the membrane

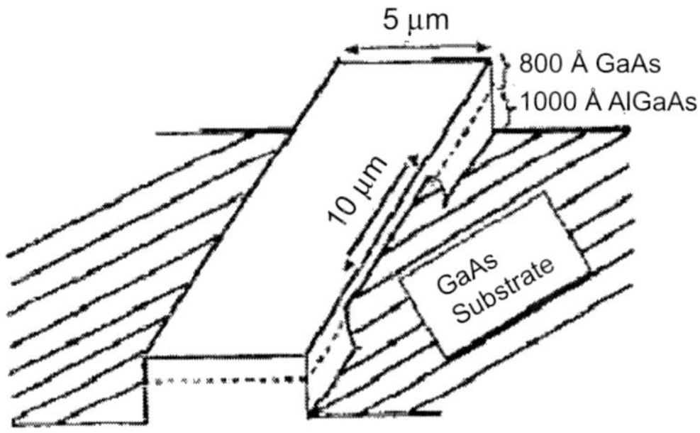

Figure 2.5. Illustration of a membrane-type GaAs CS used to grow a 200 nm-thick $In_{0.14}Ga_{0.86}As$ layer. (Ref. [12], reprinted with permission from AIP).

is about 80 nm. They grew then 200 nm-thick $In_{0.14}Ga_{0.86}As$ on the membrane, which is about 20 times of the critical thickness predicted by the Matthews' model. The X-ray diffraction (XRD) measurements revealed unrelaxed strain of 0.9% in the film on the CS, close to the designed total strain of 1%, but a more relaxed and broader peak from the film grown on an unprocessed normal substrate. Atomic force microscopy (AFM) showed a smooth surface with an average roughness of 1.6 nm compared with 13 nm measured from the reference surface.

A common problem with the growth on a membrane-type CS is the strain induced warping after re-growth of the strained layer. To overcome this problem, Jones *et al.* demonstrated another technique using the disk-shaped CS on a single pedestal [18]. They first grew a 40 nm InGaAs on a 5 μm-thick AlGaAs layer by metalorganic vapor phase epitaxy (MOVPE). Different mesa shapes and sizes were used, and selective etching was employed to etch away the AlGaAs to form a single pedestal with a diameter of 1 μm, see Fig. 2.6. The maximum overhang radius is 6 μm before the membrane collapses. The processed samples were re-grown with a lattice mismatched QW and barriers on it. Photoluminescence (PL) at room temperature was observed with an emission wavelength at 1.35 μm from the overhang structure but no signals from un-etched relaxed structures. Cathodluminescence revealed bright emission from the QW disk with a shorter wavelength in the middle part (slightly more strained) than in the ring.

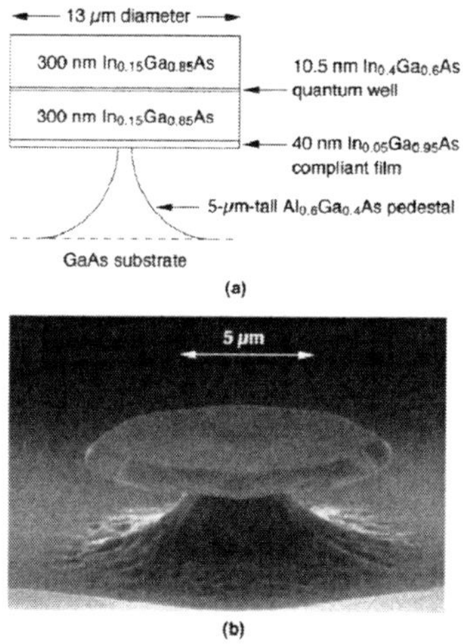

Figure 2.6. (a) Schematic and (b) an SEM image of a fully processed long wavelength InGaAs disk fabricated on a CS membrane. (Ref. [18], reprinted with permission from AIP).

2.3.2 *Amorphous Oxide Buffer*

Amorphous oxides have lower shear modulus than semiconductor crystals and thus can be used as buffers for CSs as well as dislocation sinks. These oxides can be introduced *before, during,* and *after* epitaxial growth in different materials. Powell *et al.* first introduced using silicon on insulator (SOI) as CSs [19]. The crystalline Si must be removed leaving about several tens nm Si left on SiO_2 before epitaxial growth. They prepared 10 nm-thick Si templates using separation by implant of oxygen (SIMOX) and grew $Si_{0.85}Ge_{0.15}$ with various thicknesses of 60–170 nm at 500°C on the same wafer using a shadow mask by molecular beam epitaxy (MBE). Post growth thermal annealing up to 1050°C was employed to examine strain relaxation. The thinnest as-grown SiGe shows nice Pendellösung fringes in XRD without strain relaxation, while the thickest as-grown sample reveals a broad peak in XRD with partial strain relaxation. Transmission electron microscopy (TEM) reveals no dislocations in both samples. After thermal annealing at 700°C for 1 hour, MDs are observed at the SiGe/Si interface with TDs penetrating *down into the Si layer* toward the amorphous oxide interface. The TD density is estimated to be in 10^5 and 10^8 cm^{-2} in SiGe and Si, respectively. These observations were explained by the induced tensile strain in Si that eventually led to plastic deformation due to the slippage at the Si/SiO_2 interface. In a later paper [20], the same group studied the annealing-temperature-dependent stain relaxation in 180 nm-thick $Si_{0.85}Ge_{0.15}$ on 65 nm-thick Si grown at 400°C by MBE. Similar results were found in the as-grown sample, i.e., the SiGe is fully strained without any dislocations introduced. Annealing at 700°C introduces 60° type MDs at the SiGe/Si interface. When the annealing temperature was raised to 1050°C, two types of dislocations: one usual 60° MDs at the SiGe/Si interface and one 90° pure-edge dislocations inside the SiGe layer were observed in TEM. Pure-edge dislocations are sessile but twice effective than the 60° mixed MDs for strain relaxation. According to the authors, these edge dislocations are probably formed by the reaction of two 60° MDs gliding on the (111) and the (11-1) planes via

$$\frac{1}{2}\left[10\overline{1}\right]+\frac{1}{2}\left[011\right]=\frac{1}{2}\left[110\right]$$

The inter-diffusion of Si and Ge atoms is believed to be the driving force for formation of a network of pure-edge dislocations. The annealing temperature can be lowered by modifying the chemical composition of the oxides. For example, Huang *et al.* demonstrated a reduction of the annealing temperature by 300°C using ion implantation of B_2O_3 [21]. Later Yang *et al.* demonstrated the compliant nature of the SOI by growing *in situ* relaxed $Si_{0.6}Ge_{0.4}$ at 500°C using MBE [22]. They observed a five orders of magnitude reduction of TDs (down to 10^6 cm^{-2}) in the relaxed film as compared with that grown on a conventional Si substrate and attributed it to the image force of the thin Si template that drags the TDs downward into the SOI.

Other types of materials such as GaAs and GaN have also been tested to be grown on CSs using SOIs. Seaford *et al.* reported growth of 4 μm-thick GaAs on the (511) plane of a 100 nm-thick SOI by MBE [23]. Compared with the reference sample grown on a normal (511) Si substrate, the TEM images showed a considerable reduction in the TD density from the sample grown on the SOI. Later on, the same group reported the growth of 3 μm-thick GaAs on 100 nm and 200 nm-thick SOIs [24] and demonstrated a narrow full width

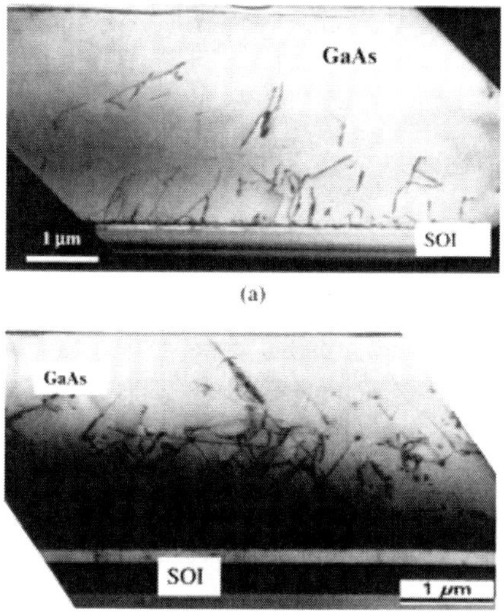

Figure 2.7. Cross-sectional TEM images of GaAs grown on (511) SOI CSs with a (a) 100 nm and (b) 200 nm Si thin membrane. (Ref. [24], reprinted with permission from AIP).

of half maximum (FWHM) of 128 arcsec for the GaAs peak in XRD. The thinner Si template resulted in a less TD density which is in the range of 10^7 cm^{-2} estimated from TEM shown in Fig. 2.7. Based on the optimized structure, they fabricated working AlGaAs/GaAs *npn* heterostructure bipolar transistors.

Yang *et al.* tested growth of GaN on SiC on SOIs by MBE [25]. They used bond and etch back SOI technology to prepare a large-size SOI wafer with a 50 nm-thick Si on top. The thin Si layer was then converted to SiC by exposing the substrate to a flux of acetylene or carbon beam at 900°C. Due to the much strong bonding strength of Si-N (9.5 eV) than that of Al-N (2.98 eV), a thin layer of Al was deposited before the introduction of N plasma. After the 10 nm-thick AlN nucleation layer followed by a 10 nm GaN, an AlN/GaN superlattice with a period of 4 nm was then grown. Finally, a 200 nm-thick GaN was grown at 500–740°C by MBE. The XRD revealed a FWHM of 28 arcmin from the thin GaN layer, while PL at 10 K showed a dominant bound exciton peak with a FWHM of 8 meV.

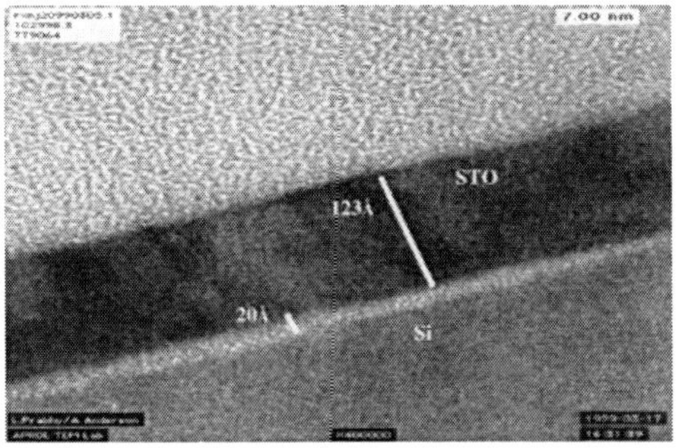

Figure 2.8. Cross-sectional TEM of STO grown on a Si (001) substrate. A thin amorphous oxide layer is clearly observed. (Ref. [28], reprinted with permission from Elsevier B. V.).

Amorphous oxides can be formed even *during* the epitaxial growth and thus be used as the buffer for CSs. A good example is the work of MBE growth of GaAs related compounds on $SrTiO_3$/Si performed by Motorola in late 1990s [26]. For a review of growth of oxides on Si, please refer to the Ref. [27]. Growth of oxides on Si is highly motivated by the need to develop high-*k* gate oxides for next generation CMOS technology. One candidate oxide is strontium titanate $SrTiO_3$ (STO) with a high dielectric constant of about 300 at room temperature. When examining the crystal structure of the STO/Si interface grown at high temperatures, Ramdani *et al.* found that a thin amorphous layer was formed as shown by TEM in Fig. 2.8 [28]. Further investigations revealed that the amorphous layer was not formed at the initial stage when the STO thickness is less than 2 nm. The thickness of the amorphous layer is constant irrespective of the STO thickness (>2 nm) but can be varied by changing the growth temperature or the oxygen flux. For the optimized growth condition, which is unfortunately not specified in the paper [28], the minimum thickness of the amorphous layer is about 1 nm. Based on these results, they argued that the amorphous layer resulted from internal oxidation of the Si substrate due to the vacancy assisted oxygen diffusion to the interface during epitaxial growth of STO. Later the same group studied the STO/Si interface grown at 300–500°C by X-ray photoelectron spectroscopy (XPS) [29]. The thickness of the amorphous layer increases with the growth temperature. The XPS results confirm that the amorphous interfacial layer is a mixture of Sr-silicates and SiO_x species, including SiO, Si_2O_3, and SiO_2.

The amorphous interfacial layer of STO/Si has been used to grow GaAs on STO/Si, functioning as a buffer in a CS [26]. $SrTiO_3$ has a simple cubic crystal structure with a lattice constant of 3.91 Å. When it is rotated 45° around the surface normal of the (001) plane, the lattice mismatch is then 1.8% (compressive) and −2.1% (tensile) with respect to Si and GaAs, respectively, and is therefore a suitable buffer for GaAs grown on Si. One of the key steps is the removal of native oxides from the Si substrate, which is realized by depositing 1–2 monolayers of Sr metal at 400–600°C followed by a thermal annealing at >750°C. The next step is to grow high-quality STO by first passivating the Si surface by 0.5 monolayer of Sr metal followed by co-deposition of Sr and Ti in the presence of oxygen up to 10^{-5} mbar at 500°C. The thin STO has a smooth surface with atomic steps and a low surface roughness value of 0.2 nm. Growth of GaAs on STO/Si was performed by two steps: initial nucleation of GaAs at a low growth temperature using a low growth rate followed by use of a high growth temperature and a high growth rate for bulk growth at temperatures >525°C. Figure 2.9 shows RHEED images after deposition of 0.5 μm GaAs on STO/Si, indicating very high surface quality and a single domain GaAs crystal structure. Figure 2.10 shows a high-resolution TEM image with a sharp GaAs/STO interface [30]. Perhaps the most interesting property is the density of TDs. Figure 2.11 depicts images

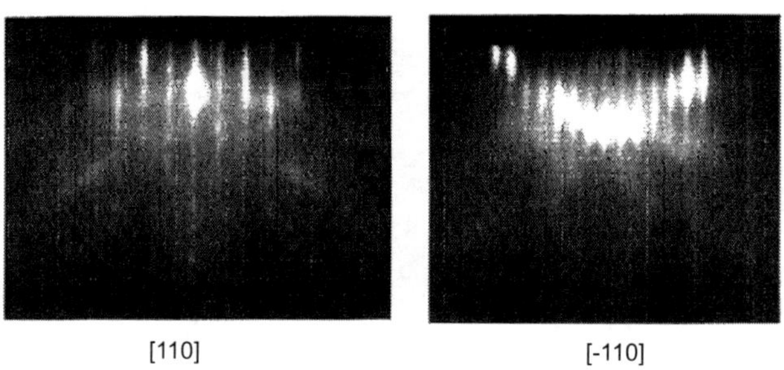

Figure 2.9. RHEED images of single domain 0.5 μm GaAs on STO/Si. (Ref. [26], reprinted with permission from IOP).

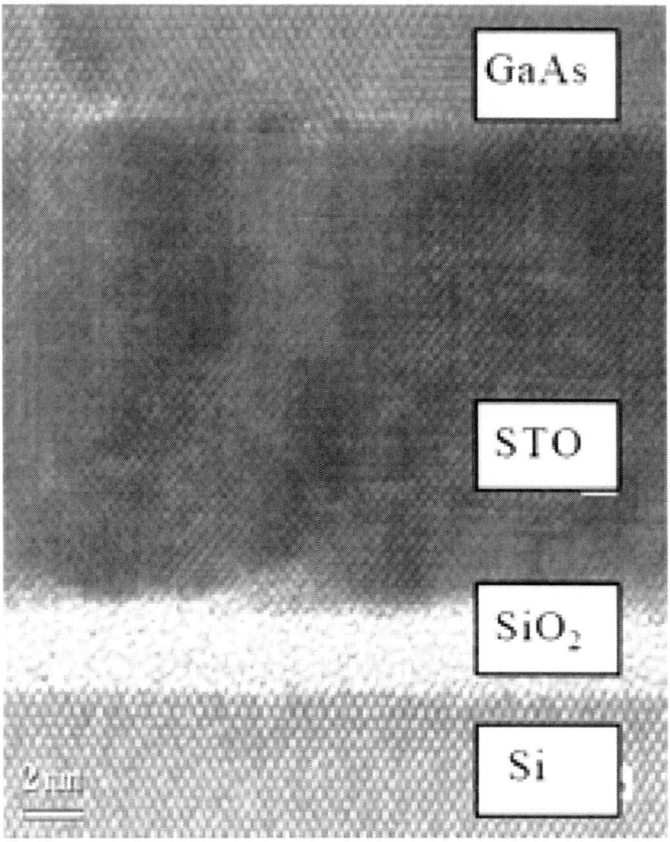

Figure 2.10. Cross-sectional TEM image showing all the layers from Si to GaAs. (Ref. [30], reprinted with permission from IEEE).

of surface by optical microscopy (left) and by plane-view TEM (right) [26]. The 2 μm-thick GaAs/STO/Si was chemically etched to reveal the TD density. The surface defect measured by the optical microscopy shows a density of 10^5 cm^{-2}, which is comparable to that grown on a commercial GaAs substrate. However, careful TEM studies suggest a high density of TDs of about 10^8 cm^{-2}, i.e., 3 orders of magnitude higher [26]. They believe that the possible explanation for the discrepancy is the difficulty of distinguishing the surface defects that are close to each other and the difficulty of counting them. One advantage of this method is the ability to grow

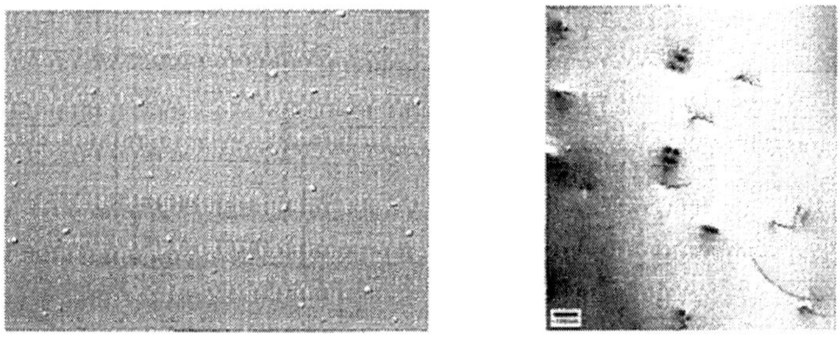

Figure 2.11. (a) Optical microscope image (2000× magnification) of an etched GaAs sample showing a defect density of $\sim 10^5$ cm^{-2}. (b) Plan-view TEM image of the same sample revealing a defect density of $\sim 10^8$ cm^{-2}. (Ref. [26], reprinted with permission from IOP).

compound semiconductors on 12" Si substrates and this was truly demonstrated by Motorola [30].

Based on the extensive works on optimizations of material quality of GaAs/STO/Si, Motorola has demonstrated impressive GaAs MESFETs on 4″ Si wafers and made comparison with the similar devices grown on GaAs substrates [30]. The MESFET structure consists of 2 μm undoped GaAs on STO to electrically isolate the GaAs active device from the Si substrate, a 150 nm n-GaAs channel doped at 8×10^{17} cm^{-3} and a 50 nm $n+$GaAs contact layer doped at 5×10^{18} cm^{-3}. The measured peak mobility was 2524 and 2682 cm^2/Vs for the MESFET grown on the STO/Si and on GaAs, respectively. Figure 2.12 shows the drain-source I–V curves for the 0.7×100 μm^2 devices at different gate biases, while Figure 2.13 shows maximum available gain from six 0.7×3000 μm^2 devices. The saturation current at a gate bias of 0.5 V is 367 mA/mm and 385 mA/mm, and the peak transconductance is 233 mS/mm and 240 mS/mm for the devices on STO/Si and GaAs, respectively. The f_{max} of the MESFETs on STO/Si is 14.5 GHz. These device results indicate that the device performance is essentially comparable when grown on STO/Si and on GaAs. As the reliability of GaAs/Si technology is of primary importance, they also performed the lifetime test. After a burn-in period, the devices are biased at a drain-source voltage of 3.5 V and at a gate bias of -0.5 V in an ambient temperature of 200°C. Both MESFETs on STO/Si and on

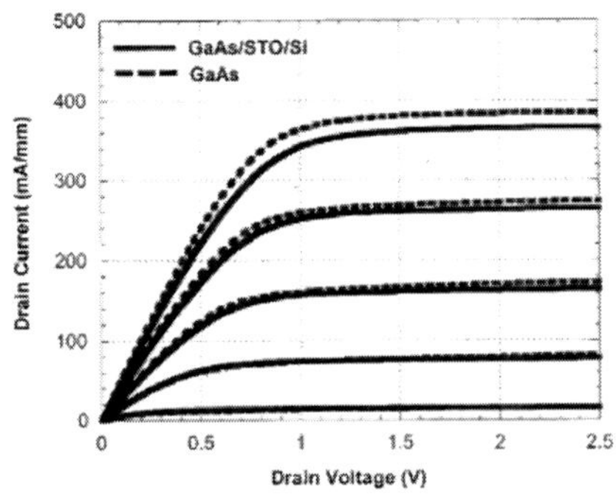

Figure 2.12. Drain current vs. drain voltage for 0.7 × 100 μm^2 MESFETs at different gate voltages up to 0.5 V with each step of 0.5 V. (Ref. [30], reprinted with permission from IEEE).

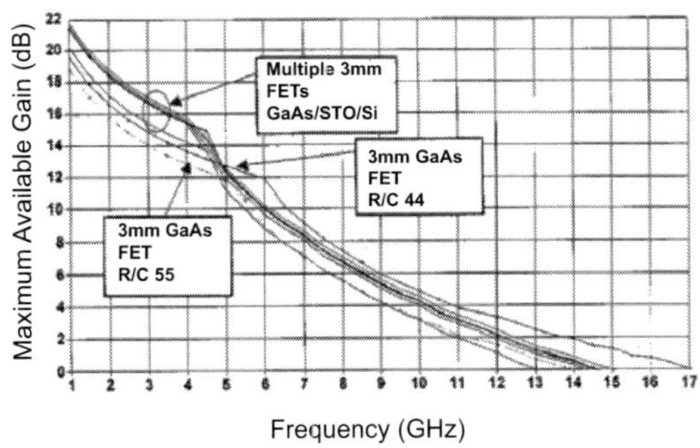

Figure 2.13. Maximum gain of six 3 mm FETs on GaAs/STO/Si and two FETs on GaAs. (Ref. [30], reprinted with permission from IEEE).

GaAs showed initial slow degradation. For a long time test, the GaAs-based device underwent a secondary degradation of 10% after 450 hours, while the degradation of the STO/Si-based device revealed a mild drop of 1.2% after 800 hours, indicating potential of this novel buffer technology.

Amorphous oxides can also be introduced *after* epitaxial growth. This is achieved by lateral oxidation after patterning of a whole epitaxial structure. Seo reported the first experiment in 1997 through the wet oxidation of AlAs [31], a method that is widely used in vertical cavity surface emitting lasers to achieve current and optical mode confinements nowadays. A 20 period 10 nm/10

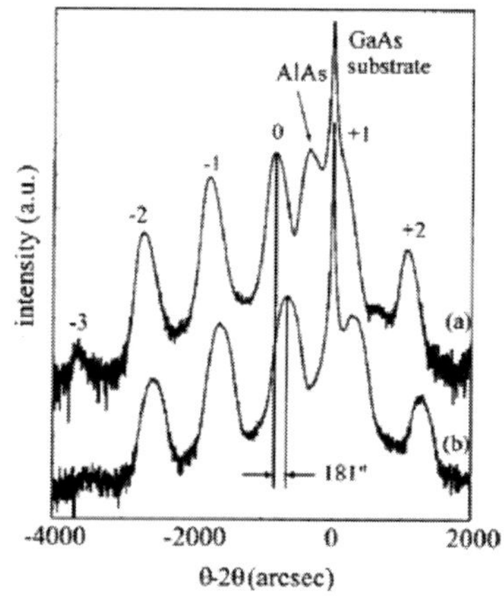

Figure 2.14. The (004) XRD rocking curves of (a) the as-grown sample with 200 nm-thick AlAs and (b) the same sample after oxidation. (Ref. [31], reprinted with permission from AIP).

nm InGaAs/GaAs superlattice with different In concentrations was grown on AlAs with a varied thickness using MOVPE. Samples were then patterned by photolithography to 200 μm strips and wet oxidized in an open quartz tube for 3–5 hours at 425°C with a flow of N_2 bubbled through the heated de-ionized water at 95°C. Figure 2.14 shows (004) XRD rocking curves from the sample with a 200 nm-thick AlAs. A peak shift of 181 arcsec toward the GaAs substrate peak is observed after wet oxidation of AlAs, indicating that the InGaAs/GaAs superlattice is partially relaxed after oxidation. It was also found that the oxide volume contracts during the conversion process from AlAs to Al_2O_3. This volume reduction is attributed as the driving force for the partial strain relaxation. They examined the effect of the In concentration in the superlattice as well as the thickness of AlAs and found that larger strain relaxation occurred in the sample with a higher In and thicker AlAs.

Later, Chavarkar *et al.* refined the experiment and gave a name "Lattice Engineered Substrate" for this type of relaxed templates residing on amorphous oxides [32]. Three $In_xGa_{1-x}As$ ($x = 0.2$, 0.3, and 0.4) layers were grown on a 50 nm-thick AlAs using MBE. Before growth of the AlAs, a GaAs buffer was grown at 250°C to facilitate the oxidation process. The InGaAs thickness was kept about 20 times of

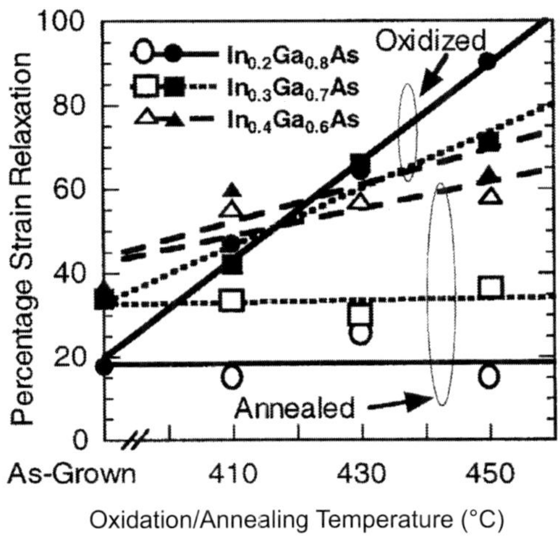

Figure 2.15. Strain relaxation of thick $In_xGa_{1-x}As$ layers on 50 nm AlAs as a function of oxidation/annealing temperature for different In concentrations. (Ref. [32], reprinted with permission from Springer Boston).

the calculated critical thickness from the Matthews' model [13]. The oxidation was performed on mesas with a size of 100 × 100 μm^2. Figure 2.15 shows the percentage of strain relaxation as a function of the oxidation/annealing temperature for the three samples. As seen, the oxidized samples show a higher strain relaxation measured by XRD than those non-oxidized but annealed samples, in particular for the samples with low In concentrations. This leads the authors to believe the change of the nature of the InGaAs/oxide interface from a diffusion limited to a reaction limited mechanism when the oxidation temperature is increased from 410 to 450°C, since the change in thermal energy is minor and not sufficient to thermally induce strain relaxation. The high oxidation temperature increases porosity of the InGaAs/oxide interface, which reduces the interaction of orthogonal MDs. From both plane-view and cross-section TEM images, they also found that strain induced contrast resulting from MDs disappeared after oxidation. This indicates the enhanced reactive removal of MD cores as a result of oxidation. Based on both effects, TDs have a less probability to be pinned when meeting a MD and can glide easily for a long distance. They also found that the initial dislocation density and control of the porosity played important roles for the final strain relaxation. In the former case, use of a low growth temperature resulted in a smooth surface and small initial strain relaxation (correspondingly a small number of TDs). The strain relaxation upon oxidation is very efficient. For the latter case, it was demonstrated that adding 10% Sb in AlAs to form AlAsSb enhanced the oxidation extent and increased strain relaxation after oxidation. Based on these observations, Romanov and Speck [33] proposed using relaxation enhancing interlayers (REIs) to reduce TDs. The key ingredient is to release local stress to eliminate blocking effect of MDs and to allow efficient gliding of TDs by introducing a REI.

2.3.3 *Viscous Interlayer*

Another method to form a CS is to use a viscous interlayer under the CS template. This interlayer can be glass, metals, or other viscous materials. The majority published works have been focused on using glass (oxides) with varied viscosity. The first demonstration of this idea was done by Tanner *et al.* in 1995 [34]. They grew 120 nm $Si_{0.86}Ge_{0.14}$ on thin bond and etch back SOIs with a 33 nm Si template using MBE following the similar approach of Powell *et al.* [19]. The samples were then thermally annealed at 700, 900, and 1100°C. It is important to note that SiO_2 becomes viscous when heated above 960°C. They monitored the dislocation related D1 peak in PL and found that the as-grown sample was nearly flat. The intensity of the D1 peak increases with the annealing temperature up to 900°C and then decreased significantly when annealed at 1100°C. This implies that the thick $Si_{0.86}Ge_{0.14}$ film relaxes upon thermal annealing by formation of MDs and TDs, and these dislocations "disappear" due to the viscous flow of the buried SiO_2 film. The XRD shows a monotonic increase of strain relaxation with the annealing temperature: <3%, 42%, 46%, and 99% for the as-grown sample and the samples annealed at 700, 900, and 1100°C, respectively. The viscosity of SiO_2 at low temperatures can be significantly reduced by adding B_2O_3 to form $(B_2O_3)_x(SiO_2)_{1-x}$ (BSO) glass [35]. Moran *et al.* grew 3 µm-thick $In_{0.44}Ga_{0.56}As$ on 10 nm GaAs templates bonded to a GaAs handle wafer coated with BSO glass at 650°C using MOVPE [36]. The mole fraction of B_2O_3 is 10% and 30%, giving a viscosity of 10^{17} Poise and 10^{12} Poise at 650°C, respectively. They found a narrower FWHM of XRD by a factor of 2 and smaller surface roughness by a factor of 4 as compared with the same structure grown on a GaAs substrate. This gives indirect evidence of improved structural quality grown on CSs. The XRD line-width is similar for the samples grown on CSs with different viscosities. This led the authors to conclude that the mechanism responsible for strain relaxation was insensitive to the viscosity in this range.

Hobart *et al.* studied relaxation of $Si_{0.7}Ge_{0.3}$ films bonded to high and low viscosity glass [37]. They combined the two established SOI fabrication processes: film separation by hydrogen implantation and bond and etch back SOI, to produce high quality Si templates with a thickness less than 5 nm. Two types of glasses: SiO_2 and borophosphorosilicate glass (BPSG) with a glass transition temperature of 1150 and 625°C, respectively, were employed. A CS wafer containing a 30 nm-thick $Si_{0.7}Ge_{0.3}$ and a 4 nm-thick Si was bonded to a MHS coated with one of the two types of glasses. The structure was then thermally annealed up to 800°C for 90 min. For the SiGe film bonded to SiO_2, no change in surface morphology and roughness (about 0.3 nm) measured by AFM was found up to 800°C. For the SiGe film bonded to BPSG, little change was observed when increasing annealing temperature to the glass transition temperature of 625°C. Above this temperature, buckling was observed as shown in Fig. 2.16, and surface roughness increases drastically. Figure 2.17 shows (004) rocking curves of XRD as a function of the annealing temperature. When the buckling occurs, surface is no longer planar, and the diffraction peak disappears. It has also been found that the buckling depends on the size of the sample and can eventually disappear for small sizes. For example, a patterned 30 × 30 μm^2 mesa reveals negligible buckling and a surface roughness of 0.1 nm, about 1000 times reduction over the un-patterned case.

In a late paper [38], they extensively examined the effect of the pattern size on strain relaxation. It was found that strain relaxation occurs via lateral relaxation at the mesa edges and surface buckling in the center of the mesa. Figure 2.18 shows the local strain distribution measured by micro-Raman for a 60 × 60 μm^2 pattern

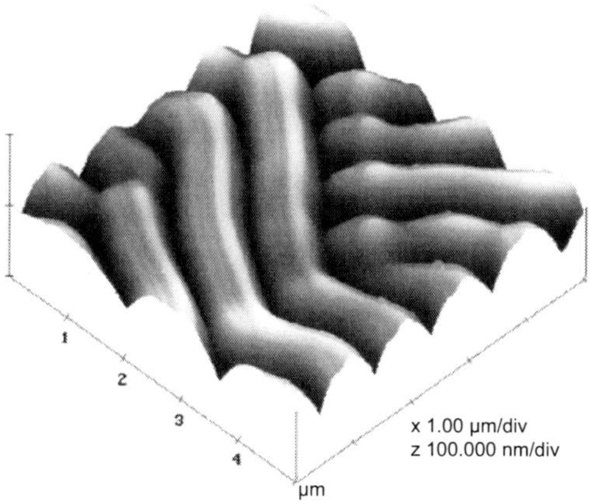

Figure 2.16. An AFM image of a buckled SiGe film grown on BPSG after annealing at 800°C for 90 min. (Ref. [37], reprinted with permission from Springer Boston).

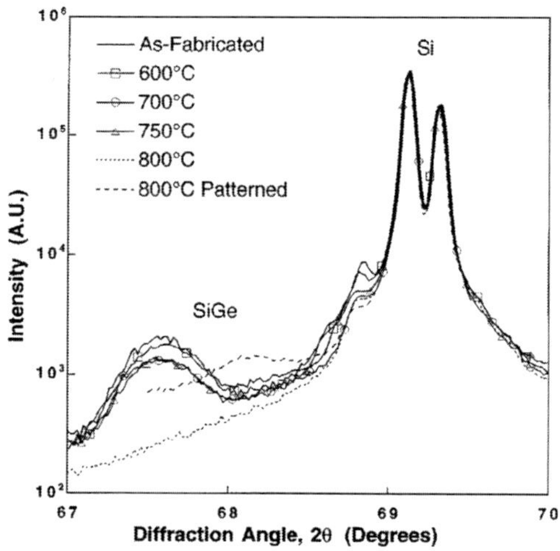

Figure 2.17. The (004) XRD rocking curves of SiGe films on BPSG under different thermal annealing conditions. The patterned sample is a 30 × 30 μm^2 mesa. (Ref. [37], reprinted with permission from Springer Boston).

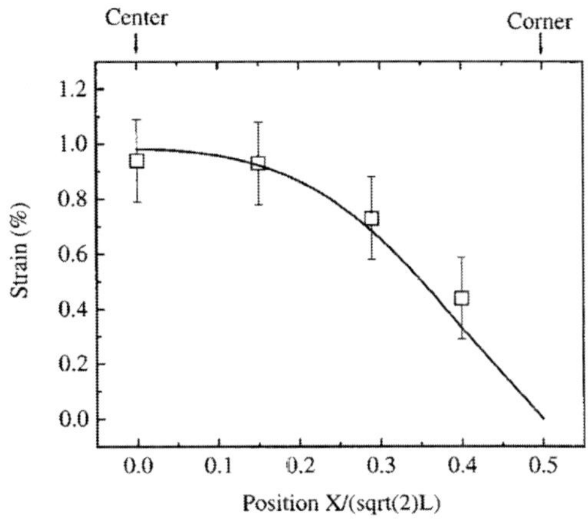

Figure 2.18. Strain distribution along the diagonal of a 60 × 60 μm^2 $Si_{0.7}Ge_{0.3}$ island after 17 min annealing at 790°C measured by micro-Raman. (Ref. [38], reprinted with permission from AIP).

after 17 min annealing at 790°C. Clearly strain relaxation gradually increases from the center to the edge. The strain relaxation at the edge is due to the slip of the template on the oxide, while the buckling results from coherent thermal expansion. This is supported by the fact that no dislocations are found inside the SiGe layer by TEM. The time scale for the lateral strain relaxation, τ_L, can be modeled as

$$\tau_L = \frac{\eta L^2}{C_{11} h_f h_g} \tag{2.10}$$

where L is the mesa edge width, η is the viscosity of the glass, h_g is the thickness of the glass, C_{11} and h_f are the elastic coefficient and thickness of the template, respectively. The buckling is nearly sinusoidal and its amplitude, A, rises exponentially with time, t, as:

$$A = A_0 \exp(t/\tau_B) \tag{2.11}$$

where A_0 is initial surface roughness and τ_B is buckling time constant which is proportional to the viscosity. Therefore, simply reducing the viscosity will enhance both the lateral relaxation and the buckling simultaneously. However, use of a long annealing time or a thick template can eventually eliminate the buckling.

Metals can also be used as viscous interlayer. Carter-Coman *et al.* [39] used thermally evaporated Al as the viscous interlayer to fabricate 300 nm-thick GaAs CSs. Various $In_{0.07}Ga_{0.93}As$ layers with thickness up to 800 nm were grown at 525°C by MBE. The XRD measurements indicate that the InGaAs layers grown on such CSs have a larger critical thickness and slower strain relaxation than the same structures grown on conventional GaAs substrates. However, the FWHM of XRD peaks is broad indicating an inferior structure quality.

2.3.4 *Structure Modification of Substrate Templates*

Another technique to fabricate CSs is to modify structures of substrate templates. Ion implantation into semiconductors can create nano-scale cavities at a certain distance below the surface [40–42]. The templates are then thermally annealed to improve the structure quality. It was believed that the generated dislocations due to the formation of over-pressurized nano-cavities induced by the implanted ions glide to interface during the annealing forming misfit segments, thus relieving strain. The distance, typically around 100 nm, can be adjusted by using different ions. This method has been applied to both SiGe and InP. Although there is some evidence of compliant nature in the templates, the crystal quality in general is inferior to equivalently relaxed structures grown on conventional substrates.

Chen *et al.* discovered that Si grown at a low temperature (LT) exhibited a tensile strain and behaved like a CS [43]. They grew 300 nm $Si_{0.76}Ge_{0.24}$ at 550°C on 50 nm Si grown at 400°C (LT-Si) using MBE. The as-grown sample was then annealed at 800°C for 20 min to release strain in the SiGe layer. It is well known that the LT growth of Si generates a high density of point defects. The tensile strain of the LT-Si confirmed by XRD provides certain strain compensation when a SiGe alloy is grown on it. Figure 2.19 shows TEM images of the as-grown and the annealed $Si_{0.76}Ge_{0.24}$ and $Si_{0.7}Ge_{0.3}$ on LT-Si templates. The as-grown $Si_{0.76}Ge_{0.24}$ film reveals TDs penetrating into the LT-Si. Thermal annealing enhances strain relaxation of the SiGe film with nearly all the observed TDs threading into the LT-Si. This indicates that the LT-Si is soft and acts like a CS. They studied the effect of the growth temperature on both the LT-Si and the SiGe film and found that the lower the growth temperature,

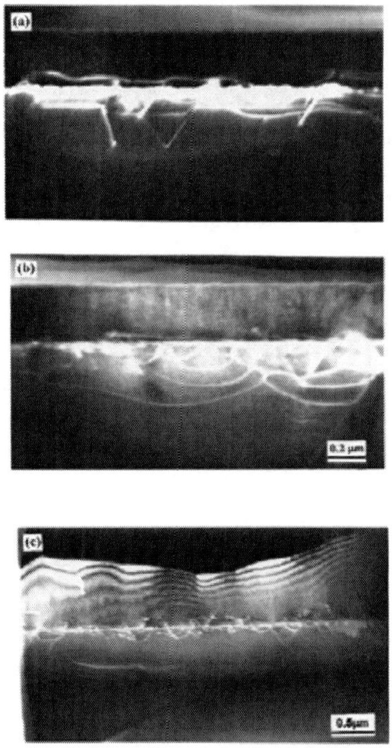

Figure 2.19. The (022) dark field TEM images of (a) as-grown, (b) annealed $Si_{0.76}Ge_{0.24}/Si$, and (c) annealed $Si_{0.7}Ge_{0.3}/Si$. (Ref. [43], reprinted with permission from AIP).

the more effective the dislocation blocking effect in the LT-Si and the fewer the TDs transmitted into the SiGe. However, for the $Si_{0.7}Ge_{0.3}$ film, dislocations are found to threading into both the LT-Si and the SiGe layer, and the surface becomes rough. Luo *et al.* studied the effect of the LT-Si thickness on surface and structure quality [44]. It was found that surface roughness decreased from 14 nm to 1 nm when the LT-Si thickness increased from 50 to 200 nm. The estimated TD density from the optimized sample is 10^7 cm^{-2}.

2.3.5 *Twist Bonding*

Twist bonding was first reported in 1997 by Ejeckam *et al.* from the group in Cornell University who proposed the concept of the CS [45]. The basic idea is to prepare a CS by bonding a thin CS layer with

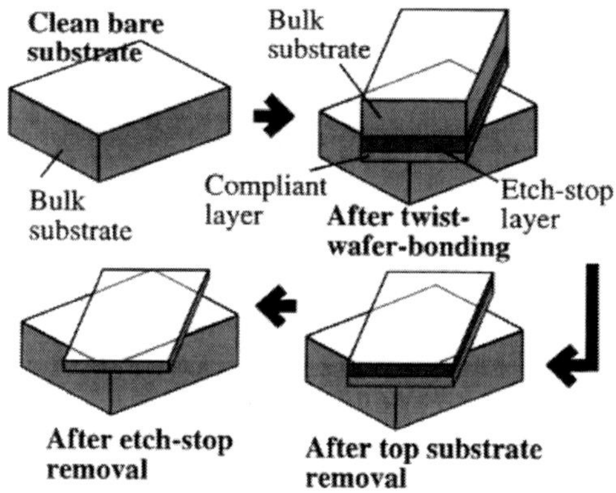

Figure 2.20. A conceptual schematic showing the process of forming a compliant universal substrate using twist bonding. (Ref. [46], reprinted with permission from AIP).

an angle to the substrate as schematically shown in Fig. 2.20 [46]. The preparation procedure is as follows. First, an etch stop layer is grown epitaxially on a dummy substrate. A thin compliant layer in a thickness of a few nanometers is then grown. The dummy substrate is then twist-bonded to a clean bare substrate. Both the dummy substrate and the etch stop layer are then carefully removed, leaving a thin CS. The bonding and etching steps are crucial for the quality of the CS template and subsequent epi-layer heterostructures [47, 48].

Ejeckam *et al.* demonstrated the proof of concept by growing 300 nm-thick $In_{0.36}Ga_{0.64}P$ (1% lattice mismatch) on the twist-bonded 10 nm-thick GaAs templates (9–32°) using MOVPE [45]. This thickness is about 30 times of the critical thickness calculated from the Matthews' model [13]. The cross-sectional TEM image shown in Fig. 2.21b reveals few dislocations in the 300 nm-thick $In_{0.36}Ga_{0.64}P$ layer, while dislocations are found in the same sample grown directly on a GaAs substrate (Fig. 2.21a). Later the same group reported successful growth of 650 nm-thick dislocation-free InSb (15% lattice mismatch) on the 3 nm-thick CS templates twist-bonded by 40° with respect to the GaAs MHS [46]. These results confirm low density of TDs in the films grown on twist-bonded CSs limited by the TEM resolution of about 10^6 cm^{-2}, compared with 10^{11} cm^{-2} found in the InSb layer grown on a normal GaAs substrate.

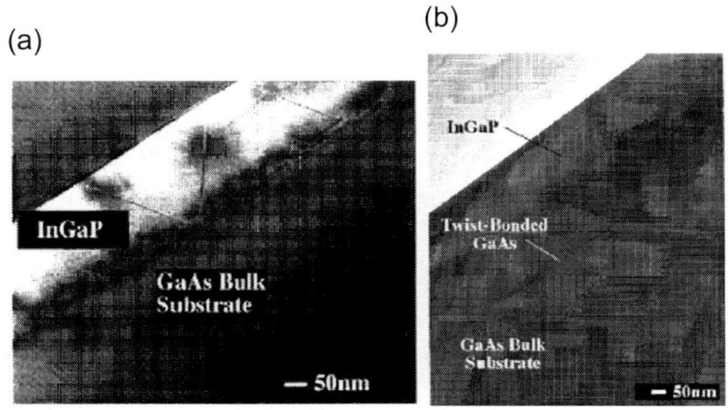

Figure 2.21. Cross-sectional bright field TEM images of an $In_{0.36}Ga_{0.64}P$ layer grown on (a) a normal GaAs substrate and (b) a CS twist bonded by 9° to the GaAs MHS. (Ref. [45], reprinted with permission from AIP).

Tomich *et al.* made comparative studies on surface, structural, and optical quality of InAs/InGaSb superlattices grown on GaAs, GaSb, and twist-bonded GaAs substrates using MBE [49]. The active structure consists of 100 periods of 4.4 nm InAs/1.7 nm GaSb superlattice intended for peak optical absorption at about 10 μm. The CS was prepared by twist-bonding a 2 nm-thick InGaAs layer with 45° to the GaAs substrate. A GaSb buffer layer was first grown before the growth of the SL. The AFM images over a scan area of 5 × 5 μm^2 show smooth surface for samples grown on the GaSb and on the compliant GaAs substrates but an undulated surface morphology for the sample grown on the GaAs substrate. The root-mean-square surface roughness value is 1.7, 0.2, and 0.18 nm for the structure grown on GaAs, GaSb, and compliant GaAs, respectively. The XRD measurements indicate a partial strain relaxation of the superlattice grown on the GaAs CS. The TEM images reveal many TDs in the structure grown directly on the GaAs substrate but very few in the sample on the CS. The cut-off wavelength is 11 and 13.9 μm for the samples grown on the GaSb and the GaAs CS, respectively, but there is no significant optical response for the identical structure grown on the GaAs substrate.

These results indicate that the twist-bonded CS is a promising technology to overcome large lattice mismatch practically. However, the compliance is not yet proven experimentally. Jesser, van der Merve, and Stoop made an extensive theoretical analysis on such a technology [15]. For a small angle of a few degrees, the interface between the CS and the MHS is modeled by narrow dislocation cores of high local atomic distortions separated by areas of lattice matching across the interface. For large twist angles, the highly distorted core regions of dislocations start to overlap and the twist boundaries degenerate into a rigid misfit vernier (MV). This hypothesis is theoretically proven. They also discussed theoretically the critical thickness in the case of weak and strong bonding between the CS and the MHS and suggested the following design criteria for twist-bonded CSs [15].

- The CS should be as thin as practical but not so thin that the interfacial dislocations at the twist-bonded interface are eliminated by attraction to its free surface.
- The twist angle of the CS should be larger than about 8° so that the homogeneous elastic strains of the CS are not hindered by the bulk substrate.
- Coincidence angles should be avoided so that the CS is not locked into a configuration that inhibits homogeneous elastic deformation of the CS.
- A CS should be selected to have a small misfit with the hetero-epitaxial film and not with the bulk substrate. It should not be of the same material as the bulk substrate, nor should it strongly interact with the bulk substrate.
- The CS should introduce layer-by-layer growth in the epitaxial film as an island growth mode of the epitaxial film is likely to lead to geometrically necessary dislocations when islands coalesce.
- A CS should be of a smaller elastic modulus than the epitaxial film so that elastic deformation and any plastic deformation with its accompanying TD generation will preferentially occur in the CS.
- The bulk substrate should be selected to have a large mismatch to the CS as accomplished by a high angle of twist and/or a high lattice mismatch and need not be of the same crystal structure as the CS.

2.4 Future Outlook

Since the first proposal by Lo in 1991, there has been great progress in both theoretical understanding and practical implementation of compliant substrates using different innovative methods described above. There are two issues that are worth for further pursuing in the future: interaction between a CS and a MHS and the CS size scaling. As addressed by Ayers [10], so far there is no clear experimental demonstration of the full compliance as most CSs reported hitherto are thicker than the critical thickness for the free-standing superlattice. Compliant substrates with a thickness of a few atomic layers are required to test this. Also there usually exists certain interaction between atoms of a CS and of a MHS at the interface, which exerts a resistance to the slippage of a CS. It is of paramount importance to reduce this interaction or even decouple the two parts. One attractive approach is to *in situ* prepare a CS by making use of different melting temperatures of epitaxial layers during an epitaxial growth. When the growth temperature approaches to the melting temperature of an epi-crystal covered by another crystal with a higher melting temperature, the underneath crystal becomes liquid-like with a very low viscosity. For example, InSb has a melting point of 527°C, which is lower than the growth temperature of many III–V compounds by MBE and MOVPE. Therefore, it would be very interesting to first grow InSb on Si, GaAs, or other conventional substrates followed by a thin CS template at a low growth temperature to maintain epitaxial nature and to suppress surface roughness. The growth temperature is then increased to close to or above the melting temperature of InSb, and the CS template will become floating then. The InSb viscosity can be adjusted by changing the growth temperature around the melting point. Subsequent epitaxial layers can be grown on the virtual free-standing CS.

Another issue is the size scaling of a CS template. It has been realized recently that a relative compliance is not enough for preparation of large-size CS templates [10]. The absolute compliance is required and its value is proportional to the template size as shown in Eq. 2.9. However, this is practically very difficult to implement for large-size CSs. The physical origin is in analog to the concept of "excess stress" introduced by Tsao *et al.* [50] when discussing dislocation kinetics in a partially relaxed heterostructure. As mentioned above, the concept of an ideal CS is based on the free slippage of the whole epi-structure together with the thin CS template on a MHS. The driving force for the slippage is the strain force in the epi-CS complex structure relative to the MHS. Initial slippage occurs at the rims of a CS. During the slippage, the lateral lattice moves toward the equilibrium value and the average strain force decreases. The driving force is then the local "excess" strain force that is the residual stress with respect to the partially relaxed local lattice constant and decreases when it approaches to the rim of the CS. This is similar to the assumption made by Kästner and Gösele [16, 17] in their model. The slippage process will eventually stop when the "excess" strain force is balanced by the resistance force at the interface. As a result, the slippage can be achieved only in the region close to the rims of a CS, while the center part will undergo elastic deformation like, e.g., buckling [37]. Therefore, it is difficult to realize absolute compliance without significantly reducing the resistance (viscosity), especially for a large template size. One approach to overcome this problem is to

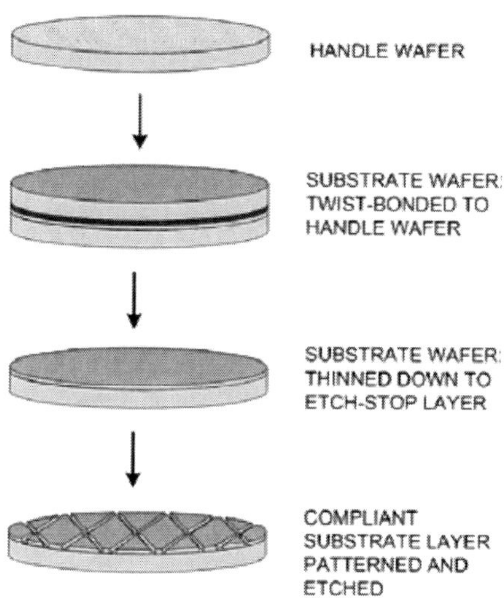

Figure 2.22. Schematics showing the fabrication process of the universal compliant trench from Ref. [10].

introduce trenches on CS templates. Ayers *et al.* introduced a new method called universal compliant trench (UCT) substrates [10]. The procedure is shown schematically in Fig. 2.22. Depending on the absolute compliance, the trenches can be in an arbitrary shape and prepared by lithography techniques. The mesa size is chosen to be suitable for any practical device applications. As mentioned in Ref. [10], this technology is under development jointly by University of Connecticut and San Jose State University in USA.

Acknowledgment

Mr. Yuxin Song is acknowledged for preparing Fig. 2.1.

References

1. Frank, F. C. and van der Merve, J. H. (1949). One-dimensional dislocations. I. Static theory, II. Misfitting monolayers and oriented overgrowth, *Proc. R. Soc. London Ser. A.*, **198**, pp. 205–225.
2. Osbourn, G. C. (1982). Strained-layer super-lattices from lattice mis-matched materials, *J. Appl. Phys.*, **53**, pp. 1586–1589.
3. Yablonovitch, E. and Kane, E. O. (1986). Reduction of lasing threshold current-density by the lowering of valence band effective mass, *J. Lightwave Technol.*, **4**, pp. 504–506.
4. Wang, S. M., Andersson, T. G., Kulakovskii, V. D., and Yao, J. Y. (1991). Critical layer thickness in $In_xGa_{1-x}As/GaAs$ quantum wells studied by photoluminescence and transmission electron microscopy, *Superlatt. Microstruct.*, **9**, pp. 123–126.
5. Stangl, J., Holy, V., and Bauer, G. (2004). Structure properties of self-organized semiconductor nanostructures, *Rev. Mod. Phys.*, **76**, pp. 725–783.
6. Bimberg, D., Grundmann, M., and Ledentsov, N. N. (1999). *Quantum Dot Heterostructures* (John Wiley & Sons Ltd, England).
7. Lo, Y. H. (1991). New approach to grow pseudomorphic structures over the critical thickness, *Appl. Phys. Lett.*, **59**, pp. 2311–2313.
8. Bourret, A. (2000). Compliant substrates: a review on the concept, techniques and mechanisms, *Appl. Surf. Sci.*, **164**, pp. 3–14.

9. Brown, A. S. and Doolittle, W. A. (2000). The status and promise of compliant substrate technology, *Appl. Surf. Sci.* **166**, pp. 392–398.
10. Ayers, J. E. (2008). Compliant substrates for heteroepitaxial semiconductor devices: theory, experiment, and current directions, *J. Electron. Mat.*, **37**, pp. 1511–1523.
11. Teng, D., and Lo, Y. H. (1993). Dynamic-model for pseudomorphic structures grown on compliant substrates — an approach to extend the critical thickness, *Appl. Phys. Lett.*, **62**, pp. 43–45.
12. Chua, C. L., Hsu, W. Y., Lin, C. H., Christensen, G., and Lo, Y. H. (1994). Overcoming the pseudomorphic critical thickness limit using compliant substrates, *Appl. Phys. Lett.*, **64**, pp. 3640–3642.
13. Matthews, J. W. (1975). Defects associated with accommodation of misfit between crystals, *J. Vac. Sci. Technol.*, **12**, pp. 126–133.
14. Hirth, J. P., and Lothe, J. (1992). *Theory of Dislocations*, 2nd Ed. (Krieger Publishing Company, USA).
15. Jesser, W. A., van der Merve, J. H., and Stoop, P. M. (1999). Misfit accommodation by compliant substrates, *J. Appl. Phys.*, **85**, pp. 2129–2139.
16. Kästner, G., and Gösele, U. (1997). Principles of strain relaxation in heteroepitaxial films grown on compliant substrates, *J. Appl. Phys.*, **88**, pp. 4048–4055.
17. Kästner, G., and Gösele, U. (2003). Reduced critical thickness for relaxing heteroepitaxial films on compliant substrates, *Appl. Phys. Lett.*, **82**, pp. 3209–3211.
18. Jones, A. M., Jewell, J. L., Mabon, J. C., Reuter, E. E., Bishop, S. G., Roh, S. D., *et al.* (1999). Long wavelength InGaAs quantum wells grown without strain induced warping on InGaAs compliant membranes above a GaAs substrate, *Appl. Phys. Lett.*, **74**, pp. 1000–1002.
19. Powell, A. R., Iyer, S. S., and LeGoues, F. K. (1994). New approach to the growth of low dislocation relaxed SiGe material, *Appl. Phys. Lett.*, **64**, pp. 1856–1858.
20. LeGoues, F. K., Powell, A., and Iyer, S. S. (1994). Relaxation of SiGe thin films grown on Si/SiO_2 substrates, *J. Appl. Phys.*, **75**, pp. 7240–7246.
21. Huang, F. Y., Chu, M. A., Tanner, M. O., Wang, K. L., U'Ren, G. D., and Goorsky, M. S. (2000). High-quality strain-relaxed SiGe alloy grown on implanted silicon-on-insulator substrate, *Appl. Phys. Lett.*, **76**, pp. 2680–2682.

22. Yang, Z., Alperin, J., Wang, W. I., Iyer, S. S., Kuan, T. S., and Semendy, F. (1998). *In situ* relaxed Si1-xGex epitaxial layers with low threading dislocation densities grown on compliant Si-on-insulator substrates, *J. Vac. Sci. Technol. B*, **16**, pp. 1489–1491.
23. Seaford, M. L., Tomich, D. H., Eyink, K. G., Grazulis, L., Mahalinham, K., Yang, Z., *et al.* (2000). Comparison of GaAs grown on standard Si (511) and compliant SOI (511), *J. Electron. Mat.*, **29**, pp. 906–908.
24. Pei, C. W., Heroux, J. B., Sweet, J., Wang, W. I., Chen, J., and Chang, M. F. (2002). High quality GaAs grown on Si-on-insulator compliant substrates, *J. Vac. Sci. Technol. B*, **20**, pp. 1196–1199.
25. Yang, Z., Guarin, F., Tao, I. W., Wang, W. I., and Iyer, S. S. (1995). Approach to obtain high quality GaN on Si and SiC-on-silicon-on-insulator compliant substrate by molecular-beam epitaxy, *J. Vac. Sci. Technol. B*, **13**. pp. 789–791.
26. Droopad, R., Curless, J., Yu, Z., Jordan, D., Liang, Y., Overgaard, C., *et al.* (2002). GaAs on Si using an oxide buffer layer, *Inst. Phys. Conf. Ser.*, **174**, pp. 1–8.
27. Yu, Z., Liang, Y., Overgaard, C., Hu, X., Curless, J., Li, H., *et al.* (2004). Advances in heteroepitaxy of oxides on silicon, *Thin Solid Films*, **462–463**, pp. 51–56.
28. Ramdani, J., Droopad, R., Yu, Z., Curless, J. A., Overgaard, C. D., Finder, J., *et al.* (2000). Interface characterization of high-quality $SrTiO_3$ thin films on Si(100) substrates grown by molecular beam epitaxy, *Appl. Surf. Sci.*, **159–160**, pp. 127–133.
29. Hu, X., Li, H., Liang, Y., Wei, Y., Yu, Z., Marshall, D., *et al.* (2003). The interface of epitaxial $SrTiO_3$ on silicon: *in situ* and *ex situ* studies, *Appl. Phys. Lett.*, **82**, pp. 203–205.
30. Eisenbeiser, K., Emrick, R., Droopad, R., Yu, Z., Finder, J., Rockwell, S., *et al.* (2002). GaAs MESFETs fabricated on Si substrates using a $SrTiO_3$ buffer layer, *IEEE Electron. Dev. Lett.*, **23**, pp. 300–302.
31. Seo, J. H., and Seo, K. S. (1998). Strain relaxation of InGaAs/GaAs superlattices by wet oxidation of underlying AlAs layer, *Appl. Phys. Lett.*, **72**, pp. 1466–1468.
32. Chavarkar, P., Mathis, S. K., Zhao, L., Keller, S., Speck, J. S., and Mishra, U. K. (2000). Strain relaxation in InGaAs lattice engineered substrates, *J. Electron. Mat.*, **29**, pp. 944–949.
33. Romanov, A. E., and Speck, J. S. (2000). Relaxation enhanced interlayers (REIs) in threading dislocation reduction, *J. Electron. Mat.*, **29**, pp. 901–905.

34. Tanner, M. O., Chu, M. A., Wang, K. L., Meshkinpour, M., and Goorsky, M. S. (1995). Relaxed $Si_{1-x}Ge_x$ films with reduced dislocation densities grown by molecular beam epitaxy, *J. Cryst. Growth*, **157**, pp. 121–124.
35. Hansen, D. M., Moran, P. D., Dunn, K. A., Babcock, S. E., Matyi, R. J., and Kuech, T. F. (1998). Development of a glass-bonded compliant substrate, *J. Cryst. Growth*, **195**, pp. 144–150.
36. Moran, P. D., Hansen, D. M., Matyi, R. J., Cederberg, J. G., Mawst, L. J., and Kuech, T. F. (1999). InGaAs heteroepitaxy on GaAs compliant substrates: X-ray diffraction evidence of enhanced relaxation and improved structural quality, *Appl. Phys. Lett.*, **75**, pp. 1559–1561.
37. Hobart, K. D., Kub, F. J., Fatemi, M., Twigg, M. E., Thompson, P. E., Kuan, T. S., *et al.* (2000). Compliant substrates: a comparative study of the relaxation mechanisms of strained films bonded to high and low viscosity oxides, *J. Electron. Mat.*, **29**, pp. 897–900.
38. Yin, H., Huang, R., Hobart, K. D., Suo, Z., Kuan, T. S., Inoki, C. K., *et al.* (2002). Strain relaxation of SiGe islands on compliant oxide, *J. Appl. Phys.*, **91**, pp. 9716–9722.
39. Carter-Coman, C., Bicknell-Tassius, R., Brown, A. S., and Jokerst, N. M. (1997). Analysis of $In_{0.07}Ga_{0.93}As$ layers on GaAs compliant substrates by double crystal X-ray diffraction, *Appl. Phys. Lett.*, **70**, pp. 1754–1756.
40. Holländer, B., Lenk, St., Mantl, S., Trinkaus, H., Kirch, D., Luysberg, M., *et al.* (2001). Strain relaxation of pseudomorphic $Si_{1-x}Ge_x$/Si (100) heterostructures after hydrogen or helium ion implantation for virtual substrate fabrication, *Nucl. Instrum. Meth. Phys. Res. B*, **175–177**, pp. 357–367.
41. Chicoine, M., Beaudoin, C., Roorda, S., Masut, R. A., and Desjardins P. (2005). III–V compliant substrates implemented by nanocavities by ion implantation, *J. Appl. Phys.*, **97**, pp. 064309–064314.
42. Chen, C., Yu, B., Liu, J., Cao, J., Zhu, D., and Liu, Z. (2005). Investigation of strain-relaxed SiGe thin film grown on ion-implanted Si compliant substrate, *Nucl. Instrum. Meth. Phys. Res. B*, **239**, pp. 433–439.
43. Chen, H., Guo, L. W., Cui, Q., Hu, Q., Huang, Q., and Zhou, J. M. (1996). Low-temperature buffer for growth of a low-dislocation-density SiGe layer on Si by molecular-beam epitaxy, *J. Appl. Phys.*, **79**, pp. 1167–1169.
44. Luo, Y. H., Wan, J., Forrest, R. L., Liu, J. L., Jin, G., Goorsky, M. S., *et al.* (2001). Compliant effect of low-temperature Si buffer for SiGe growth, *Appl. Phys. Lett.*, **78**, pp. 454–456.

45. Ejeckam, F. E., Lo, Y. H., Subramanian, S., Hou, H. Q., and Hammons, B. E. (1997). Lattice engineered compliant substrate for defect-free heteroepitaxial growth, *Appl. Phys. Lett.*, **70**, pp. 1685–1687.

46. Ejeckam, F. E., Seaford, M. L., Lo, Y. H., Hou, H. Q., and Hammons, B. E. (1997). Dislocation-free InSb grown on GaAs compliant universal substrates, *Appl. Phys. Lett.*, **71**, pp. 776–778.

47. Zhu, Z. H., Zhou, R., Ejeckam, F. E., Zhang, Z., Zhang, J., Greenberg, J., *et al.* (1998). Growth of InGaAs multi-quantum wells at 1.3 μm wavelength on GaAs compliant substrates, *Appl. Phys. Lett.*, **72**, pp. 2598–2600.

48. Zhang, C., Lubyshev, D., Jackson, T. N., Miller, D. L., and Mayer, T. S. (1999). The effect of $Al_{0.7}Ga_{0.3}As$ etch stop removal on the preparation of wafer-bonded compliant substrates, *J. Electrochem. Soc.*, **146**, pp. 1597–1601.

49. Tomich, D. H., Eyink, K. G., Grazulis, L., Brown, G. L., Szmulowicz, F., Mahalingam, K., *et al.* (2000). Comparison of InGaSb/InAs super-lattice structures grown by MBE on GaSb, GaAs and compliant GaAs substrates, *J. Electron. Mat.*, **29**, pp. 940–943.

50. Tsao, J. Y., Dodson, B. W., Picraux, S. T., and Cornelison, D. M. (1987). Critical stresses for Si_xGe_{1-x} strained-layer plasticity, *Phys. Rev. Lett.*, **59**, pp. 2455–2458.

Chapter 3

Patterned Substrate Epitaxy

Huan Zhao and Shumin Wang
Department of Microtechnology and Nanoscience, Chalmers University of Technology, 41296 Göteborg, Sweden
shumin@chalmers.se

In this chapter, we provide a brief overview of patterned substrate epitaxy (PSE) with intention to improve material quality of lattice mismatched thin films. We describe a short history, mechanisms of strain relaxation in diamond and zincblende heterostructures with small lattice mismatch when grown on large size blanket substrates and on patterned substrates, and implementation techniques.

3.1 Introduction

Patterned substrate epitaxy, as we discuss in this chapter, refers to epitaxial growth of crystals on a patterned substrate with a limited area in the scale from sub-millimeters down to hundreds of nanometers at the initial growth stage. This is in contrast to conventional epitaxial growth on wafer sizes as well as nanowire growth in nanometer scale. In some cases, the consequent lateral overgrowth may lead to coalescence of individual seeds to eventually form large crystalline templates in millimeter or centimeter scale. There

Lattice Engineering: Technology and Applications
Edited by Shumin Wang

ISBN 978-981-4316-29-3 (Hardcover), 978-981-4364-25-6 (eBook)
www.panstanford.com

have been different names in nomenclature describing such an epitaxial growth process: selective area epitaxy (SAE), epitaxial lateral overgrowth (ELO), micro- or nano-channel epitaxy (MCE or NCE), and aspect ratio trapping (ART), etc. We prefer to use the term patterned substrate epitaxy in this chapter. The first experimental demonstration of SAE dates back to almost 50 years ago when Joyce and Baldrey from Texas Instruments successfully grew Si thin films on Si (111) substrate windows formed by SiO_2 using chemical vapor deposition (CVD) in 1962 [1]. The growth selectivity was achieved at high growth temperatures and at low growth rates. The main technical advantage is that it allows for integrating multi-function devices at desired sites on the same substrate. In 1965, Tausch Jr and Lapierre III [2] and Michelitsch [3] reported single crystal GaAs overgrowth on SiO_2, the first experimental demonstration of ELO (the name was coined by Jastrzebski late in 1983 [4]). The concept of SAE was also demonstrated using molecular beam epitaxy (MBE) by Cho and Ballamy later in 1975 [5]. The difference between the SAE grown by CVD and MBE is that deposited materials are only nucleated on opened window seeds initially and overgrown on masks laterally to form single crystals in CVD, while materials are usually deposited on both seeds and masks resulting in single crystals and amorphous/poly-crystals, respectively, in MBE. The effect that the periphery of a patterned substrate can act as dislocation sinks and thus will have influences on strain relaxation was first realized by Matthews *et al.* from IBM in 1970 [6] but didn't catch much attention. Unlike a large-size blanket substrate with a well-defined crystal direction, a crystal formed on a reduced area is often faceted with different crystal orientations having different surface energy [7]. Vohl [8] discovered that growth rate of InP was anisotropic on these facets and decreased with the order of (110) > (100) > (111)A > (111)B. It is this anisotropic growth that leads to lateral overgrowth on, e.g., a (001) substrate window. In 1982, researchers in Lincoln Laboratory employed ELO for the first time to grow highly lattice mismatched GaAs on Si using a thin Ge buffer [9], a very hot topic in 1980s aiming at realizing monolithic integration of electronic and optoelectronic devices. Using this technique for GaAs/Ge/Si, a very low dislocation density of 10^4 cm^{-2} was claimed in GaAs on a Si substrate. Although such a low number is doubtful due to experimental resolution, it is perhaps the first experimental attempt to use PSE to harness the lattice mismatch problem.

Another approach employing small growth areas to harness lattice mismatch is to grow lattice mismatched thin films on etched mesa substrates proposed by Luryi and Suhir from Bell Laboratory in 1986 [10]. They theoretically investigated critical thickness as a function of both mesa (strip in their original paper) size and mismatch based on energy balance model and concluded that for fixed lattice mismatch the critical thickness increased with decrease of the mesa size. When the mesa size is below a critical value, often in nanometer scale, the critical thickness approaches to infinity, a theoretical basis for growth of coherent heterostructure nanowires with large lattice mismatch. This concept was late experimentally verified by Fitzgerald *et al.* in a variety of material systems, including InGaAs/GaAs [11–13], SiGe/Si [13, 14], and GaAs/Si [15] in late 1980s. Interestingly, the finite growth size not only enhances the critical thickness and delays initial strain relaxation by nucleation of misfit dislocations (MDs) but also has a strong impact on dislocation dynamics in thick and relaxed films, resulting in effective mitigation of dislocation density.

In this chapter, we will first summarize understandings of strain relaxation in ideal 2D diamond and zincblende thin films by formation of a network of MDs. It is well known that when the lattice mismatch is high, typically larger than 2%, the initial strain relaxation will undergo a surface morphological transition from 2D to 3D and the growth mode switches to the Sttranski–Krastanov mode. This situation will not be considered here, implying that the techniques reviewed may apply to small or moderate lattice mismatched heterostructures. We will then outline physical mechanisms showing the significance of PSE on extension of critical thickness and reduction of the dislocation density. Different techniques to realize PSE are presented with examples from selected III–V and SiGe materials. One great success of widely used ELO nowadays is growth of high-quality nitride templates, and this topic will not be touched here to avoid overlap with the nitride chapter in this book. Finally, a summary will be given with some discussions on recent applications using PSE.

3.2 Strain Relaxation in Ideal 2D Lattice Mismatched Films

For strain relaxation in an ideal 2D lattice mismatched film, two questions need to be answered. First, what is critical thickness at which a film begins to relax by nucleation of a grid of MDs at heterointerface? Second, how do residual strain and density of threading dislocations (TDs) evolve with film thickness after the critical thickness is reached? Critical thickness has been predicted theoretically [6, 16–21]. For a thin film with a thickness, h, grown on a substrate with lattice mismatch, f, the lattice mismatch is shared between elastic deformation characterized by residual strain, ε, and plastic deformation due to nucleation of MDs characterized by degree of strain relaxation, δ, by:

$$f = \varepsilon + \delta \tag{3.1}$$

The area density of elastic energy, E_e is

$$E_e = Ah\varepsilon^2 \tag{3.2}$$

where A is a constant. Assuming uniform plastic deformation caused by a grid of MDs with a component of Burgers' vector perpendicular to a dislocation line, $b_\perp$, the average distance between two neighbor MDs is $b_\perp/(f - \varepsilon)$. The area energy density of a MD network, E_d, is

$$E_d = \frac{f - \varepsilon}{b_\perp} B \ln\left(\frac{\alpha h}{b}\right) \tag{3.3}$$

where B is a constant, b is Burgers' vector, and α is the dislocation core parameter. At initial strain relaxation, the MD density is so low that the average distance between two neighbor MDs is much larger than the film thickness, and dislocation interaction can be neglected. The critical thickness is calculated by minimizing the total energy with respect to the residual strain:

$$\frac{\partial\left(E_e + E_d\right)}{\partial\varepsilon} = 0 \tag{3.4}$$

leading to

$$\varepsilon = \frac{B}{2Ahb_\perp} \ln\left(\frac{\alpha h}{b}\right) \tag{3.5}$$

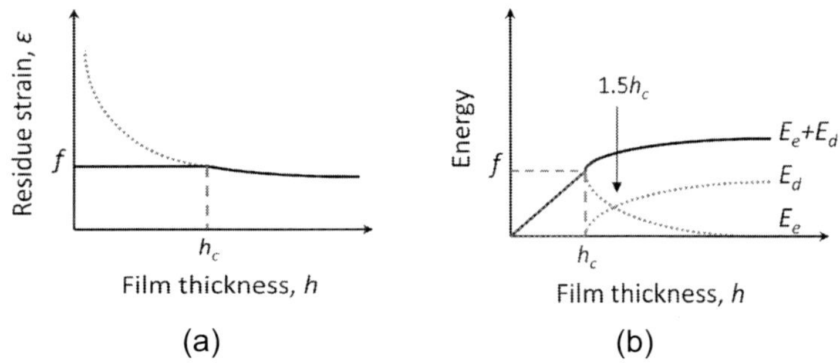

Figure 3.1. Schematic diagrams showing (a) residual strain and (b) strain energy, dislocation energy, and the sum of both as a function of film thickness.

Figure 3.1a shows residual strain as a function of film thickness. The maximum value is limited by the lattice mismatch. Critical thickness is defined when the calculated residual strain from Eq. 3.5 equals to f. The above energy minimization approach is general for any dislocation source if E_d is replaced by energy of the particular source, e.g., dislocation half loops [18]. Critical thickness can also be calculated using the force-balance model first proposed by Jesser and Matthews [22, 23] and later developed by Matthews *et al.* [6, 16, 17], where the strain force is balanced by dislocation line tension at the critical thickness. In a more delicate way, other corrections such as the tension in a surface step and the Peierls-Nabarro friction force [24, 25] can be included [16, 21, 26]. Both the energy minimization and the force balance approaches are proved to be equivalent [27], and the well-known Matthews and Blakeslee (MB) model [16] to predict the thermodynamic limit of critical thickness has been verified by numerous experiments from a wide range of metal and semiconductor heterostructures.

It is worth to mention that the MB model gives a necessary condition for strain relaxation but is not sufficient. Early experiments revealed larger measured critical thickness values compared with the MB model. This was motivated by People and Bean [19] to propose the energy balance model in which the area strain energy density is equal to the area dislocation density which is converted by assuming a fixed distance between two neighbor MDs. The People and Bean (PB) model implies that the lattice mismatch can't be shared between partial elastic strain and nucleated MDs, and all the strain energy will be transformed to dislocation energy when the critical thickness is reached. This is not the case in practice, since the strain relaxes gradually with film thickness. In fact, Eq. 3.5 also predicts how the residual strain will proceed after the critical thickness is reached. Figure 3.1b depicts thickness dependence of area energy density of strain, MDs, and the sum of both. Above the critical thickness, the strain energy density monotonically decreases, while the MD energy increases and intersects with the strain energy density at 1.5 times of the critical thickness. Therefore, the lattice mismatched layer begins to relax as a result of system energy minimization before the two energy densities become equal. The discrepancies between the measured critical thicknesses and the predicted values from the MB model can be attributed to combined effects of limited experimental resolution and initial sluggish plastic relaxation [28]. The limited experimental resolution often leads to larger critical thickness values. Initial strain relaxation is usually measured by X-ray diffraction (XRD), photoluminescence (PL), transmission electron microscopy (TEM), and cathode luminescence (CL). Among these methods, TEM provides the best resolution limited by the scanning area resulting in a minimal resolution in 10^5 cm^{-2} if none of dislocations are observed from the specimen. The initial sluggish plastic relaxation implies that strain relaxation is a kinetic process and dislocation dynamics needs to be considered [20].

There are several identified sources for nucleation of MDs as shown in Fig. 3.2: existing TDs from a substrate, homogeneous and heterogeneous formation of dislocation half-loops [17]. Dislocation multiplication is not important at the initial relaxation but can be one of mechanisms when the MD density is high. An existing TD can be bent under stress forming a MD segment. However, the TD density from a commercial substrate is in 10^2–10^3 cm^{-2} and is too little to cause any detectable strain relaxation. Dislocation half-loops can be a possible source for initial nucleation of MDs. Calculations show that this mechanism will operate only for large lattice mismatch, typically 2–6% for InGaAs on GaAs, depending on the selected dislocation core values [29]. For such high lattice

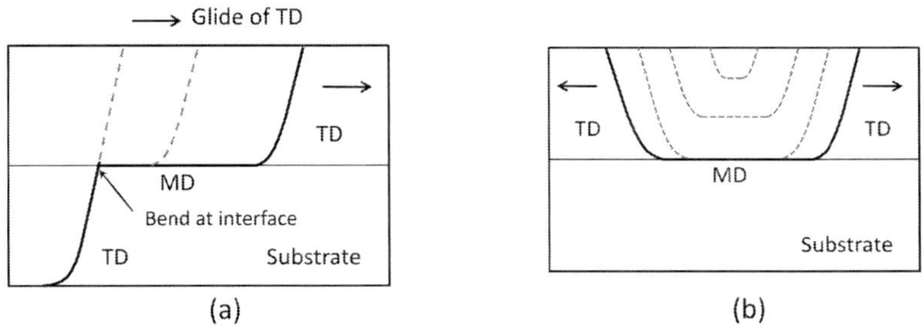

Figure 3.2. Schematic diagrams showing nucleation of a MD by (a) gliding of an existing TD and (b) gliding of a dislocation half-loop.

mismatch, initial strain relaxation may occur via formation of 3D islands rather than nucleation of MDs. Heterogeneous nucleation of half-loops is more likely to be operable as sources for nucleation of initial MDs. This can occur due to surface imperfections such as particles, impurities, surface steps, or damages, etc. The Hagen-Strunk type [30] dislocation multiplications can occur in a thin layer if two perpendicular MDs with the same Burgers' vector intersects. Fitzgerald [12] showed a hierarchy of different mechanisms due to different activation energy in the order of fixed TDs from a substrate, heterogeneous nucleation of half-loops, homogeneous nucleation of half-loops, and dislocation multiplications. For a detailed historical review of critical thickness and strain relaxation, we recommend an excellent review article by Fitzgerald [29].

Since MDs are formed by gliding of dislocation half-loops or existing TDs in diamond and zincblende semiconductors, the TD density is related to the average length of MD segments. For dislocation half-loops, one MD segment is connected with two threading components as shown in Fig. 3.2b, while for MDs caused by displacement of existing TDs, one MD segment results in one TD. Misfit dislocations are required to realize strain relaxation and usually have little influence on device performance if the device active layers are far away from MDs. However, TDs penetrate through the whole heterostructure and can trap free carriers, cause extra scattering, or act as sources of dark currents. It is imperative to eliminate or minimize the TD density for achieving good device performances. Dislocation geometry is also an important factor. In diamond or

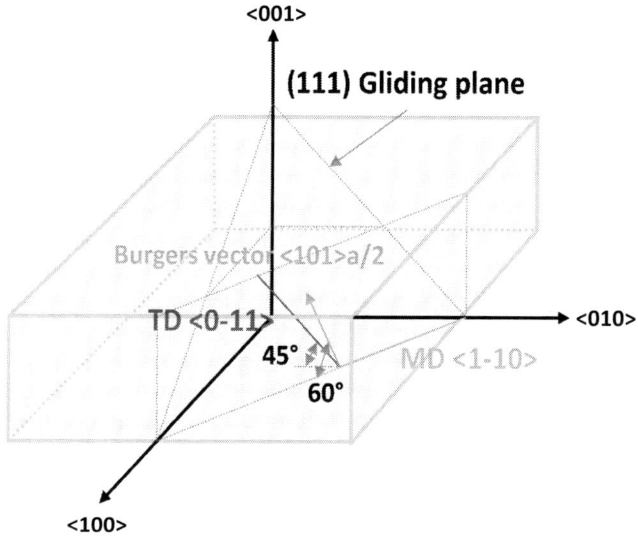

Figure 3.3. Schematic diagram showing geometry of an MD and a TD in a diamond or zincblende lattice mismatched heterostructure. See also Color Insert.

zincblende crystal structures, dislocations are usually 60 mixed type with a Burgers' vector of a $<\pm101>$ $a/2$ or $< 0 \pm 11 >$ $a/2$ gliding in the (111) plane as illustrated in Fig. 3.3. An important consequence of such geometry is that the termination point of a TD on surface moves away from the connected MD with increasing film thickness.

An ideal strain relaxation process shown in Fig. 3.2 is that if all the TDs can move to sample edges. This is rarely the case in practice due to dislocation interactions and pinning by impurities and crystal imperfection. Figure 3.4 shows schematics of different scenarios of dislocation interactions. A TD can be blocked by a perpendicular MD during gliding shown in Fig. 3.4a or simply across a perpendicular MD shown in Fig. 3.4b. Two TDs running in perpendicular directions can repel (Fig. 3.4c) or attract to annihilate (Fig. 3.4d). The threading components of two parallel MDs can also merge to form an edge dislocation if they are close enough (Fig. 3.4e) and the resulting edge dislocation is sessile. Figure 3.5 shows a TEM image where some of

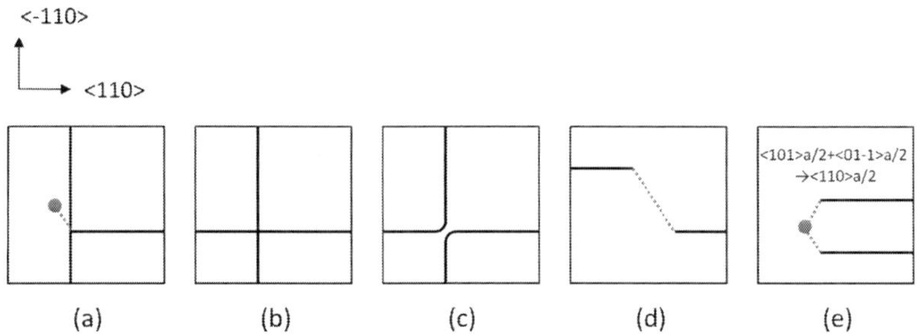

Figure 3.4. Schematic diagrams showing dislocation interaction (top view): MDs (solid lines) and TDs (dotted lines and circles).

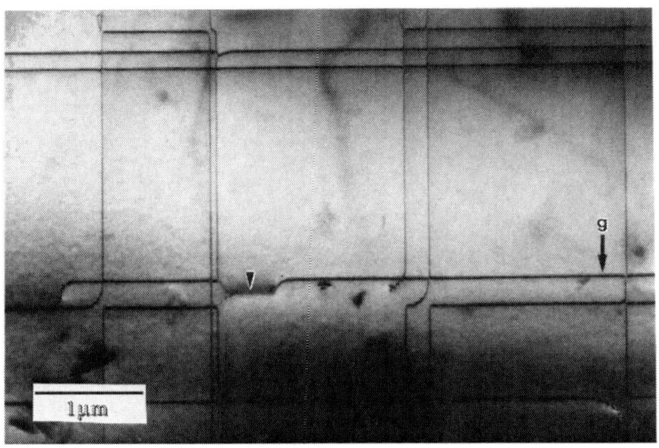

Figure 3.5. A planar-view TEM image of an InGaAs layer grown on GaAs. (Ref. [31], reprinted with permission from AIP).

the described scenarios are clearly observed [31]. Finally, TDs can also be pinned by impurities like dopants.

The TD density can be reduced as a result of dislocation interactions if two TDs are within an interaction distance. The inclined geometry of 60° mixed type TDs enables TDs to move with film thickness, and two TDs originating from two initially far separated MDs may have chances to interact when the film thickness is large. Sheldon *et al.* [32] and Tachikawa and Yamaguchi [33] found that the TD density monotonically decreases with film thickness as shown in Fig. 3.6. Three regions are identified with different physical

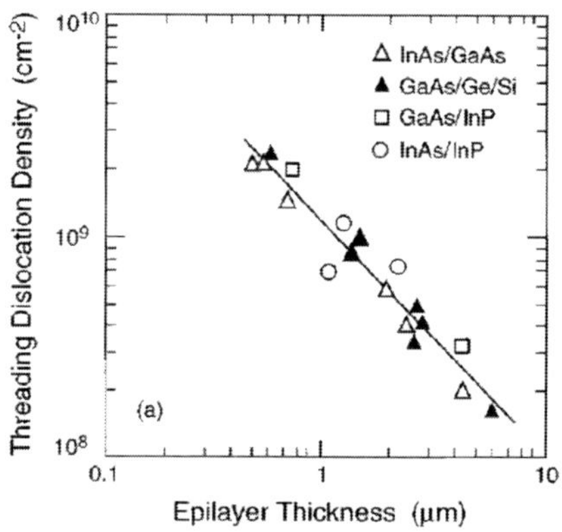

(a)

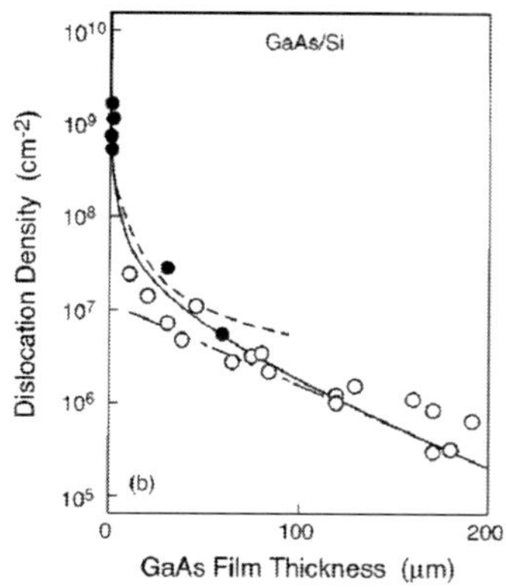

(b)

Figure 3.6. Reported experimental data of the TD density as a function of film thickness. (Ref. [32, 33], reprinted with permission from AIP).

mechanisms governing the TD density: entanglement region, $1/h$-dependence region, and exponential decay region. For a very thin film typically below 0.5 μm, there is a dense TD density in the range of 10^9 cm^{-2} or higher. In this entanglement region, new MDs are nucleated continuously, and residual strain quickly decreases generating a large amount of TDs. For film thickness in the range of 1–10 μm, the TD density inversely decreases with film thickness ($1/h$-dependence). For higher thicknesses, the TD density decreases exponentially. This film-thickness-dependent TD density was theoretically modeled [33] and numerically simulated [34]. Assuming a thickness-dependent TD density, $D(z)$, there are two contributions for the TD reduction: annihilation via dislocation interactions (proportional to $D^2(z)$) and TD deflection by impurities/imperfections or moving to edges (proportional to $D(z)$). Thus, we can write

$$\frac{dD(z)}{dz} = -\alpha D(z) - \beta D^2(z) \tag{3.6}$$

where α and β are two constants related to the two mechanisms, respectively. For a medium thickness, annihilation via dislocation interactions (the second term on the right side of Eq. 3.6) is dominant, and the solution shows $D(z) \sim 1/h$. For very thick layers, the TD density is so low, typically around 10^7 cm^{-2}, that the probability of dislocation interaction is negligible. The first term on the right side of Eq. 3.6 becomes important, and the solution shows an exponential decay. However, the TD reduction in this region is ineffective. For GaAs/Si, the TD density is decreased from 2.2×10^7 cm^{-2} at 10 μm to 7×10^6 cm^{-2} at 60 μm and to 4×10^5 cm^{-2} at 180 μm [30]. Such an inefficient strain relaxation by simply increasing film thickness calls for novel techniques to effectively suppress the TD density. The above model does not account for the lattice mismatch. A correction made by Ayers [35] showed similar $1/h$ behavior and a sub-linear dependence on the lattice mismatch without using fitting parameters $D(z) \sim f/\ln(1/4f)$.

3.3 Dislocation Reduction by Reduced Area Epitaxy

Luryi and Suhir [10] first studied critical thickness of a lattice mismatched layer grown on a mesa strip with a limited width using the energy balance approach similar to the PB model. The stress induced by a dislocation has an exponential decrease with the distance from the dislocation line. In a mesa strip, sidewalls allow for free elastic crystal deformation resulting in zero stress. The area strain energy density (integration over the film thickness) thus depends on the strip width. For a small width, the overall strain energy decreases and the critical thickness increases. When the strip width is below a critical value, the overall strain energy is so small that nucleation of MDs will be energetically unfavorable for any thickness, and the critical thickness will be infinitively large. Figure 3.7 shows calculated half of the critical strip width as a function of Ge content in Ge_xSi_{1-x}/Si. Although they used the energy balance approach to derive the critical thickness from a strip, the qualitative conclusion is correct even for a mesa, i.e., for certain lattice mismatch, it is possible to grow an infinitively thick layer without relaxation if the mesa size is small enough.

Fitzgerald *et al.* [11] demonstrated that the TD density in $In_{0.05}Ga_{0.95}As$ grown on GaAs reduces with decrease of the mesa size: from >5000 cm^{-1} for large growth area to zero for a 25 μm mesa. They argued that for small lattice mismatch (<2–6%) where homogeneous nucleation of dislocation half-loops and dislocation

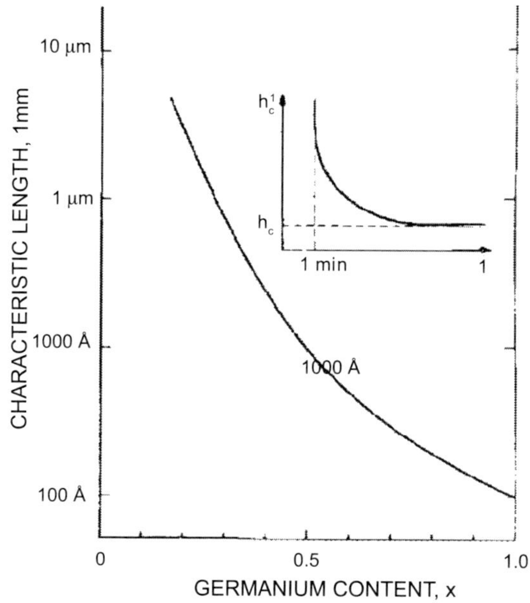

Figure 3.7. Critical strip width to reach an infinitive critical thickness as a function of Ge content in Ge_xSi_{1-x}/Si, the inset shows the critical thickness as a function of strip width. (Ref. [10], reprinted with permission from AIP).

multiplications are insignificant during initial strain relaxation, the fixed sources for MD nucleation is proportional to the growth area [13]. A small growth area generates a small amount of MDs and can tolerate a large amount of TDs. Growth on a mesa substrate not only increases the critical thickness but also enables TDs to glide toward sidewalls and thus reducing the TD density. Zhang *et al.* [36] proposed a quantitative model using the force balance approach to account for this phenomenon. For a TD segment with a distance, r, from the sidewall in a film grown on a mesa, there is an attractive image force from a mirror TD segment at $-r$ with an opposite Burgers' vector. This attractive force can cause the TD moving toward the sidewall leaving an MD segment, if it overcomes the dislocation line tension. As seen in Eq. 3.3, the line tension has a logarithmic form, and the h in Eq. 3.3 should be the minimum value between the film thickness and half of the mesa size. This implies that the line tension is either a constant or increases slowly with film thickness by $\ln(h)$. Since the image force is proportional to $1/r$, TDs close to sidewalls can easily glide to mesa edges. Critical thickness and critical mesa size are defined when the image force for the TD at the middle point of the mesa balances the line tension. Zhang *et al.* [36] considered an array of MDs that completely relaxed the film and replaced the h in Eq. 3.3 by half of the distance between two neighbor MDs leading to the critical mesa size, L, by

$$L < \frac{16h}{\ln(1/4f)} \tag{3.7}$$

The result indicates that for large lattice mismatch, the critical mesa size also becomes large, which is opposite to the predication by Luryi and Suhir [10].

Finally, the inclined geometry of 60° mixed dislocations implies that all TDs will eventually terminate at sidewalls if the film is thick enough, i.e., the ratio of thickness/mesa size is larger than a value depending on geometry. This technique is referred as epitaxial necking [15].

3.4 Technical Implementation of Patterned Substrate Epitaxy

In this section, different techniques to achieve patterned substrate growth will be reviewed. We categorize them into six variations in three different groups as sketched in Fig. 3.8. All of them employ bottom-up approaches starting with patterning on a large substrate before epitaxy. Both CVD and MBE have been used but growth mechanisms, dislocation nucleation, and dynamics and resulting film quality are very different.

3.4.1 *Epitaxial Lateral Overgrowth and Micro-Channel Epitaxy*

Micro-channel epitaxy includes both vertical and lateral crystal growth. The latter is a type of ELO that is of technological significance to form large area films. The name of MCE is coined by Nishinaga [37] and widely used by researchers in Asia. In this

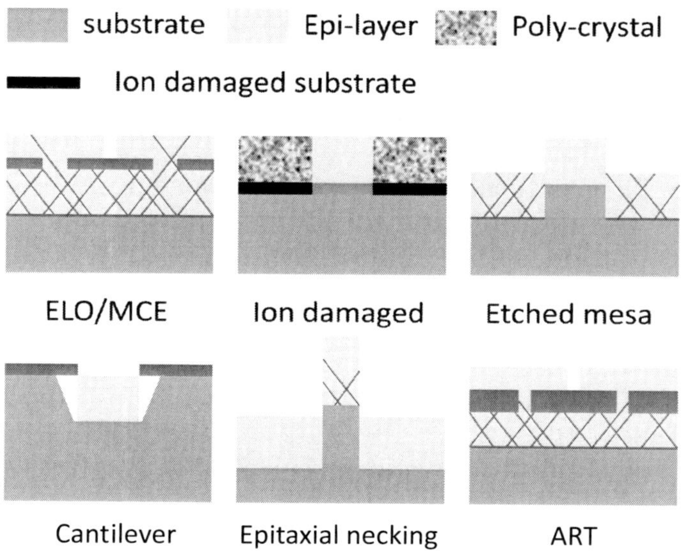

Figure 3.8. Different technical approaches to realize patterned substrate epitaxy. The orange color represents dielectric masks, while the red and blue lines are TDs and MDs, respectively. See also Color Insert.

technique, a relaxed buffer layer containing dislocations is first grown on a lattice mismatched substrate. Dielectric patterns, often SiO_2 but also SiN [38], are formed by lithography and etching, leaving substrate windows exposed for epitaxy. The window size can be in micrometers for MCE and in nanometers for NCE, and the window geometry can be quite innovative as will be shown below. Epitaxy proceeds first in the window region and then crystals grow laterally over the dielectric patterns and coalesce eventually to form a large area thin film. This type of ELO can be easily achieved by CVD and liquid phase epitaxy (LPE). For MBE, if the size of dielectric patterns is in sub-micrometers or larger, material deposition occurs on both substrate windows and dielectric patterns forming single crystals and amorphous/poly-crystals, respectively. Although the crystal-amorphous (or poly-crystal) boundary offers dislocation sinks to improve epi-material quality, the ELO is generally forbidden. The diffusion length of Ga on SiO_2 was found to be about 200 nm at 600°C [39] and 230–290 nm at 630°C [40], which is much smaller than the diffusion length on GaAs. Therefore, selective and lateral overgrowth in conventional MBE can occur only when the dielectric pattern size is below 200 nm. This poses technical challenges for cost-effective lithography technique to form such nanoscale dielectric pillars. If incident beams have small angles with respective to the sample surface in MBE, lateral overgrowth is possible [41–43]. A conventional MBE system can't meet this requirement, and a special design modification is necessary.

MCE can effectively block most TDs. Existing TDs from buffer underneath will be absorbed if they meet the sidewalls of dielectric masks or the crystal/poly-crystal boundaries. Only a small amount of TDs can penetrate through the opened windows depending on the window size. Newly generated TDs can glide to the sidewalls and boundaries. The overgrown region usually contains few dislocations, and it is this region that is of interest for making devices. For coalesced large area films, crystal defects are often formed in the coalesced regions and proper design is necessary to eliminate such defects as discussed below.

The first demonstration aiming at achieving a low TD density was reported by Tsaur *et al.* from Lincoln Laboratory in 1982 [9]. They first deposit 0.3 μm Ge on (001) Si using electron beam evaporation at 450–500°C followed by a GaAs layer grown by CVD. A growth mask of 0.1 μm-thick SiO_2 is then fabricated by lithography with 3–5 μm opening and 30–50 μm apart. In total, 2-4 μm-thick GaAs has been deposited on window strips along different crystal orientations. It is found that the maximal lateral overgrowth occurs along the direction 25° from the <110> crystal direction. No dislocations are observed in the overgrown region by TEM. They estimate an upper limit of TD density of 10^4 cm^{-2} compared with 10^7–10^8 cm^{-2} from a reference GaAs/Ge/Si sample.

Nishida *et al.* reported growth of SiGe on Si with SiO_2 windows from 1×1 μm^2 to 100×100 μm^2 using MBE [44]. They showed that MDs in 150 nm-thick $Si_{0.8}Ge_{0.2}$ on Si are eliminated when the mesa size is reduced to 2.5×2.5 μm^2. Figure 3.9a shows the linear TD density as a function of square window size for 150 and 75 nm-thick $Si_{0.8}Ge_{0.2}$ on Si. The TD density linearly increases with the window size up to about 7 μm and saturates afterward. They also performed Raman scattering spectroscopy and deduced position

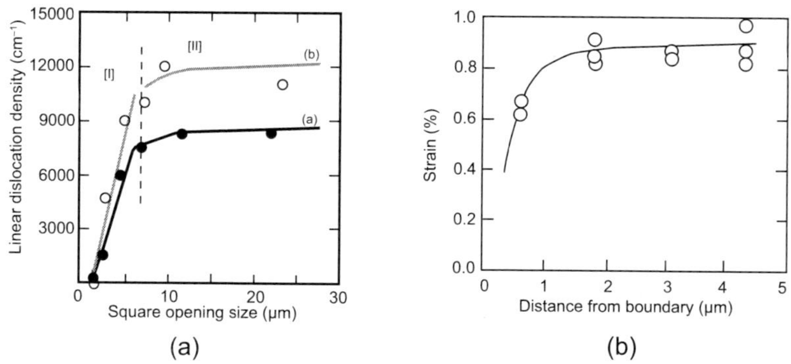

Figure 3.9. (a) Linear TD density as a function of square SiO_2 window size for 150 nm (thick line) and 75 nm (thin line)-thick $Si_{0.8}Ge_{0.2}$ films; (b) Residual strain distribution deduced from Raman scattering from 150 nm-thick $Si_{0.8}Ge_{0.2}$ film grown on a pattern with a size of $10 \times 10\ \mu m^2$. (Ref. [44], reprinted with permission from AIP).

dependent residual strain inside the 150 nm-thick $Si_{0.8}Ge_{0.2}$ on a $10 \times 10\ \mu m^2$ window as shown in Fig. 3.9b. The residual strain close to the boundary is found smaller than that in the middle of the film as expected.

In a conventional MCE approach, the window opening is usually smaller than or similar to the window separation. The useful area is small for device applications and the lack of surface planarity makes device processes rather complicated. Hull *et al.* presented an innovative idea using a 2D array of small SiO_2 pillars [45]. The pillars have a small diameter of 1–2 μm and slightly off-line from the <110> direction such that the two neighbor pillars overlap if viewed from the <011> direction, see the sketch in Fig. 3.10. A crystalline film is grown in areas between the pillars. A dislocation nucleated in the film will propagate and eventually be terminated by one of the pillars. Two sets of $Si_{1-x}Ge_x$ ($x = 0.15 - 0.2$) films with varied thickness and 84% and 96% wafer surface uncovered, respectively, were grown by MBE. The TD density is in the order of 10^5 cm^{-2} in as-grown samples which is about 10–100 times less than the same film grown by a conventional way. Post growth thermal annealing at 950°C induces more TDs but still keeps 10–100 times less in TDs compared with reference samples under the same

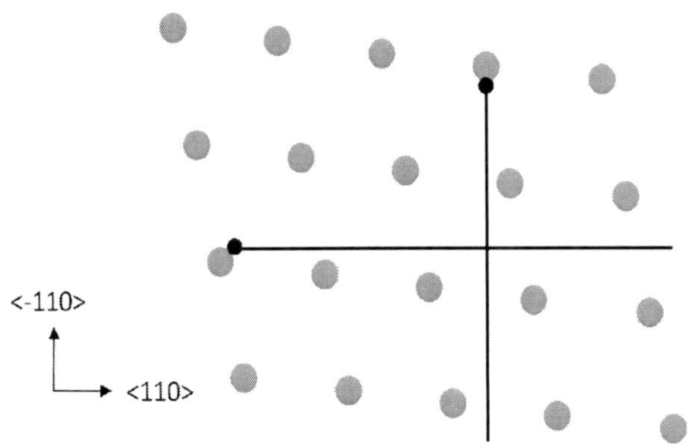

Figure 3.10. Schematic showing a set of slightly misaligned dielectric array pillars (gray circles) used to block dislocations (TDs shown by black dots).

annealing condition. They also tried to achieve a large planar film by growing thick layer. However, poly-crystals are formed on the pillars due to relatively large sizes. If the pillar size is small enough and growth temperature is high, it might be possible to have selective growth only on the opened window. In this case, the minimum TD density is limited by the pillar density.

It is possible to create selective regions for epitaxy without physically forming dielectric patterns that are non-planar in geometry. Watson *et al.* [46] suggested using ion implantation to create substrate windows. They used 300 keV Xe ions with a dose of 10^{16} cm^{-3} to damage exposed GaAs using a SiO_2 mask for protection. The damaged depth is estimated to be about 100 nm that is thinner than the SiO_2 mask. After removing the mask, epi-growth proceeds on a planar but square patterned GaAs surface with a size ranging from 25 to 800 μm. $In_{0.05}Ga_{0.95}As$ layers of 350 nm thickness were grown at 400 and 500°C, respectively, by MBE. The same structures were also grown on etched mesas for comparison. Figure 3.11 shows CL images of samples grown at 400°C on ion-damaged patterns along the <110> direction and on etched mesas along the <100> direction, respectively. Strain relaxation occurs predominant along the <1-10> direction with a linear MD density larger than 10^4 cm^{-1}. The MD density is less in large non-patterned area compared with that in

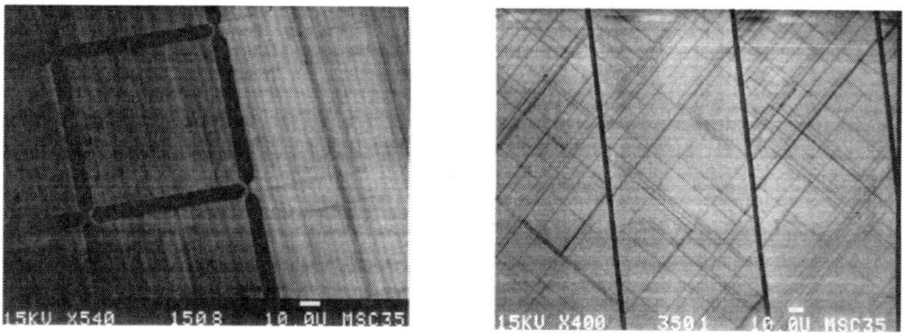

(a) (b)

Figure 3.11. CL images of 350 nm $In_{0.05}Ga_{0.95}As$ grown on GaAs using (a) ion-damaged patterns along the <110> direction and (b) etched mesa along the <100> direction. (Ref. [46], reprinted with permission from Springer).

the patterns surrounded by ion-damaged channels, suggesting that the boundary provides sources for nucleation of dislocations. Strain relaxation is more symmetric with an average linear MD density of 800 cm^{-1} along both <110> and <1–10> directions when the channels are along the <100> direction. Structures grown on etched mesas show similar results. When the growth temperature is increased to 500°C, the average linear MD density from an ion-damaged sample is 5000 cm^{-1} and 1500 cm^{-1} along the <1–10> and the <110>, respectively, and is 5000 cm^{-1} and 500 cm^{-1} along the <1–10> and the <110>, respectively, from an etched mesa sample. MDs are also observed in the ion-damaged channels, and some MDs from one square run through a channel and are extended to the neighbor square indicating a re-crystallizing process in the channels. Majority MDs are 60° mixed type but there are still about 5–10% pure-edge-type MDs. The asymmetric strain relaxation is explained by α- (along the <1–10> direction) and β-dislocations (along the <110> direction) that have gliding velocities differed by several orders of magnitudes (α-dislocations move faster than β-dislocations). Once a dislocation is nucleated, it can glide toward mesa edges and terminated at ion-damaged channels or mesa sidewalls. However, different MD densities as shown in Fig. 3.11 when comparing square size and crystal directions indicate that nucleation source is a crucial factor and it depends on crystal direction. Both ion-damaged channels and sidewalls can act as nucleation sites. When mesas are formed with <110> and <1–10> boundaries, it is easy to nucleate α-dislocations. The fact that more symmetric strain relaxation but a less average MD density found in Fig. 3.11b implies that nucleation of β-dislocations is enhanced while nucleation of α-dislocations is suppressed when mesas are formed along the <100> direction.

Both MCE and ELO have been investigated in a variety of materials and by different epitaxial technologies [37]. They provide means of growing high-quality heterostructures with large lattice mismatch. In ELO, material quality is in general superior in the region above the mask than in the region above the opening as shown in Fig. 3.8. As example, Kazi *et al.* [47] fabricated 850 nm GaAs/AlGaAs quantum well lasers on Si from the two different regions using ELO. The laser fabricated in the region above the mask revealed a lower threshold current density (0.81 vs. 1.17 kA/cm^2), a higher differential quantum efficiency (42% vs. 36%), and a longer lifetime compared to the counterpart fabricated in the region above the opening.

An important issue is the coalescence of laterally overgrown crystals to form a large area film. This has been extensively investigated in growing SOI by CVD or LPE [4, 48–50] and GaAs and InP by LPE [51, 52]. There are two modes of coalescence. If the coalescence proceeds at one point through the whole process, the two crystals will be closed by a zipper without any nucleation of dislocations. If the coalescence proceeds at two or more points, closed open areas will be formed. The open areas will diminish with crystal growth and eventually form dislocations or other types of crystal defects. Figure 3.12 shows one example of ELO growth of InP [52]. The etch pits are only observed from the seed region, and the laterally overgrown InP layer within the triangle shows no crystal defects. Since lateral overgrowth depends on facet direction, it is important to design right pattern geometry for ELO to eliminate crystal defects from a large coalesced film [52].

3.4.2 *Reduced Area Growth on Etched Mesas*

In MCE or ELO, the reduced growth area is defined by dielectric mesas or ion-damaged surface. Growth on an etched mesa is

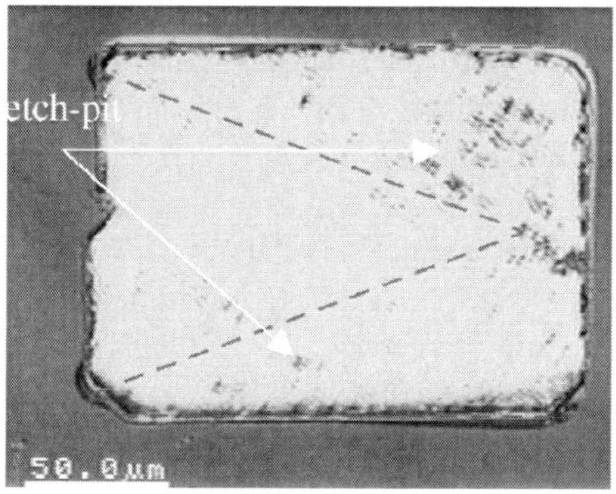

Figure 3.12. Photograph showing etch pits from the seed region. The lateral overgrown InP within the triangle reveals no crystal defects. (Ref. [52], reprinted with permission from Elsevier).

another way to realize reduced area growth. Here, micrometer-scale etched mesas are formed by photolithography followed by wet or dry etching separated by trenches with an adjustable mesa height. Epitaxial growth takes place on both mesas and trenches and the resulting film is non-planar. Mesa edges and sidewalls act as dislocation sinks for dislocations nucleated within the mesa and can also block dislocations from trenches [53]. The small number of fixed dislocation sources due to a limited growth area, the short distance for a TD to glide to mesa sidewalls, and thus the reduced probability for dislocation multiplication ensure that a lattice mismatched layer grown on such a mesa substrate will contain no or fewer MDs and TDs compared with that grown on a large size substrate.

Fitzgerald *et al.* first demonstrated this scheme by growing 350 nm $In_{0.05}Ga_{0.95}As$ on GaAs [11] using MBE. They subsequently studied $In_{0.05}Ga_{0.95}As$ with different thicknesses grown on GaAs with varied initial TD densities [12] and $Si_{1=x}Ge_x$ $(0 < x < 0.2)$ on Si [14]. MDs are observed and quantified by CL. The general findings are:

(a) The MD density along the two <110> directions are different indicating asymmetric strain relaxation. This is explained by the fact that α-dislocations glide much faster than β-dislocations, resulting in a higher linear MD density along <110> than that along <1–10>.

(b) The MD density scales with the mesa size linearly up to a certain mesa size. For small mesas, e.g. 350 nm $In_{0.05}Ga_{0.95}As$ grown on a 67 μm GaAs mesa or 280 nm $Si_{0.81}Ge_{0.19}$ grown on a 70 μm Si mesa, MDs are completely eliminated.

(c) They examined linear dependence of the MD density on the mesa size for substrates with different initial defect densities (10^5 to 10^2 cm^{-2}). The slope reflects area dislocation sources from the mesa while the intersection at zero mesa size accounts for dislocation nucleation sites from edges. For thin films grown on high initial defect density substrates, the line passes though the original point. The agreement between the slopes and the initial defect densities in a mesa in the range of 10^4–10^5 cm^{-2} led them to conclude that defects from the substrate (presumably pre-existing TDs and inhomogeneities) are the sources for nucleation of MDs. For thick films or growth on low initial defect density substrates, the line intersects with the coordinate, indicating that heterogeneous nucleation sites on edges are additional sources for nucleation of dislocations. The observed linear dependence of the MD density on mesa size excludes the dislocation multiplication mechanism that should otherwise result in a super-linear dependence.

These results imply that critical thickness for layers grown on small mesas with a low surface defect density can be enhanced by many folds of the equilibrium value predicted by the MB model and both MDs and TDs can be significantly reduced or even eliminated as compared with grown on a largesize substrate. Etch pit density (EPD) in commercial substrates is in 10^2–10^3 cm^{-2}. This implies that the average EPD for a square mesa with a size below 30 μm is less than 1 and heterogeneous nucleation sites on edges are most likely dislocation sources. Jones *et al.* [54] reported that not only the mesa size but also the mesa orientation has a strong impact on nucleation of MDs. They studied crosshatch patterns from 290 nm-thick $In_{0.08}Ga_{0.92}As$ grown on 100 μm GaAs mesa strips with different orientations using MOVPE. All the samples were annealed at 700°C for 1 hour to avoid metastable

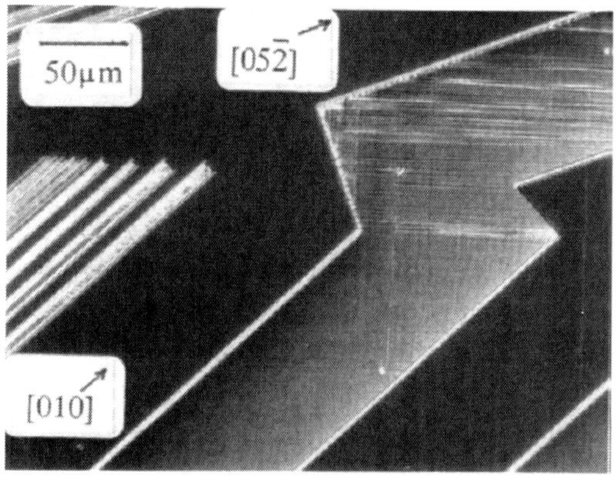

Figure 3.13. Photograph showing intersection of two mesa strips along the [05$\bar{2}$] and the [010] directions, respectively. Crosshatch patterns are clearly observed along the [05$\bar{2}$] but not along the [010]. (Ref. [54], reprinted with permission from Springer).

films. Crosshatch patterns are formed as a result of strain-induced local growth rate variation by the underneath grids of MDs and therefore reflect the network of MDs at interface. Heavy crosshatch patterns are observed on films grown on the un-patterned substrate while films grown on patterned strips showed varying degrees of crosshatch patterns. For mesa strip along [010] direction, however, no crosshatch patterns are observed. This conclusion is further substantiated by the image shown in Fig. 3.13 where two mesa strips with different orientations are merged and there is a strong demarcation between the heavily crosshatched area along the [05$\bar{2}$] direction and the crosshatch free area along the [010] direction. Since the GaAs substrates used contains an EPD density of 200 cm^{-2}, they postulated that heterogeneous nucleation sites on strip edges are dislocation sources to nucleate dislocation quarter-loops. Each TD section glides over a short distance and is terminated at the opposite side of the mesa strip. A mesa strip aligned with a certain orientation forms specific facets at the mesa edges. The activation energy for nucleation of dislocation quarter-loops on edge facets depends on facet orientation, surface step density and edge geometry etc. If the activation energy is large enough, nucleation of dislocation quarter-loops is forbidden and no crosshatch patterns are formed. It is interesting to compare this experiment with the one by Watson *et al.* [46] using ion-damaged patterning technique described above. In the latter case, strain relaxation is symmetric with MDs observed along the two orthogonal <110> directions when the patterned channel direction is along the <100>, see Fig. 3.11. The main difference is that in Watson's case, the InGaAs films are surrounded by poly-crystals deposited on ion-damaged channels without exposure of sidewalls, while in Jones' case, facets are formed at strip edges

There are several commonly used growth techniques to reduce TDs grown on large area blanket substrates such as step and continuous alloy grading [55–57], use of strained or strain compensated superlattice to filter out TDs [58–60] and post-growth thermal cycling [61, 62]. These techniques can also be combined with growth on patterned substrates to further improve material quality [63, 64]. Beam III and Kao reported growth of $In_xGa_{1-x}As$ on GaAs mesas with x up to 0.53 using both alloy grading and strained superlattices [63]. Insertion of a strained superlattice can reduce TDs for small lattice mismatch (about three times reduction up to $x = 0.2$) where the TD density is low and becomes ineffective for large lattice mismatch. Alloy grading was found more effective in TD reduction grown on both blanket and mesa substrates. For small lattice mismatch, large TD-free areas up to 20–30 μm wide bounded by heavy TD-cluster regions can be found in TEM images from a reference sample grown on a blanket substrate. This size is approximately the same as an average TD gliding length before interacting with other dislocations. The same structure grown on mesas with a smaller size than 20–30 μm would result in TD-free films and this was truly demonstrated by the authors. The step grading can work up to $x = 0.21$ while the linear grading can work for even higher In contents up to $x = 0.25$ but becomes ineffective for $x = 0.53$. Even in this case, the TD density is less than in the same structure using a step grading. TD-free $In_{0.25}Ga_{0.75}As$ layers grown on $30 \times 30\ \mu m^2$ GaAs mesas using a 1.5 μm-thick linearly graded InGaAs buffer were demonstrated.

There are several drawbacks for growth on etched mesas described above. First, etched mesa edges are often faceted and the growth rate is different on mesas and on facets leading to changes of crystal morphology on edges. Second, the material area is in principle defined by the mesa size. It is difficult to achieve large area templates like ELO. Third, the non-planarity of the final structure significantly increases complexity for device process. One solution to the last drawback is to selectively grow a lattice mismatched layer in trenches with a controlled thickness. This requires a shadow mask to select the growth area. Mechanical masks were invented in early days of development of MBE equipments to achieve selective growth [65] but the technique is impractical and not flexible. Wet etching of III–V compounds results in undercut in only one crystal orientation due to the anisotropic etching nature. Cantilever type shadow masks are more flexible to implement as shown in Fig. 3.8. One example is to use GaAs/AlGaAs [66]. After lithographic patterning and removal of GaAs, AlGaAs can be selectively etched to form a cantilever shadow mask. Another method is to pattern a surface, e.g, GaAs with dielectric films like SiO_2 and then selectively etch GaAs to form a cantilever shadow mask [67]. Using such a shadow mask, it is possible to selectively grow mismatched layers in trenches [63].

3.4.3 *Epitaxial Necking and Aspect Ratio Trapping*

Epitxial necking makes use of geometry of TDs as show in Fig. 3.3. For an inclined TD that is not parallel with the growth direction, it will eventually end at sidewalls of an etched mesa or at the mask/film interface if the ratio of the film thickness over the size of growth area is large enough. This concept was introduced by Fitzgerald and Chand [15]. TDs in diamond and zincblende semiconductors with small lattice mismatch are 60° mixed type that has an inclined angle of 45° from the $<001>$ direction. If the film thickness exceeds the lateral seed size, all TDs will be terminated at sidewalls. Langdo *et al.* reported growth of Ge on 4 inch Si (100) wafers with 120 nm-thick SiO_2 patterns using epitaxial necking by CVD [68]. 100 nm holes are formed by interferometric lithography followed by reactive ion etching. Ge films grow first inside holes and then laterally overgrown on the oxide and finally coalesce. TDs are found to be terminated at Ge/oxide interface without being deflected forming zigzag defects. The areas above nano-holes are free of dislocations. Crystal defects, predominately microtwin defects as confirmed by TEM, are found from the coalesced areas, which is common also in coalesced Si films by selective area growth. There might be other mechanisms to trap TDs at the edges in addition to epitaxial necking. The stress in Ge inside nano-holes might be different due to the edge effect. The attractive dislocation image force can also pull TDs towards the edges.

The concept of epitaxial necking was re-named as aspect ratio trapping (ART) recently [69]. Careful TEM investigations on Ge grown on Si using ART uncover that Ge crystals form facets during the growth, initially on (-113) and gradually changing to (−111) [70]. TDs are found to be approximately perpendicular to the facet planes. With the increase of the facet angle from (−113) to (−111), TDs are bent toward the Ge/SiO_2 interface. ART was also employed to grow GaAs on Si [71, 72] using MOVPE. TDs can be effectively suppressed by ART, but heavy crystal defects like microtwins exist at the coalesced area as shown in Fig. 3.14a [72]. These coalesce defects are not related to GaAs quality inside trenches but are very sensitive to growth conditions during the overgrowth on SiO_2. Lower growth pressure and higher growth temperature improve selective growth morphology [73], while higher V/III ratio and lower growth temperature facilitate to reducing microtwin defects [74]. Based on these findings, a two-step growth strategy is used, and the resulting coalesced 1.5 μm GaAs shows a very low defect density with an XRD FWHM of 190 arcsec as compared with 300 arcsec from the

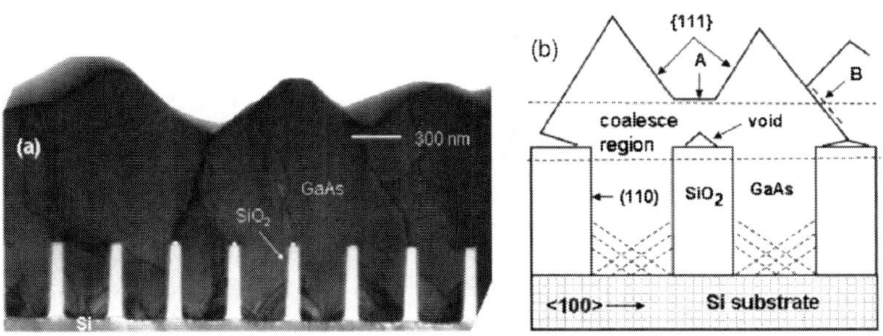

Figure 3.14. (a) Cross-sectional TEM image showing a rough surface and coalesce defects. (b) Schematic growth model of ART. (Ref. [72], reprinted with permission from AIP).

optimized GaAs film grown on a blanket Si. Ge and GaAs films grown on Si by ART are usually very rough. Chemical mechanical polishing is used for further growth of GaAs on Ge/Si [75] and for device applications [76]. It was reported that a 1 μm-thick GaAs grown on polished Ge/Si using ART showed an XRD FWHM of only 140 arcsec [75].

3.5 Future Outlook

Integration of III-Vs and Ge on Si substrates has received renewed interest recently motivated by pushing the limit beyond Moore's law in microelectronics. According to the international technology roadmap of semiconductors, high mobility channel materials are one of the potential solutions beyond the 22 nm CMOS technology. III-Vs and Ge are identified as potential candidates of high mobility channel materials due to their superior mobility for electrons and holes, respectively. Epitaxial lateral overgrowth combined with ART has been recognized as an efficient tool to harness lattice mismatch between III-Vs/Ge and Si. The major challenges are microtwin defects at the coalesced regions and rough surface for GaAs/Ge on Si, which are very sensitive to growth conditions, as well as the growth selectivity and initial nucleation within growth windows for III-Vs other than GaAs. A few examples are given below.

Araki *et al.* studied growth selectivity and initial seed nucleation of InP on Si using MCE by MBE [77]. The growth selectivity is achieved when the growth temperature is 500°C or higher. However, multiple InP seeds are formed in holes which are deleterious to form crystal defects after merging together. The square hole size needs to be reduced to $0.7 \times 0.7\ \mu m^2$ to achieve one InP seed per hole. Deura *et al.* reported detailed study of growing InGaAs on (111) Si using MCE by MOVPE [78]. Plural InGaAs seeds are nucleated in 2 μm wide holes, while single seed is formed in each hole for pure InAs. The lateral overgrowth depends on the Ga content, see Fig. 3.15. For InGaAs with a small Ga content, micro pillars are grown. These pillars contain rotational microtwins, which are believed to come from layer-by-layer growth on the (111) plane according to the authors. Increasing Ga facilitates lateral overgrowth as shown in

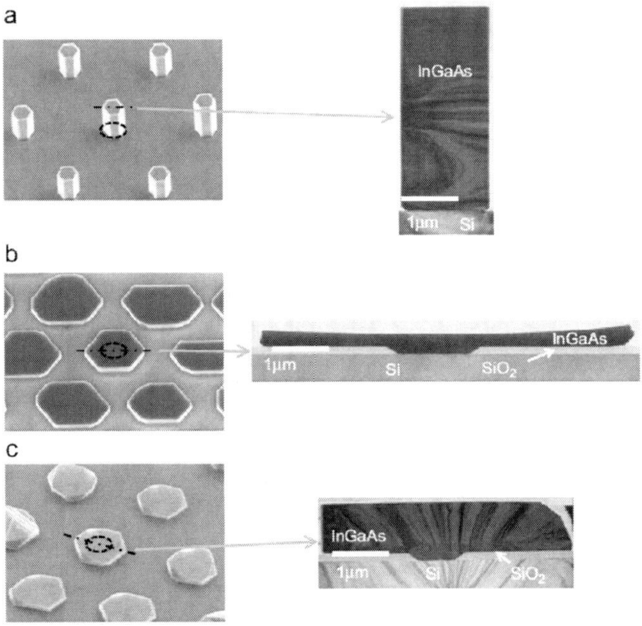

Figure 3.15. SEM images showing different morphology of InGaAs grown on 2 μm wide patterned Si holes. The Ga content increases from (a) to (c). (Ref. [78], reprinted with permission from Elsevier).

Fig. 3.15b. Further increasing Ga increases growth rate on both (111) and on side facets and also suppresses rotational microtwins. These results help authors propose a four-step growth scheme at different growth stages to optimize uniformity and crystal quality. They demonstrated a top 50 nm twin-free layer from a 380 nm-thick InGaAs film with a lateral size of 7.1 μm using the optimized growth recipe.

ART discussed above requires that the ratio of the height over the width in a trench should be larger than 1 for 60° mixed TDs to effectively trap all TDs at the interface. Yu *et al.* recently demonstrated that use of hydrogen annealing can relax this condition when growing Ge on patterned Si with 500 nm wide growth windows and also lead to a very smooth coalesced Ge surface [79]. After standard lithography and etching, a hydrogen bake at 1000°C was performed to remove all the residual oxides in the growth windows followed by a very thin Si buffer deposited at 700°C. The initial Ge layer was grown at 400°C at which the growth rate is 30 nm/min and 3 nm/min on the (100) and the (113) facets, respectively. This leads to formation of (113) facets at the growth front. The sample was then annealed at 825°C for 30 min under a hydrogen flow. After coalescence, the growth temperature is raised to 600°C for growth of 15 min, and the growth rate increases to 60 nm/min and 12 nm/min for the (100) and the (113) facets, respectively. This enhances filling in the valleys formed by neighbor (113) facets. The sample is again annealed at 825°C under hydrogen exposure. The total Ge thickness can be further increased by growth at 600°C. According to authors, use of hydrogen annealing modifies surface reconstruction and increases Ge surface diffusion, which are crucial for reduction of the TD density and enhancement of surface smoothness [80]. Defects are found only from the first 60 nm-thick

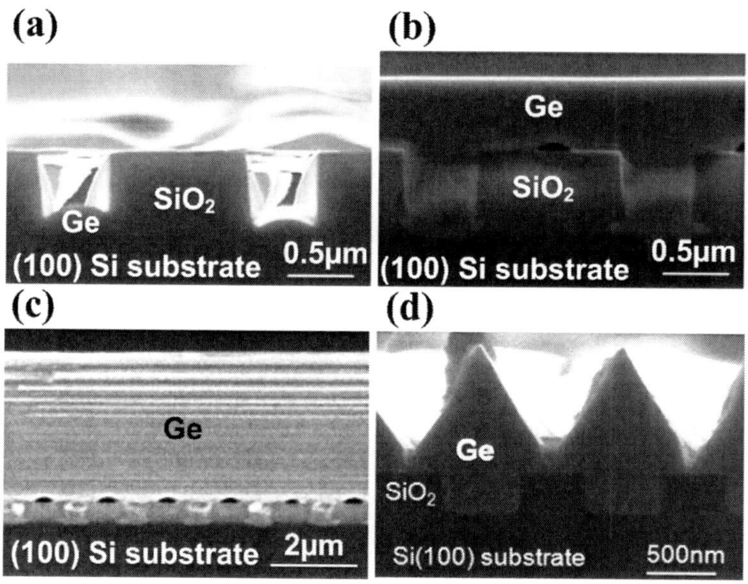

Figure 3.16. Cross-sectional SEM images of (a) initial Ge films grown at 400°C, (b) Ge films further grown at 600°C for 15 min, (c) Ge films further grown at 600°C for 35 min, and (d) Ge films grown at 400°C for 1 h. Hydrogen annealing is performed in (a)–(c) but not in (d). (Ref. [79], reprinted with permission from AIP).

Ge above the interface, and a smooth surface with a root-mean-square roughness value of 0.7 nm is demonstrated as compared with 3.5 nm from the sample without hydrogen annealing. However, the TD density is estimated to be about 4×10^6 cm^{-2}. Figure 3.16 shows cross-sectional SEM images of the Ge films at different stages with and without hydrogen annealing.

In general, patterned substrate growth is a powerful epitaxial method to harness lattice mismatch and is unlimited for innovations in the future. It can provide good quality lattice mismatched crystals with a limited size. TDs can be significantly reduced or eliminated depending on the pattern size and geometry. However, the residual strain is often non-uniform over the surface, in particular for small size crystals, and the resulting surface can be rough. It is still challenging to achieve a large-size coalesced films up to centimeter or larger with few coalesce defects and smooth surface.

References

1. Jopyce, B. D. and Baldrey, J. A. (1962). Selective epitaxial deposition of silicon, *Nature*, **195**, pp. 485–86.
2. Tausch, F. W., Jr, and Lapierre III, A. G. (1965). A novel crystal growth phenomenon: single crystal GaAs overgrowth onto silicon dioxide, *J. Electrochem. Soc.*, **112**, pp. 706–709.
3. Michelitsch, M. (1965). Epitaxial growth of GaAs through cracks in SiO_2 masks, *J. Electrochem. Soc.*, **112**, pp. 747–748.
4. Jastrzebski, L. (1983). SOI by CVD: epitaxial lateral overgrowth (ELO) process — review, *J. Cryst. Growth*, **63**, pp. 493–526.
5. Cho, A. Y. and Ballamy, W. C. (1975). GaAs planar technology by molecular beam epitaxy (MBE), *J. Appl. Phys.*, **46**, pp. 783–785.
6. Matthews, J. W., Mader, S., and Light, T. B. (1970). Accommodation of misfit across the interface between crystals of semiconducting elements or compounds, *J. Appl. Phys.*, **41**, pp. 3800–3804.
7. Shaw, D. W. (1966). Selective epitaxial deposition of GaAs in holes, *J. Electrochem. Soc.*, **113**, pp. 904–908.
8. Vohl, P. (1981). Vapor-phase epitaxy of GaInAsP and InP, *J. Cryst. Growth*, **54**, pp. 101–108.

9. Tsaur, B.-Y., McClelland, R. W., Fan, J. C. C., Gale, R. P., Salerno, J. P., Vojak, B. A., *et al.* (1982). Low-dislocation-density GaAs epilayers grown on Ge-coated Si substrates by means of lateral epitaxial overgrowth, *Appl. Phys. Lett.*, **41**, pp. 347–349.
10. Luryi, S., and Suhir, E. (1986). New approach to the high-quality epitaxial-growth of lattice-mismatched materials, *Appl. Phys. Lett.*, **49**, pp. 140–142.
11. Fitzgerald, E. A., Kirchner, P. D., Proano, R., Pettit, G. D., Woodall, J. M., and Ast, D. G. (1988). Elimination of interface defects in mismatched epilayers by a reduction in growth area, *Appl. Phys. Lett.*, **52**, pp. 1496–1498.
12. Fitzgerald, E. A., Watson, G. P., Proano, R., Ast, D. G., Kirchner, P. D., Pettit, G. D., *et al.* (1989). Nucleation mechanisms and the elimination of misfit dislocations at mismatched interfaces by reduction in growth area, *J. Appl. Phys.*, **65**, pp. 2220–2237.
13. Fitzgerald, E. A. (1989). The effect of substrate growth area on misfit and threading dislocation densities in mismatched heterostructures, *J. Vac. Sci. Technol.* B., **7**, pp. 782–788.
14. Fitzgerald, E. A., Xie, Y.-H., Brasen, D., Green, M. L., Michel, J., Freeland, P. E., *et al.* (1990). Elimination of dislocations in heteroepitaxial MBE and RTCVD Ge_xSi_{1-x} grown on patterned Si substrates, *J. Electron. Mat.*, **19**, pp. 949–955.
15. Fitzgerald, E. A., and Chand, N. (1991). Epitaxial necking on GaAs grown on pre-patterned Si substrates, *J. Electron. Mat.*, **20**, pp. 839–853.
16. Matthews, J. W., and Blakeslee, A. E. (1974). Defects in epitaxial multilayers: 1 Misfit dislocations, *J. Cryst. Growth*, **27**, pp. 118–125.
17. Matthews, J. W. (1975). Defects associated with accommodation of misfit between crystals, *J. Vac. Sci. Technol.*, **12**, pp. 126–133.
18. Maree, P. M. J., Barbour, J. C., van der Veen, J. F., Kavanagh, K. L., Bulle-Lieuwma, C. W. T., and Viegers, M. P. A. (1987). Generation of misfit dislocations in semiconductors, *J. Appl. Phys.*, **62**, pp. 4413–4420.
19. People, R., and Bean, J. C. (1985). Calculation of critical layer thickness versus lattice mismatch for Ge_xSi_{1-x}/Si strained-layer, *Appl. Phys. Lett.*, **47**, pp. 322–324.
20. Dodson, B. W., and Tsao, J. Y. (1987). Relaxation of strained-layer semiconductor structures via plastic flow, *Appl. Phys. Lett.*, **51**, pp. 1325–1327.
21. Jesser, W. A., and Matthews, J. W. (1967). Evidence for pseudomorphic growth of iron on copper, *Philos. Mag.*, **15**, pp. 1097–1106.

22. Jesser, W. A. and Matthews, J. W. (1968). Pseudomorphi deposits of cobalt on copper, *Philos. Mag.*, **17**, pp. 461–473.

23. Jesser, W. A. and Fox, B. A. (1990). On the generation of misfit dislocations, *J. Electron. Mat.*, **19**, pp. 1289–1297.

24. Peierls, R. E. (1940). The size of a dislocation, *Proc. Phys. Soc.*, **52**, pp. 35–37.

25. Nabarro, F. R. N. (1947). Dislocation in a cubic lattice, *Proc. Phys. Soc.*, **59**, pp. 256–272.

26. Price, G. L. (1991). Critical-thickness and growth-mode transitions in highly strained $In_xGa_{1-x}As$ films, *Phys. Rev. Lett.*, **66**, pp. 469–472.

27. Hu, S. M. (1991). Misfit dislocations and critical thickness in heteroepitaxy, *J. Appl. Phys.*, **69**, pp. 7901–7903.

28. Fritz, I. J. (1987). Role of experimental resolution in experiments of critical layer thickness for strained-layer epitaxy, *Appl. Phys. Lett.*, **51**, pp. 1080–1082.

29. Fitzgerald, E. A. (1991). Dislocation in strained-layer epitaxy: theory, experiment, and applications, *Mat. Sci. Reports*, **7**, pp. 91–142.

30. Hagen, W. and Strunk, J. (1987). New type of source generating misfit dislocations, *Appl. Phys.*, **17**, pp. 85–87.

31. MacPhersen, G. and Goodhew, P. J. (1996). A refined scheme for the reduction of threading dislocations densities in InGaAs/GaAs epitaxial layers, *J. Appl. Phys.*, **80**, pp. 6706–6710.

32. Sheldon, P., Jones, K. M., Al-Jassim, M. M., and Yacobi, B. G. (1986). Dislocation density reduction through annihilation in lattice-mismatched semiconductors grown by molecular-beam epitaxy, *J. Appl. Phys.*, **63**, pp. 5609–5611.

33. Tachikawa, M., and Yamaguchi, M. (1990). Film thickness dependence of dislocation density reduction in GaAs-onSi substrates, *Appl. Phys. Lett.*, **56**, pp. 484–486.

34. Speck, J. S., Brewer, M. A., Beltz, G., Romanov, A. E., and Pompe, W. (1996). Scaling laws for the reduction of threading dislocation densities in homogeneous buffer layers, *J. Appl. Phys.*, **80**, pp. 3808–3816.

35. Ayers, J. E. (1995). New model for the thickness and mismatch dependences of threading dislocation densities in mismatched heteroepitaxial layers, *J. Appl. Phys.*, **78**, pp. 3724–3726.

36. Zhang, X. G., Li, P., Zhao, G., Parent, D. W., Jain, F. C., and Ayers, J. E. (1998). Removal of threading dislocations from patterned heteroepitaxial semiconductors by glide to sidewalls, *J. Electron. Mat.*, **27**, pp. 1248–1253.

37. Nishinaga, T. (2002). Microchannel epitaxy: an overview, *J. Cryst. Growth,* **237–39**, pp. 1410–1417.

38. Lee, H. P., Wang, S., Huang, Y. H., and Yu, P. (1988). Photoluminescence studies of selective-area molecular beam epitaxy of GaAs film on Si substrate, *Appl. Phys. Lett.*, **52**, pp. 215–217.

39. Lee, S. C. and Brueck, S. R. J. (2009). Scaling of the surface migration length in nanoscale patterned growth, *Appl. Phys. Lett.*, **94**, p. 153110.

40. Lee, S. C., Malloy, K. J., and Brueck, S. R. (2001). Nanoscale selective growth of GaAs by molecular beam epitaxy, *J. Appl. Phys.*, **90**, pp. 4163–4167.

41. Bacchin, G., and Nishinaga, T. (2000). A new way to achieve both selective and lateral growth by molecular beam epitaxy: low angle incidence microchannel epitaxy, *J. Cryst. Growth*, **208**, pp. 1–4.

42. Bacchin, G., Umeno, A., and Nishinaga, T. (2000). Lateral growth in molecular beam epitaxy by low angle incidence microchannel epitaxy, *Appl. Surf. Sci.*, **159/160**, pp. 270–276.

43. Umeno, A., Bacchin, G., and Nishinaga, T. (2000). AFM analysis of sidewall formation in low angle incidence microchannel epitaxy of GaAs, *J. Cryst. Growth*, **220**, pp. 355–363.

44. Nishida, A., Nakagawa, K., Murakami, E., and Miyao, M. (1992). Elimination of misfit dislocations in $Si_{1-x}Ge_x$/Si heterostructures by limited-area molecular-beam epitaxial growth, *J. Appl. Phys.*, **71**, pp. 5913–5917.

45. Hull, R., Bean, J. C., Higashi, G. S., Green, M. L., Peticolas, L., Bahnck, D., *et al.* (1992). Improvement in heteroepitaxial film quality by a novel substrate patterning geometry, *Appl. Phys. Lett.*, **60**, pp. 1466–1468.

46. Watson, G. P., Thompson, M. O., Ast, D. G., Fischer-Colbrie, A., and Miller, J. (1990). The isolation and nucleation of misfit dislocations in strained epitaxial layers grown on patterned, ion-damaged GaAs, *J. Electron. Mat.*, **19**, pp. 957–965.

47. Kazi, Z. I., Thilakan, P., Egawa, T., Umeno, M., and Jimbo, T. (2001). Realization of GaAs/AlGaAs lasers on Si substrates using epitaxial lateral overgrowth by metalorganic chemical vapor deposition, *Jpn. J. Appl. Phys.*, **40**, pp. 4903–4906.

48. Kitajima, H., Fujimoto, Y., Kasai, N., Ishitani, A., and Endo, N. (1989). Lattice defect in selective epitaxial silicon and laterally overgrown regions on SiO_2, *J. Cryst. Growth*, **98**, pp. 264–276.

49. Subramanian, C. K. and Neudeck, G. W. (1992). Large area silicon on insulator by double-merged epitaxial lateral overgrowth, *J. Vac. Sci. Technol. B*., **10**, pp. 643–647.

50. Nagel, N., Banhart, F., Czech, E., Silier, I., Phillipp, F., and Bauser, E. (1993). The coalescence of silicon layers grown over SiO_2 by liquid-phase epitaxy, *Appl. Phys.*, **A57**, pp. 249–254.

51. Huang, W., Nishinaga, T., and Naritsuka, S. (2001). Microchannel epitaxy of GaAs from parallel and nonparallel seeds, *Jpn. J. Appl. Phys.*, **40**, pp. 5373–5376.

52. Yan, Z., Hamaoka, Y., Naritsuka, S., and Nishinaga, T. (2000). Coalescence in microchannel epitaxy of InP, *J. Cryst. Growth*, **212**, pp. 1–10.

53. Watson, G. P., Ast, D. G., Anderson, T. J., and Pathangey, B. (1993). The barrier to misfit dislocation glide in continuous, strained, epitaxial layers on patterned substrates, *J. App. Phys.*, **74**, pp. 3103–3110.

54. Jones, J. R., Jones, S. H., Zybura, M. F., Seidel, L. K., and Oh, D. K. (1991). Effects of mesa area and orientation on defects in strained $In_xGa_{1-x}As$ films grown by LPMOCVD, *J. Electron. Mat.*, **20**, pp. 709–717.

55. Abrahams, M. S., Weisbergm L. R., Buiocchi, C. J., and Blanc, J. (1969). Dislocation morphology in graded heterojunctions — $GaAs_{1-x}P_x$, *J. Mat. Sci.*, **4**, pp. 223–235.

56. Stringfellow, G. B., and Green, P. (1969). Dislocations in $GaAs_{1-x}P_x$, *J. Appl. Phys.* **40**, pp. 502–508.

57. Olsen, G. H., Abrahams, M. S., Buiocchi, C. J., and Zamerowski, T. J. (1976). Reduction of dislocation densities in heteroepitaxial III-V VPE semiconductors, *J. Appl. Phys.* **46**, pp. 1643–1646.

58. Bedair, S. M., Humphreys, T. P., El-Masrey, N. A., Lo, Y., Hamaguchi, N., Lamp, C. D., *et al.* (1986). Defect reduction in GaAs grown by molecular-beam epitaxy using different superlattice structures, *Appl. Phys. Lett.*, **49**, pp. 942–944.

59. Dodson, B. W. (1990). Dislocation filtering — why it works, when it doesn't, *J. Electron. Mat.*, **19**, pp. 503–508.

60. Watanabe, Y., Uneta, M., Kodota, Y., and Ohmachi, Y. (1990). Dislocation reduction in GaAs grown on Si by MOMBE using multiple indium atomic plane structures, *J. electron. Mat.*, **19**, pp. 219–224.

61. Lee, J. W., Shichijo, H., Tsai, H. L., and Matyi, R. J. (1987). Defect reduction by thermal annealing of GaAs-layers grown by molecular-beam epitaxy on Si substrates, *Appl. Phys. Lett.*, **50**, pp. 31–33.

62. Choi, C., Otsuka, N., Munns, G., Houdre, R., Morkoc, H., Zhang, S. L., *et al.* (1987). Effect of *in situ* and *ex situ* annealing of dislocations in GaAs on Si substrates, *Appl. Phys. Lett.*, **50**, pp. 992–994.

63. Beam III, E. A. and Kao, Y. C. (1991). The effect of reduced growth area by substrate patterning on misfit accommodation in molecular beam epitaxially grown $In_xGa_{1-x}As/GaAs$, *J. Appl. Phys.*, **69**, pp. 4253–4262.

64. Knall, J., Romano, L. T., Biegelsen, D. K., Bringans, R. D., Chui, H. C., Harris, J. S., Jr. *et al.* (1994). The use of graded InGaAs layers and patterned substrates to remove threading dislocations from GaAs on Si, *J. Appl. Phys.*, **76**, pp. 2697–2702.

65. Tsang, W. T. and Cho, A. Y. (1978). Molecular beam epitaxy writing of patterned GaAs epilayer structures, *Appl. Phys. Lett.*, **32**, pp. 491–493.

66. Demeester, P., Buydens, L., and Van Daele, P. (1990). Growth velocity variations during metalorganic vapor phase epitaxy through an epitaxial shadow mask, *Appl. Phys. Lett.*, **57**, pp. 168–170.

67. Beam III, E. A., Kao, Y. C., and Yang, J. Y. (1991). A cantilever shadow mask technique for reduced area molecular beam epitaxial growth, *Appl. Phys. Lett.*, **58**, pp. 152–154.

68. Langdo, T. A., Leitz, C. W., Currie, M. T., Fitzgerald, E. A., Lochtefeld, A., and Antoniadis, D. A. (2000). High quality Ge on Si by epitaxial necking, *Appl. Phys. Lett.*, **76**, pp. 3700–3702.

69. Park, J.-S., Bai, J., Curtin, M., Adekore, B., Carroll, M., and Lochtefeld, A. (2007). Defect reduction of selective Ge epitaxy in trenches on Si (001) substrates using aspect ratio trapping, *Appl. Phys. Lett.*, **90**, p. 052113.

70. Bai, J., Park, J.-S., Cheng, Z., Curtin, M., Adekore, B., Carroll, M., *et al.* (2007). Study of the defect elimination mechanisms in aspect ratio trapping Ge growth, *Appl. Phys. Lett.*, **90**, p. 101902.

71. Li, J. Z., Bai, J., Park, J.-S., Adekore, B., Fox, K., Carroll, M., *et al.* (2007). Defect reduction of GaAs epitaxy on Si (001) using selective aspect ratio trapping, *Appl. Phys. Lett.*, **91**, p. 021114.

72. Li, J. Z., Bai, J., Major, C., Carroll, M., Lochtefeld, A., and Shellenbarger, Z. (2008). Defect reduction of GaAs/Si epitaxy by aspect ratio trapping, *J. Appl. Phys.*, **103**, p. 106102.

73. Kamon, K., Takagishi, S., and Mori, H. (1985). Selective epitaxial-growth of GaAs by low-pressure MOVPE, *J. Cryst. Growth*, **73**, pp. 73–76.

74. Chang, Y. S., Naritsuka, S., and Nishinaga, T. (1997). Effect of growth temperature on epitaxial lateral overgrowth of GaAs on Si substrate, *J. Cryst. Growth*, **174**, pp. 630–634.

75. Li, J. Z., Bai, J., Hydrick, J. M., Park, J. S., Major, C., Carroll, M., *et al.* (2009). Growth and characterization of GaAs layers on polished Ge/Si by selective aspect ratio trapping, *J. Cryst. Growth*, **311**, pp. 3133–3137.

76. Wu, Y. Q., Xu, M., Ye, P. D., Cheng, Z., Li, J., Park, J.-S., *et al.* (2008). Atomic-layer-deposited Al2O3/GaAs metal-oxide-semiconductor field-effect transistor on Si substrate using aspect ratio trapping technique, *Appl. Phys. Lett.*, **93**, p. 242106.

77. Araki, K., Seo, J. U., Hasegawa, S., and Asahi, H. (2008). Selective growth of InP on localized areas of silicon (100) substrate by molecular beam epitaxy, *Phys. Stat. Sol. (c)*, **5**, pp. 2766–2768.

78. Deura, M., Kondo, Y., Takenaka, M., Takagi, S., Nakano, Y., and Sugiyama, M. (2010). Twin-free InGaAs thin layer on Si by multi-step growth using micro-channel selective-area MOVPE, *J. Cryst. Growth*, **312**, pp. 1353–1358.

79. Yu, H. Y., Cheng, S. L., Park, J. H., Okyay, A. K., Cengiz Onbasli, M., Ercan, B., *et al.* (2010). High quality single-crystal germanium-on-insulator on bulk Si substrates based on multistep lateral over-growth with hydrogen annealing, *Appl. Phys. Lett.*, **97**, p. 063503.

80. Nayfeh, A., Chui, C. O., and Saraswat, K. C. (2004). Effects of hydrogen annealing on heteroepitaxial-Ge on Si: surface roughness and electrical quality, *Appl. Phys. Lett.*, **85**, pp. 2815–2817.

Chapter 4

Low-Temperature Direct Wafer Bonding

Anke Sanz-Velasco,[a] Cristina Rusu,[b] Isabelle Ferain,[c] Cindy Colinge,[c] and Mark Goorsky[d]

[a]*Department of Microtechnology and Nanoscience — MC2, Chalmers University of Technology, Kemivägen 9, SE-41296 Gothenburg, Sweden*
[b]*Imego/ACREO AB, Arvid Hedvalls backe 4, SE-40012, Gothenburg, Sweden*
[c]*Tyndall National Institute, University College Cork, Cork, Republic of Ireland*
[d]*Department of Material Science and Engineering, UCLA, Los Angeles, California, USA*
anke.sanz-velasco@chalmers.se

The breakthrough of direct wafer bonding was achieved with silicon-on-insulator (SOI) allowing for high-power and high-frequency electronics, microelectronics components with low energy consumption. After this, several bonding techniques have been developed (e.g. silicon direct bonding, anodic bonding) and are being developed (e.g. low-temperature plasma-assisted direct bonding) to achieve hybrid components, as III-V semiconductors on silicon-based substrates, monolithic integration of optoelectronic devices with high-speed silicon integrated circuits, three-dimensional stacking of integrated circuits (ICs) or circuits transfer onto a variety of substrates. An overview of more recent activities on several techniques for attaining low-temperature bonding is presented.

Lattice Engineering: Technology and Applications
Edited by Shumin Wang

ISBN 978-981-4316-29-3 (Hardcover), 978-981-4364-25-6 (eBook)
www.panstanford.com

4.1 Introduction

Wafer bonding is a key process for the manufacture of micro-engineered substrates for high performance logic complementary metal oxide semiconductor (CMOS) devices, SOI, strained silicon-on-insulator (sSOI), germanium-on-insulator (GeOI) [1–3] and three-dimensional structures in micro electro mechanical systems (MEMS) [4]. In this way new three-dimensional structures and devices integrating different materials can be put together (e.g. polymers on Si, glass on Si, and either GaAs, Ge, or InP on Si, etc.).

The wafer bonding technologies can be classified as (i) *direct bonding* such as silicon direct bonding (SDB), also referred to as fusion bonding, anodic bonding (also named electric field-assisted bonding), plasma enhanced bonding, and (ii) *indirect bonding* where an intermediate layer is used in between the bonded surfaces e.g. adhesive bonding, thermo compression bonding, eutectic bonding, glass frit bonding and solder bonding. All these technologies have emerged in response to the wide range of applications utilising a multitude of materials, which have different bonding requirements [2,4–8]. An overview of wafer bonding properties is shown in Table 4.1.

Low-temperature (LT) bonding for integrating different materials is becoming an increasingly important micro-fabrication method [9–13]. The demand for faster devices will require new materials (alternative channel materials) with higher low-field carrier mobility than silicon. Compound semiconductors and germanium have higher mobility for electrons and holes, respectively, thus it is attractive to integrate these materials with traditional silicon devices. As a result, thermal properties of the different materials to be bonded, such as the coefficient of thermal expansion (CTE) shown in Table 4.2 must be considered. Additionally, bonding temperature-sensitive materials and/or components with an already processed silicon wafer will also require LT bonding [14]. The upper limit for LT bonding is regarded as 400°C originating from the CMOS technology for the production of ICs. However, each LT bonding method (Table 4.1) requires more detailed analysis in terms of mechanical, electrical and chemical characteristics, reliability and hermeticity. Such study aims at simplifying the optimisation of the

Table 4.1. Overview of bonding techniques ("Strong bonds" withstands common processing steps; "High temperature" is above 400°C)

Bonding techniques		Advantages	Disadvantage
Direct (no intermediate layer)	Anodic	Hermetic, strong bond	Requires glass with sodium and a high electric field
	Direct	Hermetic, strong bond	Surfaces must have an average roughness <1 nm
	Low temperature plasma	Low temperature	Requires direct or indirect plasma exposure
Indirect (intermediate layer)	Adhesive	Non-flat surface, flexible method	Not hermetic
	Glass frit	Hermetic	Requires large line widths when used for capping
	Thermo compression (metallic)	Hermetic, Non-planar surface	High forces applied
	Solder (metallic)	Hermetic, self-aligning	Possible solder flow
	Eutectic (metallic)	Hermetic, strong bond	Gold-silicon alloy can reflow into unwanted areas

Table 4.2. CTE for materials commonly used in bonding

Material	Linear CTE at RT ($\times 10^{-6}/^{\circ}C$)
Si	2.3–2.6
SiO_2	0.5
SiN	3.2
InP	4.6
quartz	0.59
Glass pyrex/borofloat	3.25
InP	4.6
GaAs	5.7–6.8
Ge	5.8

bonded interface quality, in terms of surface energy or bond strength and surface chemical composition.

There are several reviews of bonding techniques for MEMS, microelectronic applications [2, 4, 5, 7, 8, 15] and many proceedings volumes for the International Symposium on Semiconductor Wafer Bonding-Science, Technology, and Applications as part of the Electrochemical Society Meetings (1992–present). In this chapter, the emphasis is mainly on the more recent activities on direct LT bonding methods. An overview of a variety of characterisation methods for the evaluation of the quality of the bonded interfaces will be introduced first because they will be used for the explanation of bonding mechanisms. The following sections will show several techniques for attaining LT bonding including anodic bonding, silicon–silicon direct bonding, and plasma-assisted bonding including silicon–silicon and germanium–silicon bonding.

4.2 Characterisation Methods for Wafer Bonding

There are two ways to classify methods for the characterisation of wafer bonding: (i) *non-destructive* (non-contact), such as optical microscopy to detect voids at the interface, x-ray methods, or acoustic microscopy and, (ii) *destructive*, such as mechanical characterisation for bond strength and electrical testing. All these characterisation techniques share the same goal: the evaluation of the quality of bonded interfaces, bonded materials and to understand the bonding mechanisms.

4.2.1 *Non-Destructive Methods*

Non-destructive methods usually include optical inspection, such as infrared (IR) transmission, ultrasonic imaging (scanning acoustic microscopy, SAM), x-ray diffraction (XRD), and Raman spectroscopy.

IR measurement can detect non-bonded areas (voids) larger than 20 μm in materials that do not absorb IR. For example, materials such as Si and GaAs are transparent to IR, while Ge absorbs IR. It is thus challenging to image bonded Ge by this method.

The condition for void detection is that the interface surfaces must be apart more than one fourth of the wavelength of the light (one fourth is considered to be $\sim$0.25μm) and that the lateral resolution is larger than the numerical aperture of the optical system. There are two types of voids that can form: one due to particles and thus cannot be removed from the interface — "extrinsic" voids, and others that appear after the bond anneal step — "intrinsic" voids. Both types can be seen in Fig. 4.1.

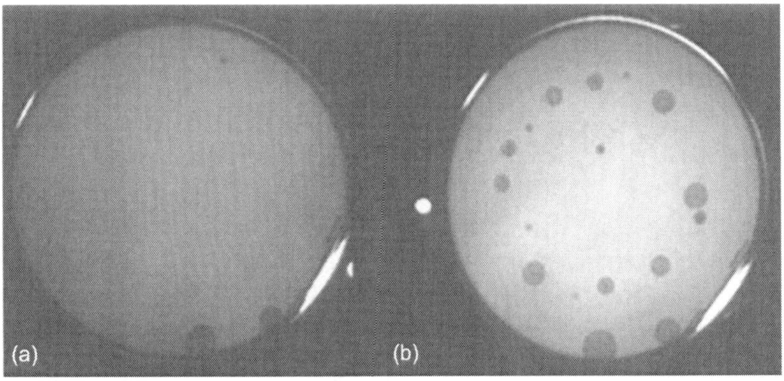

Figure 4.1. Infra-Red detection of voids for the same wafer: (a) "extrinsic" voids, due to particles; (b) "intrinsic" voids, appeared upon annealing.

4.2.1.1 Fourier transform IR spectroscopy

Fourier transform IR spectroscopy (FTIR) is used to investigate the chemistry of bonded interfaces (e.g. the absorption band of oxygen, hydrogen or O–H vibration modes bonded to semiconductors) and makes it possible to understand the bonding chemistry at the interface [1617]. A method of FTIR called multiple internal reflection configuration of FTIR (MIR-FTIR) greatly amplifies the signals of OH–Si, H–Si, and H_2O–Si at the bonded interface. With appropriate doping levels within the bonded wafers and with proper sample preparation (the wafers should be doubleside polished to prevent scattering of the IR signal [18]), prisms or sample bevelling are also considered for guiding the light to the sample. MIR-FTIR allows IR light to enter the bonded pair at one end of the sample where it will be reflected multiple times before leaving the sample at the other end An example of such a measurement is shown in Fig. 4.8.

4.2.1.2 Scanning acoustic microscopy

Acoustic microscopy gives information about interface voids (focus at interface) and micro-cracks within bulk material (focus in bulk) (e.g. Figs. 4.11 and 4.13). For example, InP and GaAs are quite fragile wafers and the stress induced during cooling can easily produce fractures in the bulk of these materials. By using SAM, the effects of the cooling rate on the cracking of bonded hetero-structures can be analysed [19]. The spatial resolution of SAM is about 1–5 μm and thin air gaps below 0.1 μm can be detected, limited by the wavelength of the acoustic beam [20, 21]. Higher spatial resolution of ca. 1 μm [22, 23] can be obtained at high frequencies but less penetration is achieved, and thus it is mainly suitable for very thin stacks; while a lower resolution at low frequency allows for greater penetration thus thicker stacks (through the wafer). The drawback is that the wafers are measured in water, thus the bond needs to be strong enough in order to avoid de-bonding of the wafers.

4.2.1.3 X-ray diffraction

X-ray diffraction has the possibility to penetrate thick samples giving physical (i.e. stress within bonded pairs) [6] and chemical information about bonded interface (i.e. bonding interface electron densities) [24] (e.g. Figs. 4.18, 4.19). The x-ray reflectivity [25] is uniquely able to determine thin film density and surface roughness. Both these parameters are critical for successful bonding and layer transfer. Further, x-ray reciprocal space mapping and double crystal x-ray topography can be used to assess the influence of thermal stress on the material properties of bonded materials with different CTEs. X-ray reflectivity and MIR-FTIR spectra are usually correlated for the understanding of the bonding interface behaviour, especially for processing and materials studies important for III-V semiconductor wafer bonding.

4.2.1.4 Raman spectroscopy

Stress in Si can be also measured by Raman Spectroscopy, since the strain in the Si layer may be calculated using the shift of the Raman peak [26, 27]. In this method, a 1–1.5 μm diameter of focused laser spot determines the spatial resolution.

4.2.1.5 Surface analysis

Surface properties (roughness, contamination, elemental composition) are commonly investigated prior to bonding after surface activation and after de-bonding. The surface roughness is usually measured by Atomic Force Microscopy (AFM) or non-contact surface profilometry (optical interferometry). Particle density on the surface can be estimated by light scattering. The surface wettability or hyrophilicity can be determined by contact angle measurements. A highly activated surface is characterised by a hydrophilic (HP) state and determined by a very low contact angle. Spectroscopic ellipsometry is used to study the structural defects in thin films and also within multilayer structures due to large wavelength range from UV to near IR [27, 28]. X-ray photoelectron spectroscopy (XPS) gives information on elemental composition and the associated chemical bonding states of the near-surface region of within ~50 Å of the outer surface.

4.2.2 *Destructive Methods*

Destructive methods are (i) cross-sectional analysis, e.g. scanning electron microscopy (SEM), transmission electron microscopy (TEM), secondary ion mass spectroscopy (SIMS) and defect etching at cross-sectioned samples, (ii) bond strength measurements, e.g. pull test, pressure burst test, tensile-shear test and crack-opening or Maszara test, and (iii) hermeticity testing, by e.g. membrane test, He-leak test, and resonator/pressure method. A comprehensive presentation of bond strength testing methods can be found in Vallin *et al.* [29]

4.2.2.1 Transmission electron microscopy

Highresolution TEM (HR-TEM) allows to see the formation of a damage layer at the bonding interface such as dislocations linked to misorientation, voids, negative crystals, and oxide-containing precipitates (e.g. Figs. 4.15 and 4.17) [4, 30].

4.2.2.2 Bond strength measurement

The fracture surface energy is measured by the crack opening method introduced by Maszara [31]: a thin blade is inserted between the bonded wafers and a crack is introduced at the interface. Equation 4.1 is used for the calculation of the fracture surface energy of two similar wafers. The length of the crack is a measure of the surface energy W of the bond,

$$W = Et^3y^2/8L^4 \tag{4.1}$$

where E is Young's modulus of the wafer material, t is the thickness of the wafers, $2y$ is the thickness of the blade, and L is the crack length. This equation holds when a pair of similar wafers is bonded. The Maszara method has the advantage of creating a well-defined load on the bonded interface. However, for a better control of the blade insertion and less uncertainties in the resulting crack length (influencing the final fracture surface energy by a fourth power of the crack length), a steady-state wedge-opening test can be used [32].

For small-size bond areas, the interfacial fracture energy at the bond interface can be analysed by indentation techniques such as Vickers test [33], micro-chevron test [34], and for even smaller surfaces as low as $5 \times 5\ \mu m^2$ the micro-indentation [35].

The bond strength can be directly measured by pull tests [36]: diced samples are glued onto stud poles that are pulled apart in a controlled manner. However, the sample mounting is critical; the obtained results depend strongly on the mounting. A better method may be burst tests [37].

Hermeticity tests of electronic devices are usually carried out using helium leak detection [38, 39] or using pressure sensor/resonators inside the cavity (quality factor of a resonator depends on the pressure) [40]. As an example, Jourdain *et al.* [41] found no fine or gross leak for Benzocyclobutene (BCB) bonds when tested according to the MIL-STD-883D method, but the function of MEMS devices inside the cavity indicated a leak. It was concluded that new standards are necessary for MEMS devices due to their small volumes. Environmental testing for humidity, vibration and shock are beginning to be used for wafer bonding [42].

4.2.2.3 Electrical characterisation

For the characterisation of the electrical properties of the bonded interfaces standard electrical methods such as current–voltage (I–V), capacitance–voltage (C–V) and resistance measurements can also be used [43, 44].

4.3 Anodic Bonding

Slightly conductive glass wafer and virtually any conducting material (semiconductor and semiconductor compounds, alloys and metals) [45, 46] can be joined. Usually an electric field in the range of 200–1000 V DC and moderate temperature ranging from 180–420°C are applied, under vacuum or atmospheric pressure (Fig. 4.2). The glass commonly used for anodic bonding has a similar CTE to Si (Table 4.2); borosilicate types such as Pyrex, Corning 7740, Schott 8330, Schott borofloat 33, lithia potash borosilicate Corning 7070, and Hoya SD-2 have a CTE behaviour with temperature that closely matches Si single crystal). This minimises the intrinsic stress generated at the Si-glass interface, while the contained high concentration of alkali metal oxides permits the bonding to take place.

The anodic bonding mechanism is based on the formation of covalent Si–O bonds and results in bond strength ranging from 10 to 40 MPa [48], stronger than the bulk strength of silicon and glass. Details of the anodic bonding mechanism have been discussed in literature [2, 49, 50].

Instead of a bulk glass wafer, thin glass layers can be used for anodic bonding, the so-called thin film anodic bonding. Its advantages are lower bonding voltage, 30–60 V [51] due to thinner glass layer 0.5–4 μm and reduced residual stress in the stack.

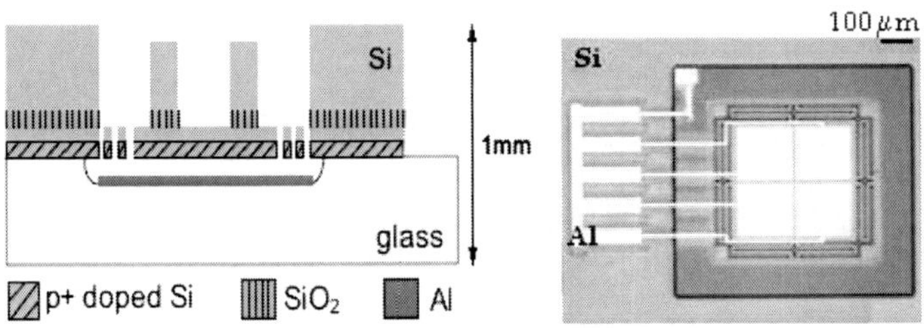

(a) (b)

Figure 4.2. Commercially available nanoindenter force sensor for materials analysis inside TEM fabricated using wafer-level anodic bonding of SOI to borofloat 33 glass (courtesy of Nanofactory Instrument AB [47]; (a) Cross-sectional sketch and (b) SEM photo through the glass.

Successful bonding was obtained with sputtered glass [38,52], evaporated glass [53], spin-on glass [54], thin layer of SiO_2 (100 nm), SiN (100 nm) [55], polySi (500 nm), and Al (1 μm) [56]. Bonding of glass to Al opens new possibilities of forming buried Al feed-through underneath a bonded Al layer. Even room temperature anodic bonding could be obtained by sputtering a thin film from Corning glass 7570 that has a softening point of 440°C (as compared to Pyrex, 821°C). The bond strength was larger than 1.5 MPa [14].

Low-Temperature Co-fired Ceramic (LTCC) is a mature component carrier system with complete infrastructure for electrical interconnects, buried passive and production that is used mainly for microwave applications. The under development low CTE LTCC material (HITK and VIA Electronic, Germany) [57] has a CTE of $3.6 \times 10^{-6}/°C$, which is much lower than the standard LTCC material ($5.5–11.6 \times 10^{-6}/°C$) [58]. This opens up for promising devices based on hybrid materials bonding (Fig. 4.3). The low CTE LTCC material contains alkali ions (Na^+ content of 1–2 wt%), which allows for direct anodic bonding to Si. The pull tests showed that the Si or LTCC material had been destroyed/broken in the joined region and not the LTCC–Si bonding area [59].

It has been shown that anodic bonding opens new possibilities for III–V photonics and electronics; epitaxial transfer of the InP-based heterostructure barrier varactor diode onto borosilicate glass substrate [60] and the measured device characteristics were similar with those of the devices on the InP substrate.

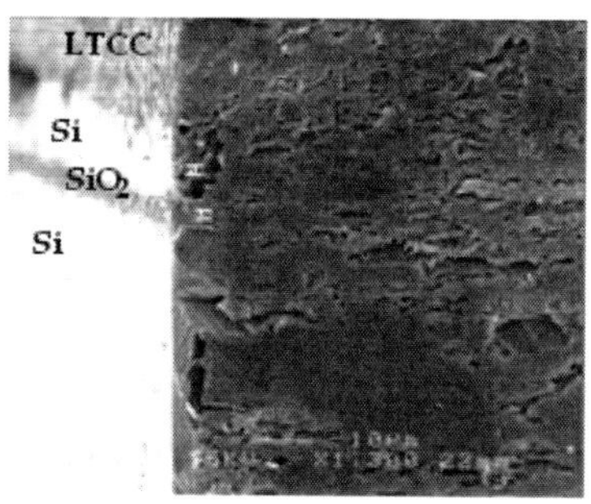

Figure 4.3. SEM cross section of LTCC-SOI bonded wafers indicating a good bond showing the corner of the chip.

Electrical compatibility of anodic bonding with MOS devices for monolithically integration of microsystems raises two main issues [61]: sodium contamination (affects both electrical and mechanical properties) and high electric field (affects electrical properties). Visser *et al.* studied the sodium distribution during thin-film anodic bonding and found that the sodium was stopped by a 1000 Å-thick film of silicon nitride [62]. By using a metal pattern deposited on the glass, it is possible to block the electric field from the circuit area [63].

The advantages of anodic bonding are reasonable low process temperature (below the sintering temperature of standard Al–Si metallisation), low residual stress, hermetic seal, and less stringent requirements on the surface quality of the wafers (as compared to SDB). However the root mean square (RMS) surface roughness must be smaller than 1 μm; the surface should be clean and dust free.

4.4 Silicon Direct Bonding

Direct bonding consists in contacting two flat, highly polished and clean surfaces [4, 64]. Spontaneous pre-bond at room temperature occurs, which is based on physical forces such as Van der Waals or hydrogen type. However, this bonding is of low strength and can be considerably improved by annealing at 600–1200°C, during which the physical bonds are converted to chemical bonds [31]. This process has been successfully exploited for fabrication of SOI wafers (used in, e.g. advanced logic applications, ASICs, microwave components, optoelectronics, MEMS, etc.) and was the main motivation for studying SDB [2]. SDB is also called silicon fusion bonding but it is misleading because the annealing temperatures are below silicon's fusion point, 1410°C.

There are two types of direct bonding: HP and hydrophobic (HB) [1, 2]. A detailed description of the SDB mechanism can be found in [1, 5, 65–67]. A HP surface is created by wet chemistry such as Piranha etch (H_2SO_4:H_2O_2) or a standard clean (RCA1, *aka* SC1) removing organic contaminants followed by another standard clean (RCA2, *aka* SC2) removing metal/alkali contaminants and particles. Such HP surface has a native oxide and a high density of OH groups which allows for immediate pre-bonding via van der Waals forces. Annealing at 800–1100°C is then performed to trigger the formation of covalent Si–O–Si bonds. The latter can achieve a bond strength equivalent to bulk Si.

Hydrophobic bonding is utilised when a bonding interface free from oxide is necessary, for example, epitaxial or hetero-epitaxial-like surfaces (e.g. p-n junction fabricated by bonding p-doped Si to n-doped Si). By using an HF dip after RCA clean, the Si wafer becomes HB, and a spontaneous pre-bonding occurs when the wafers are brought in contact at room temperature. Annealing at 300–800°C forms covalent Si–Si bonds.

The surface energy between the bonded wafers is strongly dependent on the annealing temperature [68]; HP surfaces, when annealed at temperatures above 900°C, can achieve full fracture strength of bulk Si (2.5 J/m^2). For HB wafers, the spontaneous pre-bonding energy is very low (13–26 mJ/m^2) without annealing. It increases quite fast for temperatures above 400°C and reaches a value of 2.5 J/m^2 at 600–700°C, which is a much lower temperature than with HP wafers.

The spontaneous pre-bond depends strongly on the quality of the surfaces, which should be very clean, have a RMS roughness below 1 nm, a total thickness variation 1 to 3 μm (4 inch wafer case), and wafer bow below 10 μm (4 inch wafer case).

Chemical Mechanical Polishing (CMP) is commonly used to give very smooth surfaces and can be applied also to brittle materials. Ceramics and metals [69], poly-Si up to 5 μm (doped or not), Si_xN_y up to 3 μm, Si after heavily B doping, Si after KOH etching (which results in a rough surface) [70], can all be bonded following careful surface treatment.

Direct bonding is possible between Si and many other materials, including Si to Si (Fig. 4.4) and different material combinations [68], silicon on sapphire, crystalline quartz on Si for high frequency applications, Si on fused quartz for High-Definition Television projection masks, GaAs or InP on Si for combining opto- and microelectronics, different III-V compounds for fabrication of Vertical Cavity Surface Emitting Lasers and Light Emitting Diodes, $SrTiO_3$ and $LiNbO_3$ for optical applications [71]. Technologies for fabrication of SOI materials are Separation by Implantation of Oxygen (SIMOX) [72],

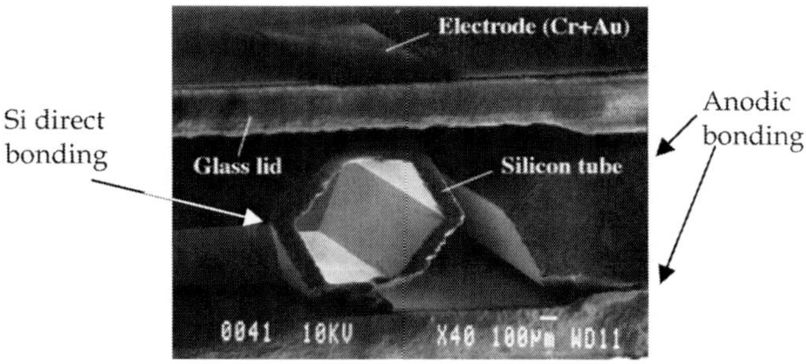

Figure 4.4. Cross-sectional SEM photo of a resonant structure (Coriolis flow sensor) (Si tube) obtained by SDB and anodic bonding at wafer level [77].

Bond and Etch back SOI (BESOI) [73], and Smart-Cut®. The Smart-Cut® process [74] is used for metallic direct bonding to transfer Si, GaAs and InP onto a Si substrate but it applies to other materials [75], SiC, InP, GaAs, Ge, diamond. Manufacture of SOI wafers with buried cavities makes it possible to take advantage of the mechanical and electronic properties of single-crystal material in the relatively thin device layer [76].

The high temperature required to achieve good bond strength has been a limitation regarding applications using Si direct bonding. However recent developments in surface preparation such as plasma activation have demonstrated that high temperatures are not of necessity. This is of great interest in MEMS design as it enables stress-free bonds to be achieved at temperatures that are compatible with metallisation layers and opens up a much larger range of applications for silicon direct bonding.

4.5 Plasma-Assisted Wafer Bonding

Plasma-assisted LT wafer bonding procedures have already been extensively documented. Plasma-assisted wafer bonding is used to facilitate strong, direct bonding at low temperature of single crystalline wafers of similar or dissimilar materials. This is an alternative approach to epitaxial growth. It opens the opportunity to combine highly latticemismatched wafers as well as wafers with different crystalline orientations.

In most plasma processes, both physical and chemical effects are important. These depend on input power, gases, substrate potential and temperature, pressure, reactor volume and geometry, gas flows, reactor material, etc. Surface damage and defects are a major drawback of plasma activation. These defects may influence the performance of materials and devices fabricated by plasma bonding [13].

A wide range of materials can be bonded using low-temperature, plasma-enhanced wafer bonding, e.g. Si, SiO_2 SiN [78], poly-Si [79], Glass [79], Sapphire [80], Quartz, InP [81–83], and Carbon-doped oxide [84].

Various plasma activation techniques are available. These are differentiated by the configuration of the plasma systems as for example barrel reactors [85, 86], Reactive Ion Etchers (RIE) [79, 83, 87, 88] RIE with Inductively Coupled Plasmas (ICP RIE) [10, 89, 90], Electron Cyclotron Resonance (ECR) systems [91], systems based on the dielectric barrier discharge (DBD) principle [3] or Sputter systems [92]. A comprehensive review of plasma process technologies may be found in the book by Vossen and Kern [93].

Different atmospheres are employed for the plasma treatment in the various plasma systems, e.g. O_2, N_2, H_2, Ar [79], or B_2H_6 [83]. After plasma exposure the wafers may be directly bonded *in situ* [11, 92, 94, 95] in the plasma chamber in vacuum. Sometimes a wet chemical treatment is used following the plasma exposure which occurs prior to *ex situ* [10, 83–85, 89, 90, 96–99] bonding in clean room environment. Typical treatments include a dip in de-ionised (DI) water or RCA1 clean.

The advantage of *in situ* bonding immediately after plasma bonding activation is that the vacuum in the plasma chamber prevents contamination from the ambient air. It is reported that some of the good bonding conditions are lost if the wafers are exposed to air after the surface treatment in vacuum [100].

However, there is a large variation in the results obtained in different systems. Some authors report surface energies above 1 Jm^{-2} after storage at room temperature [10, 88, 90, 98], while other authors report a need of low-temperature annealing to increase the surface energy [85, 87, 89, 94, 96, 101].

The results of the different investigators cannot be compared directly because experiments are made with a wide range of plasma systems, process parameters and wet chemical treatments.

Oxygen plasma is especially attractive for surface pre-treatment in wafer bonding, since it is very efficient in removing hydrocarbon and water-related species [85, 96, 102]. However, if applications require oxide-free interfaces fluorine containing plasmas [86, 103] or argon plasmas are preferred [104].

4.5.1 *Low-Pressure Plasma Processes*

Low-pressure (LP) plasma processes are dominated by the formation of high energetic ions, UV radiation and electrons and their interaction with the sample surface. In the following application examples of LP plasma processes are described in short.

Pasquariello *et al.* [105] assessed the differences between *ex situ* HP and HB surface pre-treatments for InP to Si direct wafer bonding. They found out that HP surface treatments using oxygen plasma resulted in spontaneous bonding of InP to Si at RT, resulting in high bond strength after LT annealing. These good results were not observed when using standard oxidising acids (H_2SO_4:H_2O_2:H_2O, NH_4OH, RCA1 or HNO_3) as HP surface treatment before bonding. The use of a hydrofluoric solution (HF:H_2O) (1:4) for cleaning InP before bonding to Si did not result in spontaneous bonding at room temperature. However, the bond strength started to increase after the bonded pair was annealed beyond 400°C. HB bonding followed by subsequent annealing at 400°C was thought to be the best choice regarding the electrical characteristics of the bonded InP-Si heterojunction.

Arokiaraj *et al.* [106] used *ex situ* oxygen plasma-assisted LT bonding for the integration of free-standing III-V thin-films, in particular, InP films on Si. The samples were wet chemically cleaned and the native oxide stripped in 5% HF for 1 min. Subsequently the Si was wet chemically oxidised using (H_2SO_4:H_2O_2) (5:2) for 10 min at 100°C. Only an InP epilayer of 2 μm thickness was exposed to O_2 plasma in a RIE system for 30 s using a chamber pressure of 90 mTorr, a gas flow of 100 sccm, and a bias voltage of 495 V. In the following step of the bonding, Si was placed in H_2O_2 and InP in H_2O

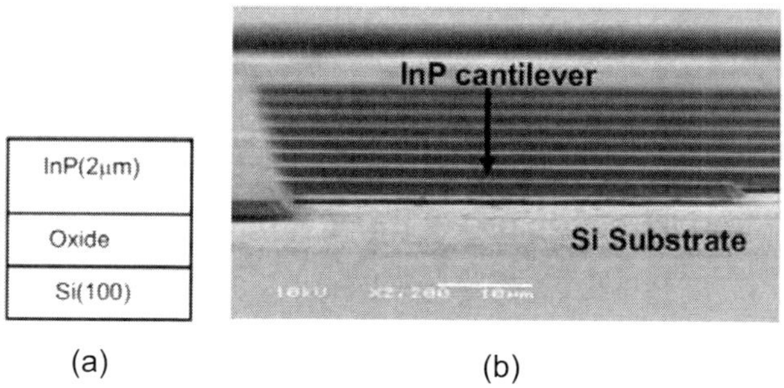

Figure 4.5. Final material structure achieved by oxidation on Si, plasma treatment of InP, epilayer and subsequent removal of InP bulk material, (b) SEM micrograph of 50 μm-long InP cantilever suspended on Si substrate by selectively etching Si substrate in KOH:H_2O (30%) [33].

for 10 min, respectively, and thereafter blow dried using N_2 gas. After initial spontaneous bonding the bonded samples were transferred to a Karl Suss wafer bonder, and the bonding was performed at 220°C for 45 min with a force of 80 N. After the bonding step, the bulk InP material was removed by lapping and wet chemical etching self-stopping at a InGaAs layer, which was removed by $FeCl_3$ etching leaving the 2 μm bonded InP material onto Si. Cantilevers are formed by the combination of lithography and subsequent dry-etching in HBr chemistry (Fig. 4.5).

Liang *et al.* [107] applied LT oxygen plasma-assisted wafer bonding to demonstrate a CMOS-compatible hybrid Si evanescent platform, which can be scaled up to 4 inch substrates. A 300°C anneal for 30 min resulted in a multiple quantum well (MQW) active region being well preserved with no threading dislocations after the epitaxial transfer to Si. A vertical outgassing channel (VOC) design was developed and employed to eliminate the intrinsic outgassing problem in LT wafer bonding. A reduction in interfacial void density is demonstrated in the order of five magnitudes. The annealing time was reduced from 15 hour to less than 30 min, increasing the process throughput.

With the help of VOCs, electrically pumped silicon distributed feedback lasers were realised on this platform for the first time. Low threshold of 23 mA with a maximum power of 5.4 mW was achieved in Continuous Wave operation. The side mode suppression ratio,

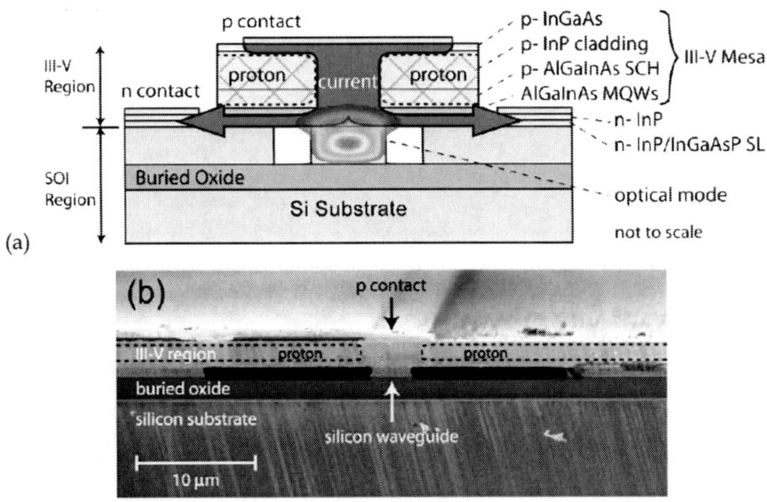

Figure 4.6. Sketch of hybrid evanescent platform, (b) SEM cross section of a polished device face [107].

line-width, and single mode operation band were 50 dB, 3.6 MHz, and over 100 nm, respectively, which are comparable to commercial III-V counterparts.

The demonstration of wafer-scale hybrid integration and on-chip light sources pave the way for realising the production of the hybrid silicon evanescent platform for the future optical interconnect system shown in Fig. 4.6ab.

Metallic Al–Al interconnects were realised by Lin and co-workers [108], using direct LT oxygen plasma-assisted bonding of silicon oxide layers with embedded Al structures. The Al–Al interconnects were realised between the bonded wafers through contact vias after the bonding process. This work, shown in Fig. 4.7, is an illustration of 3D integration.

Tan and Chong [84] used oxygen plasma-assisted LT wafer bonding to integrate low-*k* dielectric films of carbon-doped oxide (CDO) for applications in 3D integration resolving issues related to signal propagation delay, cross talk, and power consumption in advanced ICs. Dielectrics with low permittivity (low-*k*) such as CDO are already used by the semiconductor industry to replace conventional PE-TEOS as the inter-level dielectric layer.

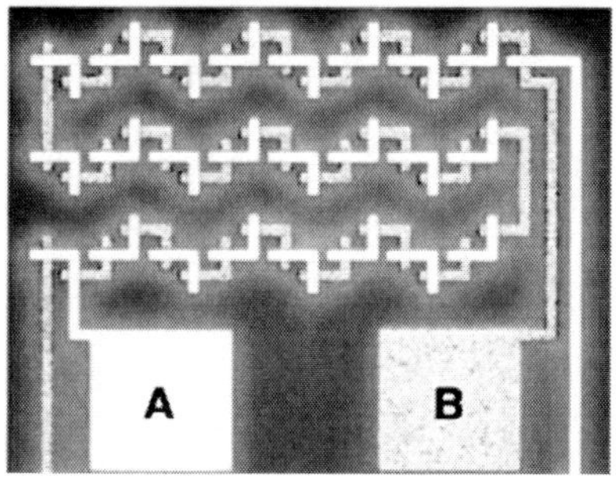

Figure 4.7. Microscope image of a contact chain after wafer thinning [108].

Kowal *et al.* [109] invented a new method of exposing semiconductor wafers to a mixture of radicals, allowing the wafer surfaces to be treated without damage from energetic ions, strong electric, fields and high UV fluxes associated with direct treatment by exposure to gas discharge plasmas, called remote plasma activation. In comparison to direct oxygen plasma activation wider process latitude, better process stability, and reduced surface roughening were obtained. Surface energies as high as 2.5 Jm^{-2} were achieved when the samples were annealed at 200°C for 2 hour. Additionally, Byun and co-workers [110] investigated oxygen and nitrogen free radicals formed using *in situ* remote plasma activation for HP and HB bonding of Si to Si in a vacuum, respectively.

As an example of *ex situ* LT plasma-assisted wafer bonding, Tong et al. [83] used B_2H_6 plasma treatment in a RIE system to bond oxide free InP to Si prior to a HF dip and room temperature initial bonding followed by subsequent annealing with temperatures up to 250°C and storage in a low vacuum chamber at 700 Pa for 24 hour. The surface energy for the B_2H_6 treated Si/InP bonded wafer pairs reached 695 mJ/m^2 at 200°C. That is greater than the fracture surface energy of [100] InP with 630 mJ/m^2. The 200°C annealed bonded wafer pair is strong enough for subsequent processing steps such as grinding, polishing, etching, and smart-cut.

4.5.2 *Ambient Pressure Plasma Processes*

Some differences between LP plasma and Ambient-Pressure (AP) plasma should be noted. In the AP plasma process, no ions are involved. The interaction of the AP plasma with the sample surface is mainly caused by UV radiation and a high concentration of electrons.

The AP plasma process is based on a DBD principle [3]. Two electrodes are coated with dielectric material and placed very close above a substrate. A continuous process gas flow is formed between the sample and the electrodes. A high frequency AC voltage is applied in order to obtain a uniform plasma discharge at AP conditions. With this method a uniform treatment of substrates up to 12″ can be realised.

The AP plasma technology was demonstrated on a range of different materials such as Si, Ge, GaAs, borofloat glass wafers and III-V substrates [3]. Prior to bonding, Gabriel and colleagues [3] used different cleaning procedure, e.g. RCA solutions or megasonic cleaning. For the AP treatment, a commercial tool, Suss NP200, was applied. Contact angles smaller than 5° were achieved for all substrates after one treatment with AP plasma. After cleaning, bonding, and annealing at 200°C, no voids were detected using IR inspection and SAM. The bond energies of AP activated wafers are analogous or higher as obtained for LP-treated wafers.

The low annealing temperatures and resulting high bond energies are comparable to those obtained with standard LP tools, e.g. RIE, AP plasma activation for the fabrication of heterogeneous material combinations. Further, advantages of this processing are the low costs for equipment and processing and the flexibility with respect to material choice and size.

4.5.3 *Oxygen Plasma-Assisted Bonding Mechanism*

The mechanism that improves the surface energy of plasma treated bonded wafer pairs, as compared to wet chemical bonded ones, is still under debate. It is known that HP silicon surfaces bond to each other spontaneously by hydrogen bonding via adsorbed water molecules [4]. The wafers therefore stay weakly bonded at room temperature even for long storage times. Si wafers treated with oxygen plasma are also HP. However, when such pre-treated wafers are bonded together, the surface energy increases (1.6 Jm^{-2}) to approximately the same value as when annealing RCA1 cleaned and bonded Si wafers at 800°C [4]. This result indicates a difference in the bonding mechanisms in the two cases. Some authors explain the difference by the increased number of OH groups due to the plasma activation [86, 102] and others by the increased number of OH groups due to the plasma activation increased in combination with enhanced water transport from the interface due to a porous oxide [10].

The following paragraph briefly presents the possible mechanism of the oxygen plasma enhanced *ex situ* bonding. It is described in more detail in Ref. [111]. When bringing two plasma treated water dipped Si wafers together, bonding first occurs spontaneously due to hydrogen bonding via adsorbed water molecules on the surfaces. If the surface energy of the plasma treated bonded wafers is measured directly after wafer contacting, the values expected for direct bonding at room temperature (about 0.2 Jm^{-2}) are obtained. During storage at room temperature, the surface energy gradually increases and eventually the wafers are bonded via covalent bonds. For this to happen, water must be transported away from the interface so that the hydroxyl groups are able to contact each other and react. This implies that the surface mobility of the reacting atoms and adsorbed molecules must be enhanced to make the reaction possible. It is speculated that the transport of water can take place through a porous plasma oxide [10]. The water molecules diffuse from the bonded interface into the porous network. As a result, the hydroxyl groups on the opposite sides of the interface become accessible to each other and a reaction (Eq. 4.2) can take place according to

$$\mathrm{SiOH} + \mathrm{HOSi} \leftrightarrow \mathrm{Si} - \mathrm{O} - \mathrm{Si} + \mathrm{H_2}. \qquad (4.2)$$

In this reaction, covalent bonds between Si and O atoms are formed as a product. This step requires temperatures higher than room temperature in conventional bonding. The theory of a porous oxide is further supported by the fact that bonding a pair of wafers when only one of the surfaces is exposed to plasma results in high surface energies as shown in [83].

When the Si surface is plasma-treated, a thin, porous plasma oxide is formed that is very reactive and attracts excess water. Since the oxide is porous, part of the water is collected in micropores. MIRS-FTIR measurements shown in Fig. 4.8 support the idea of water trapped in micropores.

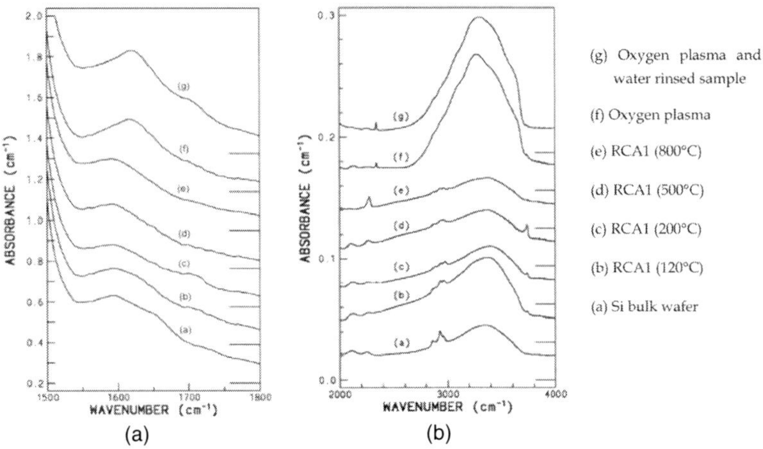

Figure 4.8. MIRS-FTIR spectra for Si to Si bonded wafer pairs: (a) the bending mode of the water region and (b) the symmetrical stretching mode of the water region.

In Fig. 4.8a, the bands at about 1600 cm^{-1} show the bending mode of water. The larger area under the bump at 1620 cm^{-1} for Fig. 4.8f,g shows that there is more water in the plasma bonded samples than in the RCA1 samples. Also, the plasma samples have more hydroxyl groups than the RCA1 cleaned samples as the region around 3400 cm^{-1} in Fig. 4.8b shows. This data is consistent with the theory that an enhanced surface area is due to a porous structure.

Enhanced bond strength of plasma pre-treated bonded wafers is attributed to the following effects [98]: (i) the removal of contaminants from the surface permits the formation of more dangling bonds and (ii) the increase in the amount of silanol groups. Water enables spontaneous bonding [111] while it also is the reason for thermally generated voids as shown in Fig. 4.9. Thermal void generation in Si to Si oxygen plasma-assisted bonded wafer pairs has been described in the literature [85, 89, 102]. The voids hamper the mass production and the reliability of the integrated devices and systems. Formation of interface voids must be reduced.

Severe void generation was observed when two Si surfaces were oxygen plasma treated in ICP RIE, bonded, and then annealed at elevated temperature in nitrogen ambient. The time needed to

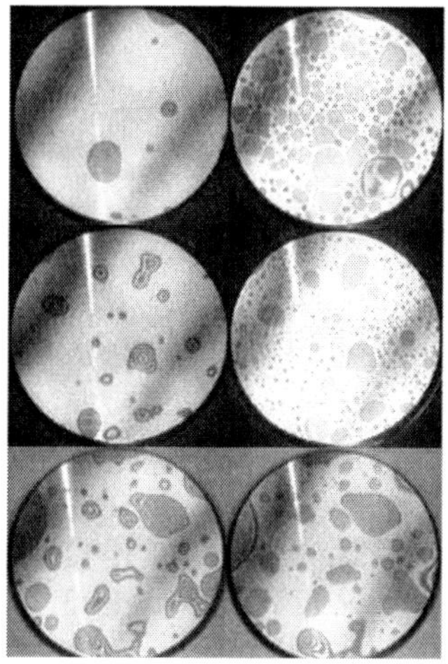

Figure 4.9. Thermally generated voids on Si to Si oxygen plasma-assisted wafer bonded interfaces. IR pictures (a), (c), and (e) are taken after the bonding step and IR pictures (b), (d), and (f) are taken after the annealing step. (a) RCA1 cleaned, 4% HF dipped, oxygen plasma treated, water dipped, spin dried, and bonded sample, (c) RCA1 cleaned, 4% HF dipped, oxygen plasma treated, and bonded sample, (b) and (d) samples from (a) and (c), respectively, annealed in nitrogen ambient ramped from room temperature to 100°C at 1°C min^{-1} and annealed for 10 min at 1000°C, (e) RCA1 cleaned, 4% HF dipped, oxygen plasma treated and heated to 100°C, in order to evaporate water, and bonded sample, (f) sample from (e) annealed in nitrogen ambient ramped from room temperature to 1000°C at 5°C min^{-1} and annealed for 20 min at 1000°C.

cause severe void generation is dependent on temperature [111]. At 400°C, voids appear after annealing for 1 hour, while annealing at 1000°C caused numerous voids in less than 10 min. These voids did not disappear during subsequent anneals at 1000°C for 24 hour or during an additional anneal at 1200°C for 1 hour. When bonding thermal SiO_2 to thermal SiO_2, no void generation is observed. Moreover, the void generation is suppressed when a thermally oxidised Si wafer is part of the bonded pair. If etched cavities are included at the bonded interface no, void generation close to the cavities is observed. In addition, the number of generated voids can be decreased by reducing the chamber pressure or by reducing the chamber pressure and the coil power simultaneously. Furthermore, no influence on the platen power was found [112].

When wafers are not dipped in water prior to bonding, there is a difference in the size, shape, and amount of generated voids after anneal, compared to a sample dipped in water and spin-dried. This is shown in Fig. 4.9a–d. The samples were annealed in nitrogen ambient, ramped from room temperature to 1000°C at 1°C min^{-1} and annealed for 10 min at 1000°C. The water dipped sample in Fig. 4.9b contains fewer but larger voids than the sample not dipped in water in Fig. 4.9d after annealing. If the amount of water at the wafer surfaces is further reduced by heating the wafers to 100°C before bonding, a drastic decrease in the amount of thermally generated voids due to annealing is observed as shown in Fig. 4.9e,f. In this case, the bonding was not spontaneous. Rather, thumb pressure was applied and a large amount of voids was generated by air enclosure during bonding at room temperature, shown in Fig. 4.9e. The sample was annealed in nitrogen ambient, ramped from room temperature to 1000°C at 5°C min^{-1} and further annealed for 20 min at 1000°C.

On the basis of the results presented above, the authors believe that the thermally generated voids are due to excess water trapped at the interface or in the porous structure of the surface layers. If the bonded interface is annealed, the water evaporates and forms a gas when the boiling point (100°C) is reached. As a consequence, the micropores are pressurised. Further increasing of the annealing temperature increases the pressure in the micropores, thus the silica skeleton begins to reorganise. Therefore, the Si can become highly deformed; micropores can join each other and voids are formed. These defects cannot be annealed out by increasing the annealing temperature, whereas in bonding RCA1 treated wafers, a distinct reduction of voids is seen after higher temperature anneal. The Si has already deformed plastically so that any further annealing has no effect on the amount or size of the imaged void.

As mentioned above, no void generation was observed when plasma treated thermal silicon oxide bonded to thermal silicon oxide bonded wafer pairs were used, as shown in Fig. 4.10. Following the above presented conclusions, this is due to the fact that the thermally grown silicon oxide is much thicker than the plasma oxide and can absorb a much greater amount of water.

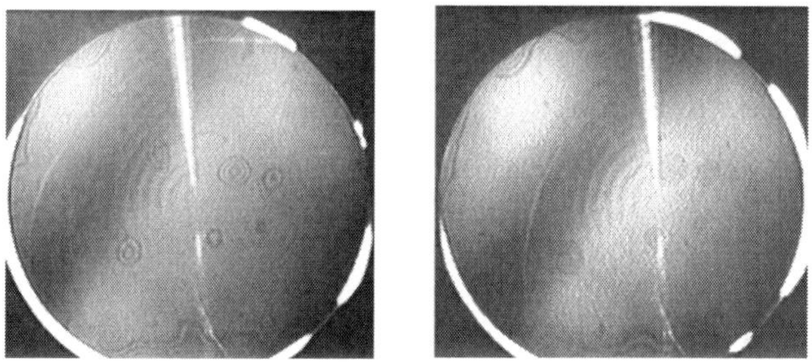

(a) (b)

Figure 4.10. IR pictures of a $SiO_2$to SiO_2 bonded wafer pair. The thermally grown oxide on both wafers is 2050 Å thick. (a) The wafer pair directly after bonding. (b) The bonded wafer pair after annealing at 1000°C for 10 min in nitrogen ambient.

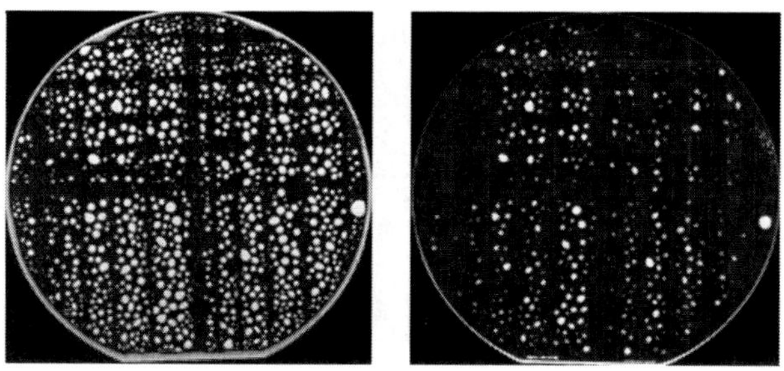

(a) (b)

Figure 4.11. SAM pictures of oxygen plasma-assisted Si to Si bonded wafers (6 inch). In one Si wafer of the bonded wafer pair, ventilation channels 10 μm wide and 10 μm deep are etched by reactive ion etching using ICP RIE. (a) After annealing ramped from room temperature to 1000°C at 5°C min^{-1} for 10 min, ramping down to 700°C and further annealing for 24 hour. (b) After further annealing at 1100°C for 2 hour.

Void generation can be reduced by controlling some plasma parameters or by bonding Si to thermal SiO_2 bonded interfaces. It can also be reduced by the amount of water at the surface of the wafer before bonding or by etching small channels in one of the Si wafers to be bonded as proposed by Kissinger [102] (see Fig. 4.11)

Zhang *et al.* [89] report a lower effective surface energy in the case of bonded wafers with patterned channels at the interface. This effect is explained by the discontinuous lateral bonding speed or contact wave velocity due to the introduction of the channels over the entire wafer surface. It is also attributed to a larger contact area between the bonded interface and the ambient. This can possibly accelerate the de-bonding process during surface energy measurements by the crack-opening method.

Besides measurement of surface energy bond strength [88, 113, 114] and leak measurements have been made on oxygen plasma bonded wafers [114]. Visser *et al.* found out that such bonding can be used to seal hermetically at low temperature (annealing at 150°C).

4.6 Low-Temperature Fabrication of Germanium-on-Insulator Using Remote Plasma Activation and Hydrogen Exfoliation

Germanium is gaining interest as a semiconductor material because bulk germanium has the highest hole mobility of all semiconductors (maximum $\mu_p = 1900\ cm^2V^{-1}s^{-1}$) and an electron mobility that is potentially twice that of Si (maximum $\mu_p = 3900\ cm^2V^{-1}s^{-1}$). Since Ge is a much rarer element than Si, it is impractical to use bulk germanium wafers. More suitable is the formation of thin germanium layers directly deposited on Si or on oxidised Si. It is also possible to produce mixed substrates in which n-channel devices are made in a thin SOI layer, and germanium p-channel transistors are made in a thin GeOI film both on the same substrate [115]. GeOI can be produced by the germanium condensation technique. In this technique, a $Si_{0.9}Ge_{0.1}$ layer is first epitaxially grown on SOI. Oxidation is performed to produce a SiO_2 layer and a Ge-enriched layer ($Si_{0.25}Ge_{0.75}$). Pure germanium is subsequently grown on the $Si_{0.25}Ge_{0.75}$ layer [116]. The epitaxial approach has the drawback of being carried out at a relatively high temperature (>650°C). Since Si and Ge have very different CTEs, the use of any high-temperature processing step results in the formation of a high density of thermal mismatch defects in the Ge epilayer [117, 118].

Ge to Si direct wafer bonding has been studied for use in high-performance photodetectors as well as high-quality epitaxial templates for GaAs growth [119–122]. However, in this section, we will focus on the characterisation of remote plasma activated Ge surfaces using oxygen and nitrogen radicals and show successful low-temperature Ge to Si direct bonding for thin-film Ge transfer. We also show that adding degassing channels at the interface decreases intrinsic voids caused by the chemical reaction at the Si–Ge interface. The chemical species on the activated Ge surfaces were investigated using angle-resolved x-ray photoelectron spectroscopy (ARXPS), while structural analysis was performed by SAM and HR-TEM.

4.6.1 *Germanium Surface Activation for Low-Temperature Bonding*

In a first experiment, 100 mm, <100>-oriented p-type Ge wafers (Ga doped with a resistivity of 0.016 Ω cm) were selected. The oxygen and nitrogen radical activated Ge surfaces were studied in a Vacuum Science Workshop Atomtech ESCA system using Al $K\alpha$ radiation ($h\nu = 1486.6$ eV). The Ge wafers were cleaned in an SC1-equivalent solution without ozone using a semi-tool spray acid tool (SAT) prior to surface activation. Wafers were then loaded into an Applied Microengineering Limited (AML) AW04 aligner bonder, and vacuum was applied. The wafers were then exposed for 10 min to either oxygen or nitrogen free radicals at a pressure of 1 mbar and a plasma power of 100 W. The radicals were generated by a remote plasma ring. A reference sample of Ge cleaned and bonded without exposure was also prepared. Wafers were bonded under a pressure of 1 kN applied for 5 min and immediately transferred for XPS analysis. The three bonded Ge to Ge samples were then de-bonded using a razor blade, cleaved into 2×2 cm^2 size samples and loaded into a high-vacuum sample holder of the XPS tool. The total exposure to ambient air was kept less than 1 min. The photoelectron peaks and chemical composition of the Ge surface were analysed at various take-off angles.

Binding state configurations of germanium atoms at the activated surfaces were evaluated by ARXPS using the de-bonded cleaved samples. Figure 4.12 shows the Ge 2p3/2, Ge 3d photoelectron features (take-off angle $= 75°$). As expected the Ge 2p3/2 signal shows two binding energy contributions at 1218.5 and 1221.0 (+/− 0.2) eV [123]. These can be assigned to zero- and quadra-valent

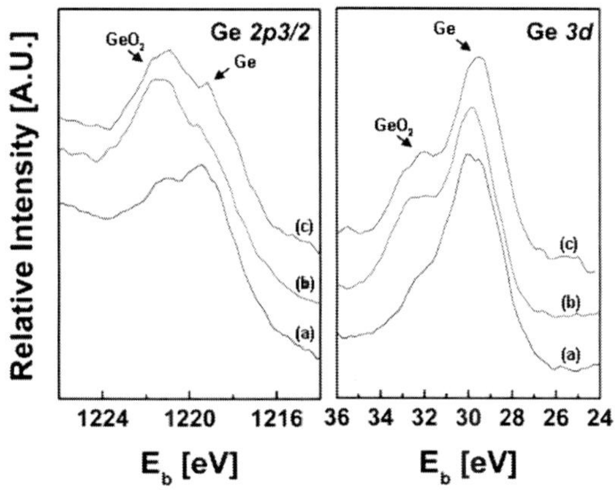

Figure 4.12. Ge 2p3/2 and Ge 3d photoelectron features (take-off angle = 75°): (a) reference cleaned sample (b) oxygen radical activated sample (c) nitrogen radical activated sample.

Ge, respectively. Zero-valent corresponds to bulk Ge, while quadra-valent indicates the presence of a GeO_2 layer. The Ge 3d peaks exhibit similar information. Two features can be resolved at 29.8 eV and 32.5 eV that once again are attributed to zero- and quadra-valent Ge, respectively [124–126]. There was no evidence in this work of the formation of detectable quantities of sub-stoichiometric oxides, which agrees with the Ge 3d signal (which is bulk sensitive) and Ge 2p signal (which is surface sensitive).

Quantitative relative contributions of the oxygen and germanium features are found by curve-fitting the XPS data. These data are reported in Table 4.3 as peak area ratios and are used to derive stoichiometric amounts using known sensitivity factors. The data collected at 75° is the most accurate representation of the actual stoichiometry close to the Ge surfaces. Curve fitting reveals a peak indicating the presence of OH^- which is located at a binding energy which is higher than the binding energy of GeO_2 [127]. This data is shown in the first column of Table 4.3 where the ratio of OH^- to GeO_2 (labelled O^-/O^{2-}) is represented. In the second column, we have derived the O/Ge ratio, which is the total O 1s peak area divided by total Ge 2p3/2 peak area. From the second column of Table 4.3,

Table 4.3. ARXPS peak area ratio. O^-/O^{2-} is the O^- signal area divided by O^{2-} signal area in O 1s spectra. O/Ge is the total O 1s peak area divided by total Ge 2p3/2 peak area

Sample/take-off angle	O^-/O^{2-}	O/Ge Total
Reference substrate 0°	0.34	0.04
25°	0.24	0.04
50°	0.09	0.08
75°	**0.00**	**0.25**
Post oxygen radical 0°	0.11	0.06
25°	0.11	0.07
50°	0.02	0.12
75°	**0.00**	**0.40**
Post nitrogen radical 0°	0.07	0.05
25°	0.09	0.06
50°	0.06	0.11
75°	**0.00**	**0.35**

it is clear that exposure of HP cleaned Ge substrates to free radicals produces some oxidation.

For samples exposed to either nitrogen or oxygen radicals, the quantification shown in Table 4.3 shows higher values of O/Ge ratio, which is more likely due to a thicker oxide film compared to the reference Ge sample. The nitrogen radical exposed Ge also shows a significant GeO_2 formation; this is similar to previous studies where Ge exposed to nitrogen plasma resulted in a more HP surface signifying formation of an oxide [128]. These quantifiable XPS data were used to estimate the film thickness as 0.39, 0.58, and 0.72 nm for the reference, nitrogen radical activated, and oxygen radical activated samples, respectively.

Most extensive hydroxylation, which is indicated in the first column (O^-/O^{2-}) of Table 4.3, was observed for the reference samples. Nitrogen radical activated samples gave the lowest –OH concentration and oxygen radical activated samples an intermediate value. The presence of hydroxyl species is important because increased hydroxyl groups can give rise to more HP reactions at bonded interfaces, which can then induce intrinsic void generation due to trapped reaction by-product. The signal due to O^- (OH^-) decreases towards grazing emission for all samples, which is due to a sub-surface hydroxide or sub-stoichiometric GeO_x.

Following this experiment, blank Ge wafers were bonded directly to Si using the same recipe and bonder used for the Ge bonded to Ge, i.e. 10 min oxygen or nitrogen radical activation at low temperature. <100>-oriented n-type Czochralski grown prime grade bare Si wafers with a diameter of 100 mm were bonded with the Ge wafers. Prior to bonding, the Ge and Si wafers were cleaned in an SC1-equivalent solution with ozone for Si and without ozone for Ge. After loading into the AML wafers were activated and bonded *in situ* under a pressure of 1 kN applied for 5 min at a chamber pressure ranging from 5 to 10 mbar. The wafers were annealed *in situ* at 100°C for 1 h with an applied pressure of 500 N in vacuum followed by an *ex situ* anneal at 200°C for 24 h in order to increase bond strength. The bonded pairs were then annealed again at 300°C for 24 h. The ramp-up rate was set to 0.5°C/min in both cases. After the anneal, Ge–Si bonded pairs remained intact despite their CTE mismatch, owing to the slow ramp-up rate. Bonded interfaces were imaged by SAM. Then Ge was directly bonded to a patterned Si wafer with 2 μm deep channels at a pitch of 400 μm using oxygen radical activation. The bond strength of the patterned Ge–Si wafer pair was measured using the crack opening method. Structural analysis of buried interfaces was studied by SAM and HR-TEM inspection.

The buried interfaces of bonded pairs were inspected using SAM after annealing at 200°C 24 h and additionally at 300°C for 24 h. Figure 4.13 shows scanning acoustic microscope images of bonded pairs: (a) pre-cleaned by SC1-equivalent solution, (b) post oxygen radical exposure, and (c) post nitrogen radical exposure. The HP reaction at the interfaces can be described by the following equations:

$$\mathrm{Ge - OH + OH - Si = Ge - O - Si + H_2O} \tag{4.3}$$

$$\mathrm{Ge(Si) + 2H_2O = GeO_2(SiO_2) + 2H_2} \tag{4.4}$$

Figure 4.13 shows that HP reactions, Eq. 4.3 and Eq. 4.4 at the interfaces create trapped gas like water vapour or hydrogen following the HP chemical reaction, which appears as a different contrast in the SAM images. Some of the water molecules resulting in covalent bonding of Ge–O–Si of oxygen and nitrogen exposed samples can diffuse through the nanometre range oxide layer and

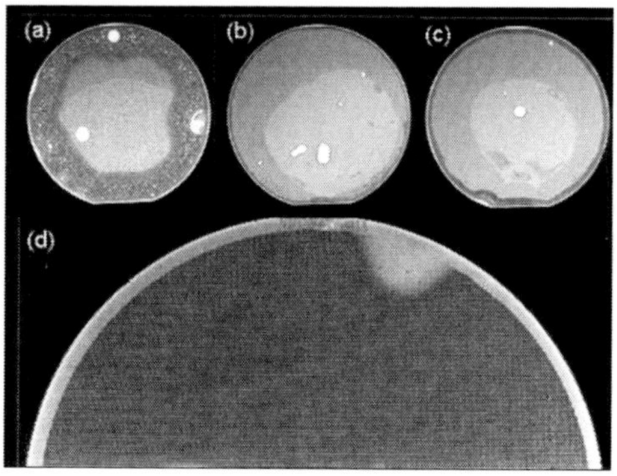

Figure 4.13. Scanning acoustic images of Ge/Si bonded wafer pairs after annealing at 200°C for 24 h, and additionally at 300°C for 24 h: (a) cleaned in an SC1-equivalent solution, followed by (b) O_2 radical 10 min exposure, (c) N_2 radical 10 min exposure, and (d) formed 2 μm depth channel in Si wafer prior to oxygen radical activated (10 min) bonding creates an exhaust path for trapped gas.

react with the bulk Ge to form Ge oxide and hydrogen. In Fig. 4.13a, however, the Ge surface is composed of a Ge suboxide, which results in an unstable interface prone to GeOx volatisation during anneal. Radical activated samples in Fig. 4.13b,c show minimal intrinsic voids due to the stability of the Ge surface (GeO_2) following exposure to free radicals. For the nitrogen radical activated sample in Fig. 4.13c, we can clearly see initiation of intrinsic voids caused by the reaction by-products near the wafer edge area. The formation of interface voids during low-temperature anneal depends on the thickness of the stoichiometric oxide film and the diffusivity of the reaction by-products at the interface, which is coincident with Si wafer bonding [129]. Thus, the SAM result agrees with the oxide thickness extracted from ARXPS as well as the diffusivity of water and hydrogen in GeO_2.

In another experiment, Ge was bonded to a Si wafer with 2 μm-deep channels formed in the surface. In Fig. 4.13d, we can clearly see that the radical activated bonding and anneal create an escape path for trapped gas or moisture, allowing high bond strength to be achieved. Additionally, the SAM image does show a uniformly coloured SAM image indicative of a well-bonded sample. The SAM

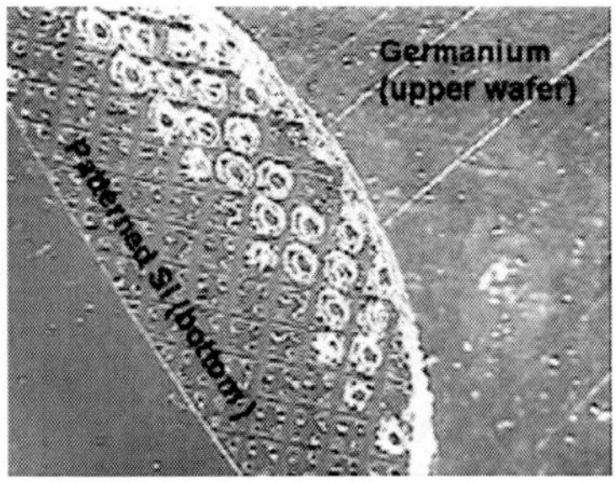

Figure 4.14. Optical microscope observation of razor blade tested sample after annealing at 200°C for 24 h and additional 300°C for 24 h. After crack opening test, bulk of the germanium crack (no de-bonding), revealing pieces of germanium strongly bonded to the thin buried oxide.

image suggests that HP reaction at the interface generates a gas phase by-product, which can easily diffuse to the channel. This result is similar to the result of Ge bonded to sapphire with patterned channels [130].

In Fig. 4.14, after low-temperature anneal at 200 and 300°C, the bonding energy of the channelled Si bonded to Ge is so high that insertion of a razor blade results in bulk fracture of the Ge, which is consistent with the respective fracture stress of single crystal Si (62 MPa) and Ge (49 MPa) [130]. The full bond strength is attributed to the oxygen radical activation.

Figure 4.15 shows a cross-sectional HR-TEM image of the buried interface using oxygen radical activation after 200°C and an additional 300°C 24 h anneal. The 2 nm-thick buried interface appears defect-free and smooth. The oxide interface corresponds with the calculated thickness (0.7 nm) measured by ARXPS of oxygen radical exposed Ge and to 1 nm-thick SiO_2 previously found on radical exposed Si. Additionally, compared to the Ge condensation technique [131], the TEM image shows no defects generated during the low thermal budget ($\leq$300°C) anneal. This low-temperature process using oxygen radical activation is the way to achieve high-quality interfaces without generating stacking fault or inducing surface damage by direct plasma activation.

Chemical and structural analyses show that remote plasma treatment using oxygen and nitrogen radical facilitates complete

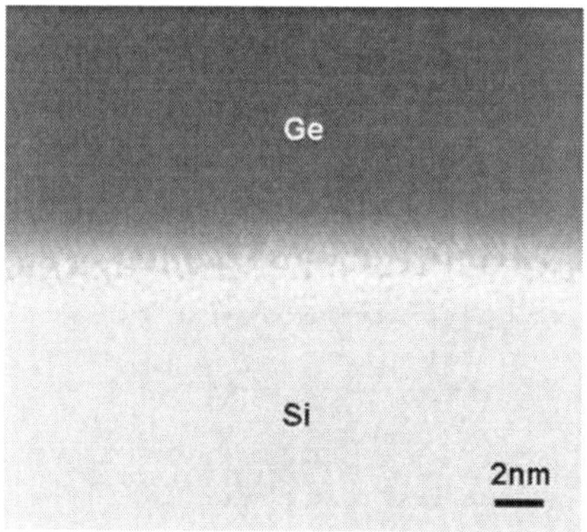

Figure 4.15. HR-TEM observation of buried interfaces after annealing at 200°C for 24 h and additional 300°C for 24 h.

oxidation on Ge surface to create a stable GeO_2 surface which increases the chemical reaction at the bonded interface. Thus, radical activation can enhance the bonding reaction allowing strong bond strength at low temperature while preserving the crystalline quality of the Ge. The radical activated bonding process is very suitable for a low-temperature process as needed for thermally mismatched materials such as those used for advanced logic or photonic devices.

4.6.2 *Low-Temperature Exfoliation of Germanium Layers*

Transfer of thin semiconductor layers by exfoliation has received a lot of attention since its first use for the fabrication of SOI substrates [127]. The implantation of hydrogen or inert gas into single crystalline semiconductor substrates leads to formation of a defective region below the surface. Under high-temperature treatment, usually in the range of 400 to 500°C, hydrogen molecules tend to be trapped in these defects and form pockets of gas at the projected range, commonly referred to as "bubbles". As temperature and/or anneal time increase, the internal pressure inside the bubbles increases and results in the formation of micro-cracks, which triggers the splitting of a thin semiconductor layer [132, 133]. The mechanisms which govern defect formation in H-implanted semiconductors and the creation of micro-cracks have already been extensively characterised in Si, Ge [134], and III-V compounds [135]. It is often addressed from a wafer bonding perspective and targets a range of applications varying from the fabrication of low defective substrates for CMOS compatible applications (SOI and GeOI substrates). The thermal budget required to generate micro-cracks is a sensitive matter for direct wafer bonding. The temperature range that is commonly considered lies above 400°C. However, such a high temperature is expected to induce significant modification of the bonded interface in heterogeneous substrates due to thermal expansion mismatch. This may result in poor bond strength and in degraded quality of the bonded interface [119, 136].

Here, we will investigate the feasibility of transferring hydrogen-implanted Ge to Si with a reduced thermal budget. Recently, co-implantation of hydrogen and helium for LT (300°C) exfoliation of Ge has been successfully demonstrated [137]. This approach presents a relatively long time-to-blister anneal (40 min) at a temperature of 300°C. In this paper, an exfoliation process which does not require any helium co-implant is investigated, which significantly reduces the time required for exfoliation at temperatures near 300°C. This process is based on a long defect nucleation step at low temperature ($\leq$150°C), followed by a very short time anneal (STA) at higher temperature (300°C). With this technique, complete exfoliation has been successfully demonstrated for hydrogen-implanted III-V materials such as InP and InAs [138, 139]. The benefits expected from this two-step process for direct wafer bonding are twofold: the low defect nucleation anneal enhances bond strength without degrading the bonded interface morphology and the STA induces minimum strain within bonded materials with dissimilar thermal expansion coefficients.

<100>-orientated n-type Ge wafers with a diameter of 100 mm (Sb doped, 0.03 Ω cm) were used for this experiment. Prior to hydrogen implant, a 100 nm-thick layer of PECVD SiO_2 was deposited and densified at 600°C. Ge substrates were implanted at room temperature with H_2^+ at a dose of 5×10^{16} cm^{-2} at 180 keV without active chuck cooling. The projected ion and Ge vacancy ranges that are expected are 650 and 590 nm below the Ge surface, respectively. Following the implant, the SiO_2 layer was completely removed in a dilute HF solution. One of the implanted substrates

Table 4.4. Ge surface roughness, as measured by AFM, following long-time anneals at LT ($\leq$200°C)

	RMS roughness (nm)	Scan area
As implanted	1.5	50 μm × 50 μm
After 100°C anneal — 22 h	0.4	10 μm × 10 μm
After 130°C anneal — 22 h	1.7	50 μm × 50 μm
After 150°C anneal — 22 h	0.4	10 μm × 10 μm

was diced into small samples (1 cm × 1 cm) prior to anneal. Strain profile within as-implanted Ge samples has been assessed by high resolution XRD measurements. The $\omega/2\theta$ diffraction patterns provide information about strain involved by the H^{2+} implant [140, 141]. Negligible strain profile variation was observed across the wafer, which suggests excellent implant uniformity among the 1 cm^2 samples. In addition, XRD measurements suggest that nucleation is already initialised during hydrogen implantation due to lack of wafer cooling during hydrogen implantation [141].

Following XRD measurements, a set of different anneal conditions was then considered in order to estimate the optimum LT thermal process to induce exfoliation. Implanted samples were sequentially annealed according to conditions described in Table 4.4. The onset of blistering was determined by tapping mode AFM and by optical microscopy. Prior to surface morphology characterisation, samples were cleaned in de-ionised (DI) water. Cross-sectional TEM (X-TEM) was performed to characterise the evolution of cracks created by hydrogen implant, as a function of thermal budget.

In addition to blister tests, Ge exfoliation was tested after direct bonding of an H_2^+ implanted Ge wafer to a single side polished 100 mm p-type [100] Si wafer. Prior to direct bonding, 100 nm of PECVD SiO_2 was deposited on the Si wafer. After oxide densification on the Si, 2 μm-deep channels were patterned through the oxide and the Si in order to facilitate the release of by-products generated during wafer bonding and annealing. The Si wafer was cleaned in a SC1-equivalent solution. The Ge wafer involved in the direct bonding experiment was implanted with implant conditions comparable to those used for blister tests. The implanted Ge wafer was cleaned in a 1:1 NH_4:H_2O solution dispensed by a SAT followed by four cycles alternating 1200:1 HF:H_2O and DI water cleans. Both Si and Ge wafers were then subjected to a Megasonic clean with DI water. Their surfaces were HP prior to the bonding. Wafers were loaded in an AML bonding chamber, which was pumped down to 10^{-5} mbar. The wafers were exposed for 10 min to free oxygen radicals generated by a remote plasma ring [110]. Wafers were bonded under a force of 1000 N applied for 5 min. The wafers were annealed *in situ* at 100°C for 1 h with an applied force of 500 N followed by an *ex situ* anneal at 130°C for 24 h in order to enhance bond strength and induce hydrogen platelet nucleation. The ramp-up rate was set to 0.5°C/min in both cases in order to minimise any wafer cracking due to CTE mismatch and to minimise the formation of thermally generated voids at the bonded interface [142]. The exfoliation was triggered by a 5 min STA at 300°C.

4.6.3 *Defect Nucleation and Oswald Ripening Mechanism*

Defect nucleation and hydrogen coalescence in implant-generated defects lead to the formation of hydrogen platelets and cracks in the bulk of the implanted semiconductor. These hydrogen-filled cavities have been reported to be located at a depth between the hydrogen projected range and the Ge vacancy range below the Ge surface [134]. Lattice deformation induced by these cavities generates surface blisters, which are clearly visible (Fig. 4.16). The evolution of the nucleation process can thus be monitored using either optical microscopy or AFM.

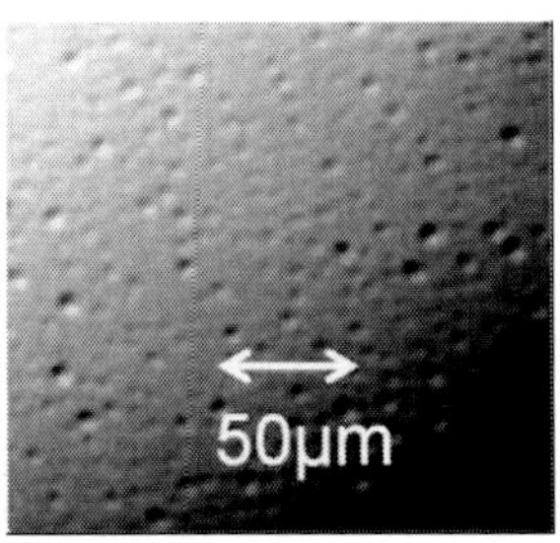

Figure 4.16. Optical micrograph showing the formation of blisters.

Significant hydrogen coalescence in III-V materials such as InP and GaAs has already been reported after long anneals at 150°C [138, 139]. In addition, dependence between the lowest temperature required to trigger the nucleation process and the melting point of the implanted material has also been highlighted in previous work [135]. Since the melting point of Ge (937°C) is close to the melting point of InP (1060°C), defect nucleation would be expected to occur in hydrogen-implanted Ge at 150°C or below the nucleation temperature in InP.

To study hydrogen nucleation, surface roughness measurements (AFM) after long (22 h) anneals at 100, 130, or 150°C have been performed. RMS roughness values are detailed in Table 4.4. As compared to as-implanted Ge, these long anneals at 100, 130, and 150°C do not modify surface roughness significantly. However, these anneals promote some migration of the hydrogen, as shown by X-TEM and XRD. X-TEM micrographs confirm that low-temperature annealing promotes the nucleation of small platelet defects. These nano-cracks are parallel to the substrate surface and are located close to the implant projected range (Fig. 4.17). The formation of these nano-cracks is known to be limited by the breaking of Ge–H lattice bonds and by hydrogen diffusion. The length of most of these cracks does not exceed 50 nm, and their propagation causes minor lattice deformation. Consequently, these nano-cracks do not modify the surface morphology significantly, when compared to as implanted Ge. It must be noted that this result does not hold for implanted Ge annealed at 200°C for 19 h, suggesting that the minimum thermal budget needed to trigger Ge exfoliation lies between 150 and 200°C. This result is in agreement with previously reported time-to-blister for H_2^+-implanted Ge samples (implant energy and dose are 160 keV and 5×10^{16} cm^{-2}, respectively): at a temperature equal to 200°C, the time-to-blister is estimated at 12 h [132].

The sample subjected to the long temperature anneal at 100°C was subsequently annealed at 200°C for 5 min. RMS roughness of this sample is not impacted by this STA. Consistently, ω-2θ diffraction patterns measured on this sample indicate very little relaxation of the strain created by the hydrogen implant (minor reduction of diffraction fringes), as compared to as-implanted Ge

Figure 4.17. (a) TEM picture of the defective region created by H_2^+ implant, in as-implanted Ge. A 5×10^{16} cm^{-2} H_2^+ dose implanted at 180 keV generates a 600 nm-thick implant-damaged region (region II) below a damage-free 150 nm-thick region under the germanium surface (region I). (b) TEM picture of hydrogen-implanted Ge following a 22 h anneal at 150°C showing the formation of nano-cracks at a depth close to the projected range, parallel to the substrate surface. (c) TEM picture of hydrogen-implanted Ge following a 22 h anneal at 150°C and a 5 min anneal at 300°C showing the transformation of nano-cracks in an almost continuous and thick micro-crack line at 645 nm below the germanium surface.

shown in Fig. 4.18a samples as-implanted and sample annealed at 100°C, 22 h followed by 200°C for 5 min. This result points towards a limited hydrogen diffusion in the implanted region and insufficient temperature (T = 200°C for 5 min) to trigger the blistering.

Additional samples annealed at 100, 130, and 150°C for 22 h were subjected to a STA at 300°C for 5 min. Surface roughness measurements suggest the formation of large hydrogen-filled cavities along the cracks and subsequent Ge exfoliation (Table 4. 5).

It should be noted that the height of surface blisters correlates well with thermal budget of the defect nucleation process: the higher the nucleation temperature, the larger the blisters. The ω-2θ diffraction patterns for these samples confirm this enhanced hydrogen diffusion after completion of such two-step anneals, as most of the diffraction fringes induced by the hydrogen implant are strongly reduced as compared to ω-2θ diffraction patterns prior to STA at 300°C (Fig. 4.18b). Such modification of the diffraction pattern suggests that significant hydrogen diffusion occurs after a

Table 4.5. Ge surface roughness, as measured by AFM, following long-time anneals at low temperature ($\leq$200°C) and STA at 200 or 300°C. Scan area is 50 μm × 50 μm unless specified

Anneal sequence	RMS roughness (nm)	Scan area
After 100°C anneal — 22 h, followed by 200°C anneal — 5 min	0.4	10 μm × 10 μm
After 100°C anneal — 22 h, followed by 300°C anneal — 5 min	7.6	10 μm × 10 μm
After 130°C anneal — 22 h, followed by 300°C anneal — 5 min	14.6	50 μm × 50 μm
After 150°C anneal — 22 h, followed by 300°C anneal — 5 min	28.1	50 μm × 50 μm

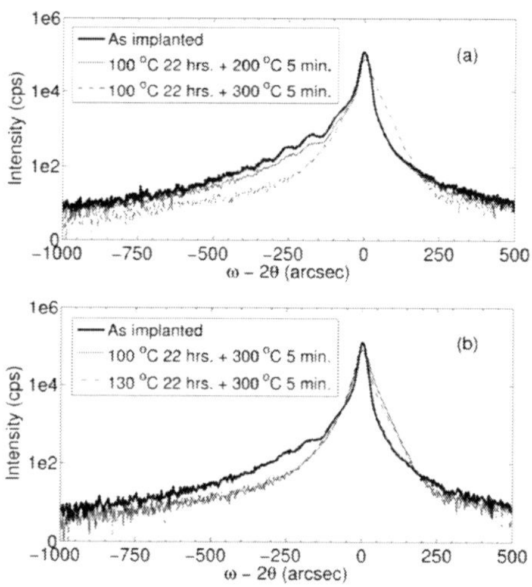

Figure 4.18. X-ray diffraction (ω-2θ) patterns: (a) impact of STA on strain relaxation in hydrogen implanted Ge after long anneals at 100°C; (b) impact of thermal budget during the nucleation process on implant-induced strain after STA at 300°C.

combined long-time anneal at temperatures as low as 100°C and a short-time anneal at 300°C.

In addition, the broadening of the Ge feature on triple-axis ω diffraction patterns suggests an increase of local deformation (Fig. 4.19a). The Full Width at 0.001 Height (FW0.001M) increases

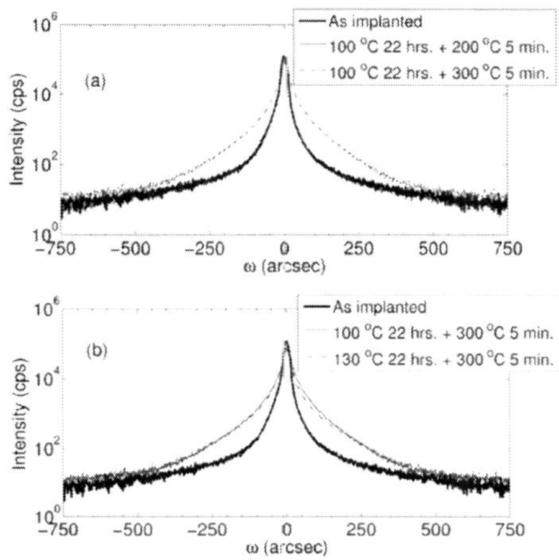

Figure 4.19. X-ray diffraction (ω) patterns: (a) impact of STA on lattice deformation in hydrogen implanted Ge after long-time anneals at 100°C; (b) impact of thermal budget during the nucleation process on the lattice deformation caused by the STA at 300°C.

from 240 arcsec in as-implanted Ge to 600 arcsec after a long-time anneal at 100°C and a short-time anneal at 300°C. It is attributed to local lattice deformations due to hydrogen Oswald ripening [141]. The latter is observed after the STA at 300°C, irrespective of the anneal temperature considered for completing the defect nucleation (Fig. 4.19b). This is a key result since it shows that a long-time anneal at a temperature as low as 100°C reduces significantly the time-to-blister at 300°C, compared with the state-of-the-art data [132, 137].

Additional roughness measurements show evidence of blistering, which is confirmed by X-TEM analysis and suggested by x-ray diffraction patterns. On the sample annealed at 150°C for 22 h and subsequently annealed at 300°C for 5 min, the formation of micro-cracks is observed by TEM. The latter is a result of the merger of nano-cracks created at LT, which form longer cracks and cause Ge exfoliation (Fig. 4.17c). The formation of these micro-cracks proceeds in a similar way to Si: during STA, hydrogen diffuses along the defect lines and forms large gas pockets (diameter >5 nm) at the expense of smaller ones [142]. The internal pressure in large gas pockets increases and ultimately leads to the extension of small cracks into micro-cracks.

4.6.4 *Low-Temperature Formation of Ge-on-Insulator*

A bonded sample made of a hydrogen-implanted Ge wafer directly bonded to a Si wafer was processed in order to demonstrate the feasibility of transferring a thin Ge layer at LT. The thermal treatment which was considered starts with a long (24 h) anneal at 130°C and ends with a 5 min anneal at 300°C. The purpose of the initial long time, LT anneal is twofold: (i) it strengthens the bonds created at the Ge/oxide interface during the bonding reaction and (ii) it promotes hydrogen platelet nucleation within the Ge substrate without modifying its morphology at the bonded interface. A 680 nm-thick layer of Ge was transferred onto 100 nm of SiO_2 deposited on the host Si wafer. High bond strength was achieved, as suggested by the fact that the transferred Ge layer follows closely the pattern printed in the oxide layer prior to bonding. The Ge surface roughness as measured by AFM is 15 nm (Ge surface roughness after exfoliation is illustrated in Fig. 4.20). This value is consistent with the surface roughness value (14.6 nm) measured on bare implanted Ge after 22 h-long anneal at 130°C (Table 4.5).

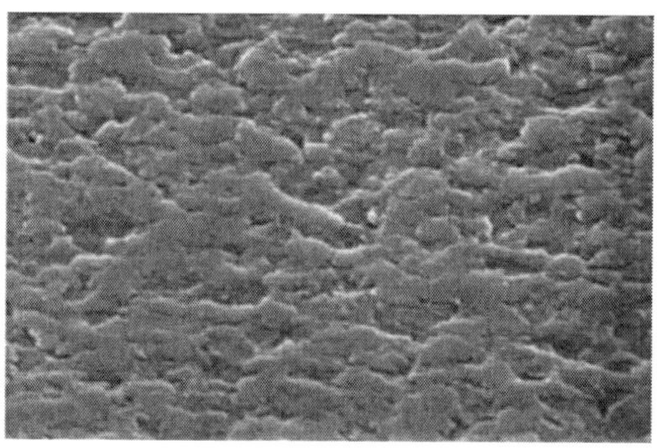

Figure 4.20. SEM graphs of the GeOI-insulator sample resulting from an exfoliation carried out after a 24 h-long anneal at 130°C and a short-time anneal at 30°C for 5 min: top-down tilted view of the Ge surface exposed after complete exfoliation.

A LT process for bonding and exfoliation of Ge has been developed using hydrogen-implanted Ge layers and surface remote plasma activation prior to bonding. It has been demonstrated that proper engineering of the annealing sequence can promote hydrogen platelet formation and allow for bonding and complete Ge exfoliation after annealing at temperatures no higher than 300°C. These results also demonstrate that the lowest thermal budget required for defect nucleation is similar for Ge and III-V materials such as InP. Enhanced bond strength in directly bonded heterojunctions, like GeOI or bonded III-V material for photonics, is the main benefit expected from such LT exfoliation process.

4.7 Surface-Activated Bonding

Another method, surface-activated bonding (SAB), realises high bond strength at LT [100]. However, by this method a considerable number of defects and damage are introduced due to the heavy Ar beam sputtering of the surfaces. Howlader *et al.* [92] present the development of SAB equipment. The developed equipment is designed for research and development purposes with a flexible fixture facilitating die-to-die and wafer-to-wafer bonding. The SAB bonder is facilitated by both Ar fast atom beam (FAB) sources and an Ar ion beam source [143]. Preliminary results are reported on Si to Ge bonding and Si to GaAs bonding performed with the following parameters: background vacuum 1.4 10^{-9} Torr, Ar-FAB 1.45 kV 48 mA, Ar irradiation 180–300 s and a bonding load of 1500 N. The bonding was successful at room temperature without void formation at the interface.

Wang *et al.* [97] tested the system for bonding of Si wafers and gold coated Si wafers (100 nm-thick Au) in low vacuum condition, nitrogen atmosphere, and ambient air successfully. Kurayama *et al.* [144] used the system to fabricate a two-layer Ge structure with a sharp doping transition at the interface on a wafer scale. Two 2 inch Ga-doped p-type, epi-ready, and <100>-oriented Ge wafers with an RMS roughness of 0.3 nm were used. Both wafers were activated by an Ar FAB source with a voltage of 1.5 kV and a current of 60 mA, resulting in an etch rate for Ge of 2.7 nm.

After activation, the wafers were bonded under 1.2 MPa at room temperature. The wafers bonded were almost void free and strong with a reported tensile strength of 14 MPa. TEM pictures revealed an 8 nm-thick amorphous interlayer, reportedly as a result of the Ar-FAB irradiation. A SIMS analysis indicated a sharp Ge concentration gradient across the bonded interface.

4.8 Summary

LT bonding is gaining renewed interest as it might enable 'More than Moore', i.e. the integration of different functions (sensors, optics, RF, etc.) with standard CMOS electronics. This will require different semiconductor materials on the same substrate. LT bonding will be necessary due to the different CTEs of the materials of interest. There are ways to increase the bonding quality and strength by introducing remote plasma or direct plasma surface modifications prior to bonding. Some examples of thin film Ge transferred to a Si substrate show that LT bonding along with exfoliation promises a way to integrate high mobility materials with standard Si CMOS.

References

1. Ristic, L. and Shah, M. (1996). Trends in MEMS Technology, Wescon '96, pp. 64–72.
2. Alexe, M. and Gösele, U. (2004). *Wafer Bonding: Applications and Technology* (Springer, ISBN 3 540-21049-0).
3. Gabriel, M., Johnson, B., Suss, R., Reiche, M., *et al.* (2006). Wafer direct bonding with ambient pressure plasma activation, *Microsystem Technol.* **12**, pp. 397–400.
4. Tong, Q.-Y. and Gösele, U. (1999). *Semiconductor Wafer Bonding Science and Technology* (John Wiley & Sons, Inc. New York).
5. Tong, Q.-Y. (2001). Wafer bonding for integrated materials, *Mater. Sci. Eng. B*, **87**, pp. 323–328.
6. Plößl, A. and Kräuter, G. (1999). Wafer direct bonding: tailoring adhesion between brittle materials, *Mater. Sci. Eng.*, **25**, pp. 1–88.

7. Najafi, K., Harpster, T. J., Kim, H., Mitchell, J. S., *et al.* (2007). Wafer bonding *Comprehensive Microsystems*, vol 1, ed. Gianchandani, Y. B., Tabata, O. and Zappe, H. (Amsterdam: Elsevier) pp. 235–270.
8. Gilleo, K. (2005). *MEMS/MOEMS Packaging — Concepts, Designs, Materials, and Processes* (McGraw-Hill).
9. Okuno, Y., Uomi, K., Aoki, M., and Tsuchiya, T. (1997). Direct wafer bonding of III-V compound semiconductors for free-material and free-orientation integration, *IEEE J. Quan. Electron.*, **33**, pp. 959–969.
10. Amirfeiz, P., Bengtsson, S., Bergh, M., Zanghellini, E., *et al.* (2000). Formation of Silicon by Plasma-Activated Wafer Bonding, *J. Electrochem. Soc.*, **147**, pp. 2693–2698.
11. Farrens, S. N. and Roberds, B. E. (2001). *In situ* plasma bonding method, *United States Patent.*
12. Tong, Q.-Y. (2003). A method of epitaxial-like wafer bonding at low temperature and bonded structures, *US 2003141502 (A1).*
13. Pasquariello, D. and Hjort, K. (2002). Plasma-assisted InP-to-Si Low temperature bonding, *IEEE J. Sel. Top. Quan. Electron.*, **8**, pp. 118–131.
14. Akatsu, T., Plössl, A., Scholz, R., Stenzel, H., *et al.* (2001). Wafer bonding of different III–V compound semiconductors by atomic hydrogen surface cleaning, *J Appl. Phys.*, **90**, pp. 3856–3862.
15. Gösele, U., Bluhm Y., Kästner G., Kopperschmidt P., *et al.* (1999). Fundamental issues in wafer bonding, *J. Vac. Sci. Technol. A*, **17**, pp. 1145–1152.
16. Feijoo, D., Chabal, Y. J., and Christman, S. B. (1994). Silicon wafer bonding studied by infrared absorption spectroscopy, *Appl. Phys. Lett.*, **65**, pp. 2548–2550.
17. Weldon, M. K., Chaba, Y. J. l., Hamann, D. R., Christman, S. B., *et al.* (1996). Physics and chemistry of silicon wafer bonding investigated by infrared absorption spectroscopy, *J. Vac. Sci. Technol B,* **14**, pp. 3095–3106.
18. Reiche, M., Hopfe, S., Gösele, U., and Tong, Q. Y. (1997). Characterization of buried interfaces by multiple internal reflection spectroscopy (MIRS), *Microchimica Acta,* **25**, pp. 367–373.
19. Farrens, S. and Sood, S. (2008). Wafer level packaging: balancing device requirements and materials properties, *41st International Symposium on Microelectronics - IMAPS*, Providence, Rhode Island, USA, 2–6 November 2008.

20. Lemons, R. A. and Quate, C. F. (1974). Acoustic microscope-scanning version, *Appl. Phys. Lett*, **24**, pp. 163–165.
21. Yu, Z. and Boseck, S. (1995). Scanning acoustic microscopy and its applications to materials characterization, *Rev. Mod. Phys.*, **67**, pp. 863–891.
22. www.sonoscan.com. *www.sonix.com.*
23. www.sonix.com.
24. Moriceau, H., Rieutord, F., Di Cioccio, L., Gueguen, P., *et al.* (2010). Low temperature by CMP or plasma activated direct bonding, *Proc. 2nd Int. IEEE Workshop on Low Temperature Bonding for 3D Integration*, p. 59.
25. Hayashi, S., Bruno, D., Sandhu, R., and Goorsky, M. S. (2003). X-ray scattering techniques for assessment of III–V wafer bonding, *J. Phys. D: Appl. Phys*, **36**, pp. A236–240.
26. DeWolf, I. (1999). Stress measurements in Si microelectronics devices using Raman spectroscopy, *J. Raman Spectrosc.*, **30**, pp. 877–883.
27. Nguyen, N. V., Maslar, J. E., Kim, J.-Y., Han, J.-P., *et al.* (2004). Crystalline quality of bonded silicon-on-insulator characterized by spectroscopic ellipsometry and Raman spectroscopy, *Appl. Phys. Lett*, **85**, pp. 2765–2767.
28. Boher, P., Piel, J. P., and Stehle, J. L. (1998). Real time spectroscopic ellipsometry: a new tool to characterize SOI wafers. The Electrochemical Society, *Semiconductor Wafer Bonding: Science, Technology and Applications IV.* pp. 249–256.
29. Vallin Ö., Jonsson, K., and Lindberg, U. (2005). Adhesion quantification methods for wafer bonding, *Mater. Sci. Eng. R*, pp. 109–165.
30. Kopperschmidt, P., Senz, S., and Scholz, R. (2001). Interface defects in integrated hybrid semiconductors by wafer bonding, *Phys. B*, **308–310**, pp. 1205–1208.
31. Maszara, W. P., Goetz, G., Caviglia, A., and McKitterick, J. B. (1988). Bonding of silicon wafers for silicon-on-insulator, *J. Appl. Phys*, **64**, pp. 4943–4950.
32. Bertholet, Y, Iker, F., Raskin, J.-P., and Pardoen, T. (2004). Steady-state measurement and modeling of wafer bonding failure resistance *Sens. Actuators, A*, **110**, pp. 157–163.
33. Anstis, G. R., Chantikul, P., Lawn, B. R., and Marshall, B. B. (1981). A critical evaluation of indentation techniques for measuring fracture toughness i: direct crack measurements, *J. Am. Ceram. Soc.*, **64**, p. 533.
34. Bagdahn, J., Plöfll A., Wiemer, M., and Petzold, M. (1999). Measurement of local strength distribution of directly bonded silicon

wafers using the micro-chevron-test, The Electrochemical Society, *Semiconductor Wafer Bonding: Science, Technology and Applications V.* pp. 218–223.

35. Bring, M., Sanz-Velasco, A., and Enoksson, P. (2006). Method for measuring fracture toughness of wafer-bonded interfaces with high spatial resolution, *J. Micromech. Microeng.,* **16**, pp. 68–74.
36. Köhler, J., Jonsson, K., Greek, S., and Stenmark, L. (2000). Weibull fracture probability for silicon wafer bonding evaluation *J. Electrochem. Soc.,* **147**, pp. 4683–4687.
37. Shimbo, M., Furukawa, K., Fukuda, K., and Tanzawa, K. (1986). Silicon-to-silicon direct bonding method, *J. Appl. Phys,* **60**, pp. 2987–2988.
38. Brooks, A. and Donovan, R. (1972). Low temperature electrostatic si-to-si seal using sputtered borosilicate glass, *J. Electrochem. Soc.*, **119**, pp. 545–546.
39. Tilmans, H. A. C., Van de Peer, M. D. I., and Beyne, E. (2000). The intend reflow sealing (IRS) technique — a method for fabrication of sealed cavities for MEMS devices, *J. Microelectromech. Sys.*, **9**, pp. 206–217.
40. Corman, T., Enoksson, P., and Stemme, G. (1997). Gas damping of electrostatically excited resonators, *Sens. Actuators, A*, **61**, pp. 249–255.
41. Jourdain, A., De Moor, P., Pamadighantam, S., and Tilmans, H. A. C. (2002). Investigation of the hermeticity of BCB-sealed cavities for housing (RF-) MEMS devices, *Proc. 15th IEEE International Conference on Micro Electro Mechanical Systems*, pp. 677–680.
42. Liao, G., Zhang, X., Lin X., Ma, C., *et al.* (2000). Ultraviolet exposure enhanced silicon direct bonding, *Front. Mech. Eng. China*, **5**, pp. 87–92.
43. Raeissi, N. B., Amirfeiz, P., and Engström, O. (2005). Electrical properties of low-temperature bonded unipolar Si/Si junctions, *Proc. 7th International Symposium on Semiconductor wafer bonding: Science, Technology, and Applications, Electrochem. Soc.*, p. 205.
44. Raeissi, B., Sanz-Velasco, A., and Engström, O. (2010). The role of mobile charge in oxygen plasma-enhanced silicon-to-silicon wafer bonding, *Electrochem. Solid-State Lett.*, **13**, pp. H179–H181.
45. Wallis, G. and Pomerantz, D. (1969). Field assisted glass-metal sealing, *J. Appl. Phys*, **40**, pp. 3946–3949.
46. Obermeier, E. (1995). Anodic wafer bonding, *Proc. 3rd Int. Symposium on Semiconductor wafer bonding: Science, Technology, and Applications, Electrochem. Soc.*, **95-7**, p. 212.

47. Nafari, A., Danilov, A., Rödjegård, H., Enoksson, P., *et al.* (2005). A micromachined nanoindentation force sensor, *Sens. Actuators, A*, **123–124**, pp. 44–49.
48. Cozma, A. and Puers, R. (1995). Characterization of the electrostatic bonding of silicon and Pyrex glass, *J. Micromech. Microeng.*, **5**, p. 98.
49. Anthony, T. R. (1983). Anodic bonding of imperfect surfaces, *J. Appl. Phys.*, **54**, pp. 2419–2428.
50. Nitzsche, P., Lange, K. Schmidt, B. Grigull, S. *et al.* (1998). Ion drift processes in pyrex-type alkali-borosilicate glass during anodic bonding, *J. Electrochem. Soc.*, **145**, pp. 1755–1762.
51. Esashi, M., Nakano, A., Shoji S., and Hebiguchi, H. (1990). Low-temperature silicon-to-silicon anodic bonding with intermediate low melting point glan, *Sens. Actuators, A*, **23**, pp. 931–934.
52. Hanneborg, A. Nese, M. Jakobsen, H., and Holm, R. (1992). Silicon-to-thin film anodic bonding, *J. Micromech. Microeng.*, **2**, p. 117.
53. Weichel, S., De Reus, R., and Lindahl, M. (1998). Silicon-to-silicon wafer bonding using evaporated glass, *Sens. Actuators, A*, **70**, pp. 179–184.
54. Quenzer, H. J., Schulz, A. V., Kinkopf, T., and Helm, T. (2001). Anodic Bonding on Glass Layers prepared by a Spin-On Glass Process: preparation process and experimental results, *11th International Conference on Solid-State Sensors and Actuators*, pp. 230–233.
55. Weichel, S., De Reus, R., Bouaidat, S., Rasmussen, P. A, *et al.* (2000). Low-temperature anodic bonding to silicon nitride, *Sens. Actuators, A*, **82**, pp. 249–253.
56. Schjølberg, H. K., Poppe, E., Moe, S., Storås, P., *et al.* (2006). Anodic bonding of glass to aluminium, *Microsys Technol.*, **12**, pp. 441–449.
57. Müller, E., Bartnitzek, T., Bechtold, F., Pawlowski, B., *et al.* (2005). Development and processing of an anodic bondable ltcc tape, *IMAPS EMPC 2005*.
58. Imanaka, Y. (2005), Multilayered low temperature cofired ceramics (LTCC) technology, (NY: Springer).
59. Khan, M. F., Ghavanini, F. A., Haasl, S., Löfgren, L., *et al.* (2010). Methods for characterization of wafer-level encapsulation applied on Si to LTCC Anodic bonding, *J. Micromech. Microeng*, **20**, p. 064020.
60. Dastjerdi, M. H. T., Sanz-Velasco, A., Vukusic, J., Sadeghi, M., *et al.* (2010). Transfer of InP-based HBV epitaxy onto borosilicate glass substrate by anodic bonding, *Electron. Lett*, **46**, pp. 1013–1014.

61. Schjølberg-Henriksen, K., Plaza, J. A., Rafí, J.-M., Esteve, J., *et al.* (2002). Protection of MOS capacitors during anodic bonding *J. Micromech. Microeng*, **12**, pp. 361–368.

62. Visser, M., Weichel, S., Storås, P., Shi, P. X., *et al.* (2001). Sodium Distribution in thin-film anodic bonding, *Sens. Actuators, A,* **92**, pp. 223–228.

63. Chavan, A. V. and Wise, K. D. (2002). A monolithic fully-integrated vacuum-sealed CMOS pressure sensor, *IEEE Trans. Electron Devices*, **49**, pp. 164–169.

64. Lasky, J. B. (1986). Wafer bonding for silicon-on-insulator technologies, *Appl. Phys. Lett*, **48**, pp. 78–80.

65. Bäcklund, Y., Hermannsson, K., and Smith, L. (1992). Bond-strength measurements related to silicon surface hydrophilicity, *J. Electrochem. Soc.*, **139**, pp. 2299–2301.

66. Ventosa, C., Rieutord, F., Libralesso, L., Morales, C., *et al.* (2008). Hydrophilic low-temperature direct wafer bonding *J. Appl. Phys*, **104**, p. 123524.

67. Spierings, G. A. C. M., Haisma, J., and Michelsen, T. M. (1995). Surface-related phenomena in the direct bonding of silicon and fused-silica wafer pairs, *Philips. J. Res.*, **49**, pp. 47–63.

68. Gösele, U., Tong, Q.-Y., Schumacher, A., Kräuter, G., *et al.* (1999). Wafer bonding for microsystems technologies, *Sens. Actuators, A,* **74**, pp. 161–168.

69. Haisma, J., Spierings, G. A., Michielsen, T., and Adema, C. (1995). Surface preparation and phenomenological aspects of direct bonding, *Philips J. Res.*, **49**, pp. 23–46.

70. Gui, C., Albers, H., Gardeniers, J. G., Elwenspoek, M., *et al.* (1996). Fusion bonding of rough surfaces with polishing technique for silicon micromachining, *Microsys Technol*, **3**, pp. 122–128.

71. Haisma, J., Spierings, B. A. C. M., Biermann, U. K. P., and Gorkum, A. A. V. (1994). Diversity and feasibility of direct bonding — a survey of a dedicated optical technology, *Appl. Opt*, **33**, pp. 1154–1169.

72. Kajiyama, K. (1995). Buried-oxide layer formation by high-dose oxygen-ion implantation into Siwafers: SIMOX (separation by implanted oxygen), *Appl. Surf. Sci*, **85**, pp. 259–264.

73. Maszara, W. P. (1991). Silicon-on-insulator by wafer bonding: a review, *J. Electrochem. Soc*, **138**, pp. 341–347.

74. Bruel, M., Aspar, B., and Auberton-Herve, A. J. (1997). Smart-cut: a new silicon on insulator material technology based hydrogen implantation and wafer bonding, *Jpn. J. Appl. Phys.,* **36**, pp. 1636–1641.
75. Tauzin, A., Akatsu, T., Rabarot, M., Dechamp, J., *et al.* (2005). Transfer of two-inch gan film by the Smart Cut^{TM}technology, 207th ECS Meeting, *8th International Symposium on Semiconductor Wafer Bonding: Science, Technology, and Applications*, p. Abstract #487. 1091–8213 (print) 2151–2043 (online).
76. Noworolski, J. M., Klaassen, E., Logan, J., Petersen, K., *et al.* (1996). Fabrication of SOI wafers with buried cavities using silicon fusion bonding and electrochemical etchback, *Sens. Actuators, A*, **54**, pp. 709–713.
77. Enoksson, P., Stemme, G., and Stemme, E. (1997). A silicon resonant sensor structure for Coriolis mass-flow measurements, *J. Microelectromech. Syst.*, **6**, pp. 119–125.
78. Hayashi, S., Sandhu, R. Wojtowicz, M., Sun, Y., *et al.* (2005). Determination of wafer bonding mechanisms for plasma activated SiN films with x-ray reflectivity, *J. Phys. D: Appl. Phys*, **38**, pp. A174–A178.
79. Luoto, H., Suni, T., Henttinen, K., and Kulawski, M. (2006). Direct bonding of thick film polysilicon to glass substrates, *J. Electrochem Soc*, **153**, pp. G981–GG985.
80. Farrens, S. and Smith, J. (1995). Chemical free wafer bonding of silicon to glass and sapphire. The Electrochemical Society, *3rd International Symposium on Semiconductor wafer bonding: Physics and Applications*, p. 595.
81. Pasquariello, D. and Hjort, K. (2001). Low temperature plasma assisted InP-to-silicon wafer bonding: an alternative to heteroepitaxial growth, *200th Meeting of The Electrochemical Society*, p. 1454.
82. Pasquariello, D. (2001). Plasma assisted low temperature semiconductor wafer bonding (Uppsala University Press, Sweden).
83. Tong, Q.-T., Gan, Q., Hudson, G., Fountain, G., *et al.* (2004). Low temperature InP/Si wafer bonding, *J. Appl. Phys. Lett*, **84**, pp. 732–734.
84. Tan, C. S. and Chong, G. Y. (2010). Low temperature wafer bonding of low-k carbon-doped oxide for applications in 3D integration, *Electrochem. Solid-State Lett.*, **13**, pp. H27–H29.
85. Wiegand, M., Reiche, M., and Gösele, U. (2000). Time-dependent surface properties and wafer bonding of o2-plasma-treated silicon (100) surfaces, *J Electrochem Soc*, **147**, pp. 2734–2740.

86. Reiche, M., Gutjahr, K., Stolze, D., Burczyk, D., *et al.* (1997). The effect of a plasma pre-treatment on the Si/Si bonding behavior, *Proc. 4th Int. Symposium on Semiconductor wafer bonding: Science, Technology, and Applications, Electrochem. Soc.*, **97–36**, p. 437.
87. Milekhin, A., Friedrich, M., Hiller, K., Wiemer, M., *et al.* (2000). Characterization of low-temperature wafer bonding by infrared spectroscopy, *J. Vac. Sci. Tech, B*, **18**, pp. 1392–1396.
88. Weichel, S. (2000). Silicon to silicon wafer bonding for microsystem packaging and formation (Academic Press, Technical University of Denmark).
89. Zhang, X. X. (2004). Investigation on the uniformity of surface energy in silicon direct-bonding technique, *J. Electrochem. Soc*, **151**, pp. G568–G573.
90. Bengtsson, S. and Amirfeiz, A. (2000). Room temperature wafer bonding of silicon, oxidized silicon, and crystalline quartz *J. Electron. Mater*, **29**, pp. 909–915.
91. Yokoyama, M., Yasuda, T., Takagi, H., Yamada, H. M., *et al.* (2009). Thin body III-V Semiconductor-On-Insulator Metal-Oxide-Semiconductor Field-Effect Transistors on Si fabricated using direct, *Appl. Phys. Exp*, **2**, p. 124501.
92. Howlader, M., Shintani, H., Suga, T., Shigeto, A., *et al.* (2010). Development of a new SAB equipment for room temperature bonding, *2nd International IEEE Workshop on Low Temperature Bonding for 3D Integration*.
93. Vossen, J. L. and Kern, W. (1991). *Thin film processes.* 1st (Academic Press, San Diego).
94. Pasquariello, D., Hedlund, C., and Hjort, K. (2000). Oxidation and induced damage in oxygen plasma in situ wafer bonding, *J. Electrochem. Soc*, **147**, pp. 2699–2703.
95. Roberds, B. E. and Farrens, S. N. (1997). Low temperature, *in situ* plasma activated wafer bonding, The Electrochemical Society, *4th International Symposium on Semiconductor wafer Bonding: Science, Technology and Applications*, p. 2466.
96. Zucker, O. (1993). Application of oxygen plasma processing to silicon direct bonding, *Sens. Actuators, A*, **36**, pp. 227–231.
97. Wang, Y-H., Oshikawa, K., and Suga, T. (2010). Low temperature bonding of si wafer in ambient air and nitrogen, *Proc. 2nd Int. IEEE Workshop on Low Temperature Bonding for 3D Integration*, p. 95.

98. Tan, C. M., Yu, W., and Wei, J. (2006). Comparison of medium-vacuum and plasma-activated low-temperature wafer bonding, *Appl. Phys. Lett*, **88**, p. 114102.

99. Bahk, J.-H., Zeng, G., Zide, J. M. O., Lu, H., *et al.* (2010). High-temperature thermoelectric characterization of III-V semiconductor thin films by oxide bonding, *J. Electron. Mater*, **39**, pp. 1125–1132.

100. Takagi, H., Meada, R., Chung, T. R., and Suga, T. (1998). Low-temperature direct bonding of silicon and silicon dioxide by the surface activation method, *Sens. Actuators, A*, **70**, pp. 164–170.

101. Suni, T., Henttinen, K., Suni, I., and Mäkinen, J. (2002), Effects of plasma activation on hydrophilic bonding of Si and SiO_2, *J. Electrochem. Soc.*, **149(6)**, pp. G348–G351.

102. Kissinger, G. and Kissinger, W. (1993). Void-free silicon-wafer-bond strentghening in the 200–400°C range, *Sens. Actuators, A,* **36**, pp. 149–156.

103. Reiche, M., Gösele, U., and Wiegand, M. (2000). Modification if Si(100)-Surfaces by SF6 plasma etching — application to wafer direct bonding, *Cryst. Res. Technol*, **35**, pp. 807–821.

104. Bower, R.W. and Chin, F. Y.-J. (1997). Low temperature direct silicon wafer bonding using argon activation, *Jpn. J. Appl. Phys*, **36**, pp. L527–L528.

105. Pasquariello, D., Camacho, M., Hjort, K., Dózsa, L., *et al.* (2001). Evaluation of InP-to-silicon heterobonding, *Mater. Sci. Eng*, **B80**, pp. 134–137.

106. Arokiaraj, J., Vicknesh, S., and Ramam, A. (2006). Integration of Indium Phosphide thin film structures on silicon substrates by direct bonding, *J. Phys: Conf. Ser*, **34**, pp. 404–409.

107. Liang, D., Fang, A. W., and Bowers, J. E. (2008). 100 mm Integration of III-V and Silicon-on-Insulator wafers for the realization of distributed feedback silicon evanescent lasers, *SPIE. Optoelectron. Mater. Dev. III*, **7135**, p. 71351O.

108. Lin, H., Stevenson, J. T. M., Gundlach, A. M., Dunare, C. C., *et al.* (2008). Direct Al-Al contact using low temperature wafer bonding for integrating MEMS and CMOS devices, *Microelectron. Eng.*, **85**, pp. 1059–1061.

109. Kowal, J., Nixon, T., Aitken, N., and Braithwaite, N. St. J. (2009). Surface activation for low temperature wafer fusion bonding by radicals produced in an oxygen discharge, *Sens. Actuators, A*, **155**, pp. 145–151.

110. Byun, K. Y., Ferain, I., and Colinge, C. (2010). Effect of free radical activation for low temperature Si to Si wafer bonding, *J. Electrochem. Soc*, **157**, pp. H109–H112.

111. Sanz-Velasco, A., Amirfeiz, P., Bengtsson, S., and Colinge, C. (2003). Room Temperature Wafer Bonding using oxygen plasma treatment in reactive ion etchers with and without inductively coupled plasma, *J Electrochem Soc*, **150**, pp. G155–G162.

112. Weinert, A., Amirfeiz, P., and Bengtsson, B. (2001). Plasma assisted room temperature bonding for MST, *Sens. Actuators, A*, **92**, pp. 214–222.

113. Jonsson, K., Köhler, J., Hedlund, C., and Stenmark, L. (2001). Oxygen plasma wafer bonding evaluated by the Weibull fracture probability method, *J. Microm. Microeng.*, **11**, pp. 364–370.

114. Visser, M. M., Weichel S., Reus R. D., and Hanneborg A. B. (2001). Strength and leak testing of plasma activated bonded interfaces, *Transducers'01*.

115. Ireland, National University of (2010). A process for manufacturing a hybrid substrate, PCT/IE2010/000049. Inventor: Colinge Cynthia Anne.

116. Vinet, M., Le Royer, C., Batude, P., Damlencourt, J. F., *et al.* (2010). Germanium on insulator and new 3D architectures opportunities for integration, *Int. J. Nanotechnol.*, **7**, p. 204.

117. Kim, M. J. and Carpenter, R. W. (2003). Heterogeneous silicon integration by ultra-high vacuum wafer bonding, *J. Electron. Mater.*, **32**, p. 849.

118. Tezuka, T., Moriyama, Y., Nakaharai, S., Sugiyama, N., *et al.* (2006). Lattice relaxation and dislocation generation/annihilation in SiGe-on-insulator layers during Ge condensation process, *Thin Solid Films*, **508**, p. 251.

119. Kanbe, H., Miyaji, M., and Ito, T. (2008). Ge/Si heterojunction photodiodes fabricated by low temperature wafer bonding, *Appl. Phys. Exp.*, **1**, p. 072301.

120. Chen, L., Dong, P., and Lipson, M. (2008). High performance germanium photodetectors integrated on submicron silicon waveguides by low temperature wafer bonding, *Opt. Express*, **16**, p. 11513.

121. Byun, K. Y., Ferain, I., Fleming, P., Morris, M., *et al.* (2010). Low temperature germanium to silicon direct wafer bonding using free radical exposure, *Appl. Phys. Lett.*, **96**, pp. 102110–102112.

122. Ferain, I. P., Byun, K. Y., Colinge, C. A, Brightup, S., *et al.* (2010). Low temperature exfoliation in hydrogen-implanted Germanium layers, *J. Appl. Phys.*, **107(5)**, p. 054315.
123. Pelissier, B., Kambara, H., Godot, E., Veran, E., *et al.* (2008). XPS analysis with an ultra clean vacuum substrate carrier for oxidation and airborne molecular contamination prevention, *Microelectron. Eng.*, **85**, p. 155.
124. Tabet, N., Faiz, M., Hamdan, N. M., and Hussain, Z. (2003). High resolution XPS study of oxide layers grown on Ge substrates, *Surf. Sci.*, **523**, p. 68.
125. Molle, A., Bhuiyan, M. N. K., Tallarida, G., and Fanciulli M. (2006). *In situ* chemical and structural investigations of the oxidation of Ge(001) substrates by atomic oxygen, *Appl. Phys. Lett.*, **89**, p. 083504.
126. Signamarcheix, T., Allibert, F., Letertre, F., Chevolleau, T., *et al.* (2008). Germanium oxynitride (GeOxNy) as a back interface passivation layer for Germanium-on-insulator substrates, *Appl. Phys. Lett.*, **93**, p. 022109.
127. Bruel, M. (1995). Silicon on insulator material technology, *IEEE Electron Device Lett.*, **31**, p. 1201.
128. Ma, X., Chen, C., Liu, W., Liu, X., *et al.* (2009). Study of the Ge wafer surface hydrophilicity after low-temperature plasma activation, *J. Electrochem. Soc.*, **156**, p. H307.
129. Vincent, S., Radu, I., Landru, D., Leterte, F., and Rieutord, F. (2009). A model of interface defect formation in silicon wafer bonding, *Appl. Phys. Lett.*, **94**, p. 101914.
130. Komanduri, R., Chandrasekaran, N., and Raff, L. (2003). Molecular dynamic simulations of uniaxial tension at nanoscale of semiconductor materials for micro-electro-mechanical systems (MEMS) applications, *Mater. Sci. Eng.*, **A340**, p. 58.
131. Le Royer, C., Damlencourt, J.-F., Romanjek, K., Lecunff, Y., *et al.* (2011). High mobility CMOS: first demonstration of planar GeOI pFETs and SOI nFETs, *Solid-State Electron.*, **59**, pp. 2–7.
132. Tong, Q. Y., Gutjahr, K., Hopfe, S., and Gösele, U. (1997). Layer splitting process in hydrogen-implanted Si, Ge, SiC, and diamond substrates, *Appl. Phys. Lett.*, **70**, p. 1390.
133. David, M. L., Pailloux, F., Babonneau, D., Drouet, M., *et al.* (2007). The effect of the substrate temperature on extended defects created by hydrogen implantation in Germanium, *J. Appl. Phys.*, **102**, p. 096101.

134. Zahler, J. M., Fontcuberta, A., Morral, I., Griggs, M. J., *et al.* (2007). Role of Hydrogen in hydrogen-induced layer exfoliation of Germanium, *Phys. Rev. B*, **75**, p. 035309.

135. Hayashi, S., Goorsky, M., Noori, A., and Bruno, D. (2006). Materials issues for the heterogeneous integration of III-V compounds, *J. Electrochem. Soc.*, **153**, p. G1011.

136. Yu, C. Y., Lee, C. Y., Lin, C. H., and Liu, C. W. (2006). Low-temperature fabrication and characterization of Ge-on-insulator structures, *Appl. Phys. Lett.*, **89**, p. 101913.

137. Hurley, R. E., Wadsworth, H., Montgomery, J. H., and Gamble, H. S. (2009). Surface blistering of low temperature annealed hydrogen and helium co-implanted germanium and its application to splitting of bonded wafer substrates *Vacuum*, **83**, p. S29.

138. Hayashi, S., Bruno, D., and Goorsky, M. S. (2004). Temperature dependence of hydrogen-induced exfoliation of InP, *Appl. Phys. Lett.*, **85**, p. 236.

139. Hayashi, S., Noori, A. M., Sandhu, R., Cavus, A., *et al.* (2006). InAs on insulator by hydrogen implantation and exfoliation, *ECS Trans.*, **3–6**, p. 129.

140. Christensen, D. H., Hill, J. R., Hickernell, R. K., Matney, K., *et al.* (1997). Evaluating epitaxial growth stability, *Mater. Sci. Eng. B*, **44**, p. 113.

141. Miclaus, C. and Goorsky, M. S. (2003). Strain evolution in hydrogen-implanted silicon, *J. Phys. D.*, **36**, p. A177.

142. Fournel, F., Moriceau, H., and Beneyton, R. (2006). Low temperature void free hydrophilic or hydrophobic silicon direct bonding, *ECS Trans.*, **3–6**, p. 139.

143. Suga, T. (2010). Surface Activated Bonding — an overview and Cu Direct Bonding, *2nd International IEEE Workshop on Low Temperature Bonding for 3D Integration.*

144. Kurayama, R., Higurashi, E., Wang, Y., Suga, T., *et al.* (2010). Low Temperature Bonding of Ge for Infrared Detectors, *2nd International IEEE Workshop on Low Temperature Bonding for 3D Integration.*

Chapter 5

Heterostructures and Strain Relaxation in Semiconductor Nanowires

Frank Glas
CNRS — Laboratoire de Photonique et de Nanostructures, Route de Nozay, 91460 Marcoussis, France
frank.glas@lpn.cnrs.fr

Associating different semiconductor materials in a single crystalline nanowire (NW) opens a rich field of investigations and applications. In this respect, the one-dimensional (1D) geometry of NWs offers several advantages over planar structures. The range of possible heterostructure geometries is much larger. We briefly review how each type of heterostructure may be fabricated. Moreover, it is often necessary to assemble materials with different lattice parameters. The sidewalls of the NWs are free surfaces that allow a very efficient elastic relaxation of the misfit strain. This allows for a much larger range of defect-free structures than in planar heterostructures or even in quantum dots (QDs). In axial heterostructures, this deeply modifies the competition between elastic and plastic relaxation. Calculations of elastic and plastic strain relaxation show that the critical layer thickness beyond which dislocations appear is much higher than in planar structures for NWs of standard radii. There also exists a critical NW radius below which an arbitrarily thick misfitting layer can be grown. In core-shell heterostructures,

Lattice Engineering: Technology and Applications
Edited by Shumin Wang

ISBN 978-981-4316-29-3 (Hardcover), 978-981-4364-25-6 (eBook)
www.panstanford.com

the sidewalls also allow for an efficient strain relaxation, and appropriate critical dimensions (shell thickness and core radius) can be calculated. Finally, strain can be used to engineer the physical properties of heterostructres in order to obtain certain electronic or optical properties.

5.1 Nanowires and Heterostructures

5.1.1 *Nanowire Growth Modes*

Following the pioneering work of Wagner and Ellis, Holonyak et al. and Givargizov on free-standing epitaxial whiskers with millimetric down to micrometric diameters made either of silicon [35, 121] or of III-V semiconductors [60], Hiruma developed in the 1990s the growth of NWs of compound semiconductors [58]. Nanowires are nowadays commonly fabricated from a large range of elemental and compound semiconductor materials with diameters ranging from hundreds down to a few nanometers [22, 26, 53, 70, 76, 96, 122]. In this chapter, we consider NWs grown epitaxially on a bulk substrate in a direction not parallel (and preferably, perpendicular) to the latter, which distinguishes them from *quantum wires*. These 1D nanostructures have remarkable physical properties and many potential applications [18, 47, 62, 69, 76, 114].

Depending on the NW material, growth may be realized either by first forming or depositing seed nanoparticles on the substrate, or without this expedient. In the former case, growth is termed *catalyzed* if the nanoparticles are foreign to the NW (metals, gold in particular, are often used) and *self-catalyzed* if the seed is made of one of the NW constituents (e.g., the group III atom for III-V NWs).[1] In the vapor-liquid-solid (VLS) mode, the nanoparticles remain liquid during growth [121], but growth from solid nanoparticles (the vapor-solid-solid (VSS) mode) has also been reported [24, 96, 124]. It is usually possible to set growth conditions such that, whatever its physical state, a nanoparticle of constant size remains on the top facet of the NW. If solid nucleation is confined to this facet, which

[1] The precise nature of the catalytic process is still debated and is probably not unique.

then also retains a constant size throughout growth, one obtains NWs with a nearly ideal cylindrical shape: this is *axial growth*. For some systems, NW formation may proceed without any seed particle, either because growth is restricted to nano-size openings made in a mask [80, 82] or, even spontaneously, as happens for III-N semiconductors. This is termed *catalyst-free* growth. In addition to axial growth, it is also possible to change the growth mode to initiate nucleation on the sidewalls of the NW. If nucleation is followed by rapid step flow along the NW sidewalls [98], this will give rise to cylindrical NWs with increasing diameter: this is *radial growth*.[2]

To form NWs, these bottom-up processes have several advantages over top-down methods. In particular, the NW is kept free of the defects generated when such structures are formed by etching and very small radii (below 10 nm) are often achievable.

5.1.2 *Heterostructures in Nanowires*

Axial and radial growth are two non-exclusive growth regimes that are at the root of the formation of the two main types of heterostructures encountered in NWs, namely *axial* and *radial* heterostructures. If, during growth, one switches from material A to material B while remaining in the axial growth mode, one obtains an axial heterostructure, where the composition changes along the NW axis (Fig. 5.1a). Conversely, if one initiates radial growth while at the same time switching from A to B, one forms a radial (or core-shell) heterostructure, with B surrounding A (Fig. 5.1b).[3] In these two types of heterostructures, the NW itself is composed of (at least) two different materials. A third type of heterostructure occurs when an homogeneous B NW is grown on substrate A. The heterointerface is then located at the base of the NW (Fig. 5.1c). Any combination of these three basic types may be considered.

This chapter is devoted to NWs with heterostructures involving materials that present a *lattice mismatch*. In this context, *lattice*

---

[2]Various conical or pencil-like morphologies have also been observed.

[3]Whereas it is often possible to achieve purely axial growth, it may be difficult to have radial growth without at the same time maintaining some amount of axial growth. It is, however, often possible to favor strongly the former at the expense of the latter.

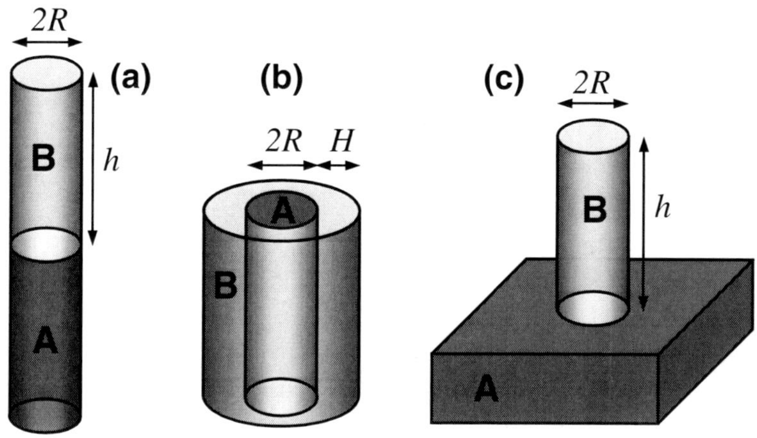

Figure 5.1. Three types of heterostructure in a nanowire and their characteristic dimensions: (a) axial, (b) radial, (c) nanowire on substrate.

engineering has two meanings. The phrase can refer to the increased latitude that the NW geometry offers for fabricating heterostructures involving highly mismatched couples. Alternatively, it denotes that the geometry and material parameters of the NW may be tailored to produce heterostructures with specific strain-related physical properties. This chapter is mainly devoted to the first aspect of lattice engineering and gives a few examples of the second.

We first recall a few general elements about the methods used for growing axial and radial heterostructures in NWs, irrespective of lattice-mismatch (Section 5.2). Section 5.3 introduces the question of mismatch accommodation in planar and non-planar systems. In the next three sections, we consider how the NW geometry permits an accommodation of this internal strain that is different from, and generally more efficient than, that which prevails in planar or QD systems. We review in turn axial heterostructures (Section 5.4), NWs on substrates (Section 5.5) and radial heterostructures (Section 5.6). Finally, we briefly consider how strain affects the electronic and optical properties of semiconductor NWs and how it can be used to produce NWs with unusual physical properties (Section 5.7).

Before proceeding further, we make two remarks. Crystalline NWs have faceted sidewalls. For the sake of simplicity, the models addressing strain relaxation in cylindrical NWs very often assume a circular cross section, rather than a polygonal one. Second, NWs of III-V compounds with bulk zinc blende cubic structure (this is the general case, apart from III-N compounds) often adopt the hexagonal wurtzite structure [41, 53, 64, 70, 96, 107]. The alternation of zinc blende and wurtzite segments creates a specific kind of heterostructure, where the crystalline phase and the lattice change in a material of fixed composition. This will not be considered here, for two reasons. First, these NWs most often grow along a [111] cubic direction, and a wurtzite/zinc blende (111) boundary then creates very little strain. Second, the phase changes tend to occur spontaneously. Although very promising [1, 108], the control of the wurtzite/zinc blende polytypism is probably not yet advanced enough to be considered as a method of lattice engineering.

5.2 Fabricating Heterostructures in Nanowires

Modern epitaxial growth techniques are ideally suited to the fabrication of planar (2D) heterostructures, consisting of layers of quasi-infinite lateral extension stacked along the growth axis. The atoms are delivered to the epilayer by vapor fluxes of the pure elements [in the case of molecular beam epitaxy (MBE)] or of simple molecules containing the latter [in the case of chemical vapor deposition (CVD) methods, such as metalorganic CVD (MOCVD)]. It is usually possible to control growth, and hence epilayer thicknesses, at the submonolayer level and, by switching from one elemental flux to another, to realize heterostructures with precisely controlled thicknesses or composition profiles. When NWs are grown from seeds, this flexibility may however be marred. Let us examine why.

5.2.1 *Axial Heterostructures*

In VLS growth with a foreign catalyst, the seed is a liquid alloy which contains not only this element but also a high concentration of the single NW constituent (for elemental semiconductors) or of the cation (for compound semiconductors) [43, 64]. This is necessary to achieve a supersaturation ensuring the repeated nucleation of NW monolayers at the catalyst/NW interface [38]. The same is true of self-catalyzed growth. Thus, a shift from one element to another (for instance from Si to Ge, or from a group III atom to another) cannot usually be realized abruptly, since it is necessary to first empty the seed particle from the first element, which is at least partly done by depositing more of this element even when its supply from the gas phase has been terminated. The interface broadening that results from this *reservoir effect* has been observed in many systems, such as (Si,Ge) [16], (Ga,In)As [57, 103], and (Ga,In)(As,P) [17]. This problem may however be overcome in various ways. For III-V compound semiconductors, it is possible to build an heterostructure by changing the group V species instead of the group III species and, for II-VI compounds, the group VI species. Since the group V (resp. VI) elements are present at very low concentration in the liquid droplet [38, 43], sharp heterointerfaces may be achieved. Indeed, the majority of axial heterostructures in III-V NWs are based on a commutation (or change of concentration) of group V atoms, whereas a single group III element is present: Ga(P,As) [8, 47, 81, 109, 119], In(P,As) [6, 34, 112], Ga(As,Sb) [20, 21, 48, 63], In(As,Sb) [12, 30].

Sharp axial heterostructures have however also been achieved in the VLS mode by changing the group III flux in the systems (Al,Ga)As [88], (Ga,In)As [54, 71, 87], or (Al,Ga,In)As [111] or the group II flux in the (Zn,Cd)Se system [4, 61]. This may be attributed to a quick replacement of the first species by the second one in the droplet or to a very rapid decrease of the supersaturation of the first element with its concentration in the liquid. The effect may be asymmetric: Paladugu et al. observed that the GaAs/InAs interface is diffuse whereas the InAs/GaAs interface is atomically flat [93] and attributed this asymmetry to differences of affinity of the various group III elements with the gold catalyst. Procedures have also been developed to minimize interface smearing: for instance, Bauer et al. used growth interruptions and played on the V/III flux ratio to sharpen the interfaces in the (Ga,In)As system [3].

However, another reason may plead for switching the group V rather than the group III species in III-V NWs. In MBE, it has been observed that the growth temperature window for NWs is dictated by the group III element: the growth temperature ranges are similar for GaAs and GaP, or InAs and InP, but the ranges for Ga-based and In-based materials have virtually no overlap [52]. MOCVD is more flexible in this respect: AlAs, GaAs, and InAs have a common range of growth temperature [23, 25, 128], and so do GaP and InP [23], or GaSb and InSb [10, 12, 48].

For certain materials, sharp interfaces may also be obtained by using VSS instead of VLS to benefit from the much lower solubility of the NW elements in a solid seed, as was demonstrated in the (Si,Ge) system [124]. Finally, catalyst-free growth is of course not concerned by the reservoir effect and also readily yields abrupt axial interfaces [9, 104].

5.2.2 *Radial Heterostructures*

Interface smearing is not expected in radial heterostructures, since the atoms that build the shell do not have to transit via the catalyst droplet, if any. Core-shell heterostructures have indeed been fabricated in III-V systems by switching the group III element [92, 110, 130], the group V element [112], or both [120, 128], as well as in the Si/Ge system [77].

The mechanisms that induce the shift from axial to radial growth depend on material and growth method. In CVD, the terminal droplet acts as a true chemical catalyst, in that it favors the decomposition of the incoming molecules. *Raising* the temperature may however allow the molecules to decompose on the NW sidewalls, which have a much larger area than the droplet (except at the very beginning of growth). A higher growth temperature may thus be used to grow the shell, as was observed for both elemental [118] and compound semiconductors [86, 105, 120]. On the contrary, in MBE, the shift from axial to radial growth may be obtained by *lowering* the temperature. This is explained as follows: in MBE, the group III species are supplied as atoms and the group V as simple molecules that may decompose thermally on the semiconductor surfaces. Axial growth is conditioned by the diffusion of the incoming species from the NW sidewalls (and substrate) to the droplet [27, 28, 51, 98, 113], which acts as a physical catalyst favoring nucleation [42]. When the temperature is reduced, the corresponding diffusion lengths decrease sharply, and a significant fraction of the atoms collected by the sidewalls cannot reach the droplet anymore: axial growth slows down or ceases and nucleation on the sidewalls generates radial growth [97, 112]. The same effect operates in catalyst-free MOCVD growth, where a shell can be formed by lowering the temperature [100].[4] In some systems, changing the material deposited suffices to change the growth mode from axial to radial, at constant temperature, as observed for Si(core)/Ge(shell) [94] and GaAs(core)/(Al,Ga)As(shell) [130] NWs; in the latter case at least, the change of growth mode is also linked to a change of diffusion regime.

5.3 Strain Relaxation in Heterostructures

5.3.1 *Modes of Strain Relaxation*

We consider mismatched heterostructures formed by depositing material B onto material A, both being single crystals. In continuum elasticity, where the atomic content of the unit cell is ignored, one may always consider that, in its intrinsic stress-free state, B results from an elastic distortion of A, the so-called *stress-free strain* [83]. We shall consider the case where A and B have the same structure and differ only by the magnitude of their lattice parameters; the stress-free strain is then a pure isotropic dilatation.[5] This can be extended to the case where the structures are different but closely related and oriented so that the interface plane may be considered as belonging either to A or B.[6] The interface is *coherent* if the lattice planes not parallel to the interface are equally spaced in A and B and continuous across the interface.

Material A is often a bulk substrate presenting a well-defined crystallographic plane to the incoming fluxes. If the deposit B

---

[4]This is particularly well demonstrated in the latter reference, since one of the multishell layers is made of the same material as the core.

[5]When B is the product of a phase transformation of A, more complex stress-free strains may occur, such as different dilatations along different axes or shear strains.

[6]such as diamond and hexagonal diamond for elementary semiconductors, or zinc blende and wurtzite for compound semiconductors, or any combination of these (see Section 5.1.2).

extends infinitely parallel to the planar interface (2D layer), the lattice mismatch makes it impossible for A and B to retain their intrinsic bulk stress-free state if the interface is coherent, simply because the densities (or spacings) of the lattice planes are different for stress-free A and B. Hence the necessity of an *accommodation* of the lattice mismatch, which can take two extreme forms. Either the coherence at the interface is preserved thanks to a deformation of one or both materials; the accommodation is then purely *elastic*. If the deposit is much thinner than the substrate, one can safely consider (ignoring possible curvature effects) that only the former is strained while the latter retains its bulk lattice constants; the deposit then adopts the lattice parameter(s) of the substrate in the interface plane and extends or contracts in the normal direction. This strain affects uniformly the whole layer. Conversely, accommodation may be realized *plastically*, via the formation of a network of *misfit dislocations* at the interface, which thereby becomes *incoherent*. The density of dislocations of a given type that accommodates a given mismatch is fixed and inversely proportional to the relative mismatch [56]. If the mismatch is fully accommodated by dislocations, both crystals are strained in the vicinity of the dislocation cores but rapidly tend to their respective stress-free state away from the latter. If the density of dislocations is insufficient, the interface is only partially coherent and the entire thin layer remains elastically strained, albeit less than in the fully coherent case.

One of the major questions regarding strain accommodation in heterostructures is to find out which factors determine the mode of relaxation (elastic or plastic). It is usually observed that dislocations do not form until the growing layer reaches some *critical thickness* h_c. Basically, such a critical thickness exists because, as seen above, the energy in the coherent state is uniformly distributed in the layer volume (so that the energy per unit area scales with layer thickness h) whereas the energy in the plastically relaxed state is concentrated in the interface region (the energy per unit area depends only weakly on h). This is at the origin of the most widely used criterion for calculating h_c for a given couple of materials, which consists in comparing the total energy of a given heterostructure in the coherent and plastically relaxed states as a function of thickness, and to define h_c as that thickness at which the energy in the former becomes larger than in the latter. This is an *equilibrium* criterion, equivalent to finding the thickness at which the misfit-induced force acting on a preexisting dislocation tends to pull and extend it into the A/B interface [33, 79, 95]. Other criteria are of a kinetic nature and deal with the nucleation of the misfit dislocations or with their motion toward the interface.

In many applications of heterostructures, in particular those based on the electrical and optical properties of semiconductor materials, dislocations have a deleterious effect, so that many methods have been devised to eliminate them, or at least to reduce their density. One obvious strategy is to keep the deposit thickness below h_C. This is however not always possible since h_C decreases rapidly when the misfit increases.

5.3.2 *Playing on Dimensionality: Quantum Dots, Quantum Wires and Nanowires*

Other methods play on the dimensionality and dimensions of the deposit (and sometimes of the substrate) to prevent or hinder dislocation formation. Indeed, when the constraint of infinite lateral extension is lifted, the deposit may recover its intrinsic (stress-free) state even if the interface remains coherent. The lattice planes may then deform continuously from the spacing of the substrate towards the intrinsic spacing of the deposit over some distance from the interface. This is realized in QDs. Formation of QDs usually starts by depositing coherent islands of a highly mismatched semiconductor onto a bulk substrate[7] [44]. All dimensions of the islands are nanometric, hence the often-used name *0D heterostructures*. Experiments demonstrate that, for a given mismatched A/B couple, the interface A/B may retain its coherence up to much larger deposit thicknesses when B forms as QDs rather than as a 2D layer [29, 40]. The reason is that the elastic energy stored in a given volume of B may be much less when B is a QD than when it is a 2D layer. This in turn stems from the fact that elastic lattice relaxation is more efficient in QDs because these have additional lateral free surfaces as compared with a 2D

[7] The islands may subsequently be capped by another material.

layer. A similar but weaker effect occurs for quantum wires parallel to the substrate.

The same kind of effect occurs in NWs. For a misfitting layer at the top of a NW (Fig. 5.1a), lateral relaxation should be even easier than for a QD, since the effective "substrate" has the same finite diameter instead of being laterally infinite. Similarly to QDs, NWs have lateral dimensions ranging from a few nanometers to a few tens of nanometers. At variance with QDs, which have inclined facets and nanometric heights, free-standing NWs may extend perpendicularly to the substrate with a uniform diameter and over potentially unlimited lengths (in practice, several microns are easily reached).

5.4 Axial Heterostructures in Nanowires

5.4.1 *Elastic Relaxation*

We consider a cylindrical heterostructured NW of radius R, constituted by a foot of material A and an epilayer of material B (Fig. 5.1a). We assume that the foot is much longer than the NW diameter, thereby constituting an effective substrate of semi-infinite extension along the NW axis. In this case, one may forget about the actual bulk substrate which may support the NW foot, since it is located far away from the interface (the reverse case of a B NW deposited directly on a bulk A substrate (Fig. 5.1c) will be treated in Section 5.5).

The easy coherent accommodation of mismatch by this axial heterostructure can be understood qualitatively by considering the variation of the strain state of the system as the thickness h of the deposit increases (Fig. 5.2). When $h \ll R$ (Fig. 5.2, left), the system is close to a 2D B layer deposited on a semi-infinite A substrate and the elastic relaxation is the same as in a laterally infinite system, with no gain in energy, except at the very edge of the NW. When h increases (Fig. 5.2, center), the region around the interface may relax in all directions (because of the top and lateral free surfaces). As a rule of thumb, this happens when $h \sim 2R$, although calculations (see Section 5.4.2.2) show that elastic relaxation becomes significant for aspect ratios $\rho = h/(2R)$ well below 1. In this case, when their

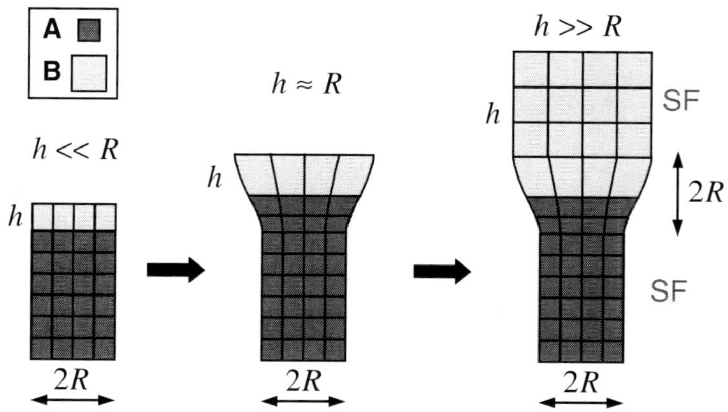

Figure 5.2. Schematics of elastic relaxation in an axial heterostructure as a function of the aspect ratio ρ. The inset gives the stress-free elementary volumes. "SF" denotes a portion of material that is nearly in its stress-free state.

elastic constants are comparable, substrate and epilayer share the strains even though the NW substrate is semi-infinite. Because of this efficient lateral relaxation, the top of the B layer nearly achieves its intrinsic stress-free state; hence, when h increases further, any additional B material is also in its stress-free state, so that the elastic energy does not increase anymore (Fig. 5.2, right).

If one is interested in the mechanisms of strain relaxation during growth, the most pertinent system involves a B layer with a free top surface. On the other hand, investigating the effect of strain on the electrical and optical properties of NWs usually requires considering structures where the surfaces of the active region are not exposed, namely capped epilayers and mixed axial/radial heterostructures.

5.4.2 *Calculation of the Elastic Relaxation in Axial Heterostructures*

5.4.2.1 Methods

The elastic state of axial heterostructures has been calculated by several authors. The differences between these calculations concern the particular type of axial heterostructure considered (capped or uncapped epilayer, single or multiple heterostructure), the elastic properties of the materials (isotropically elastic or not, with identical or different elastic constants for substrate and epilayer), and the method of solution of the equations of elasticity (analytical, numerical for continua, atomistic).

The elastic problem is solved by imposing that the forces on any free surface cancel. In a NW, this applies at least to the radial and tangential forces on the sidewalls. This is the only condition in the case of an infinitely long NW, which has no free surface normal to the axis. The axial heterostructure then consists of two semi-infinite materials A and B ($h \to \infty$). This is however not relevant for growth, where one is interested in the evolution of thin epilayers with time. Such infinite A/B NWs were treated by Ertekin *et al.* with a variational method, consisting in positing analytical forms for the elastic strains in the system, defined by a few parameters, calculating the total elastic energy and minimizing it as a function of the parameters [31, 32]. This gives reasonable results if the functional forms are well chosen. However, since the trial strain fields are not solutions of the equations of elasticity, this cannot be used quantitatively. A more general problem was actually treated long before and rigorously by Barton [2, 115]. Working in 1941, Barton was not interested in NWs. However, in the framework of continuum linear elasticity (where the strain state depends not on the absolute dimensions of the system but only on their ratios), his solution for an infinitely long cylindrical bar subjected to an excess pressure along part of its length describes perfectly well an infinite NW A, including a misfitting section B.[8] Barton gave an analytical solution in integral form, valid for elastically-isotropic materials with identical elastic constants. The same problem was recently treated by other authors, who overlooked Barton's work [66, 129].

The case of a an epilayer B of finite height with a free surface was treated by Glas under the same hypotheses about materials [37]. The strains and the elastic energy were determined numerically in two ways, either by finite element (FE) calculations or by using Barton's solution. In the latter case, to find the solution for the A/B heterostructure with a free top surface, an infinitely long NW is

[8]Barton considered a finite section, but the $h \to \infty$ limit of his solution can be taken.

considered, which includes the A/B heterostructure of interest and a small number of additional misfitting layers of finite thicknesses placed above layer B. Superimposing Barton's solutions for each layer of this infinite composite NW insures that the lateral surfaces remain free. The number, positions, lengths, and misfits of the additional inclusions are optimized to cancel as best as possible the forces on the top of layer B (which is now an interface internal to the system). The two methods give very similar results, except for the thinnest epilayers [37]. The second one has the advantage of providing analytical expressions for the elastic fields. FE calculations for heterostructures involving specific materials have also been performed [50, 126, 127].

Several authors have used atomistic methods to compute strains in NW heterostructures. Only such methods can provide the positions of every atom in the structure, as requested by some techniques of calculation of the electronic and optical properties, such as the tight binding method. The materials, the geometry, and the crystallography of the system have to be specified for each new structural calculation. Keating's valence force field (VFF) method for elementary semiconductors [68] and its extension to compound semiconductors [78] have been used since the 1980s to calculate static atomic displacements and strains. The relative simplicity of the calculations makes it possible to apply the VFF method to systems of several millions of atoms [36]. Niquet and coworkers used the VFF method to compute strains in axial GaAs/InAs and InAs/InP heterostructures (or more exactly superlattices, since periodic boundary conditions were applied along the NW axis) [84, 85]. These authors assumed a cubic zinc blende structure although III-V compounds of bulk zinc blende structure very often adopt the hexagonal wurtzite structure when grown as NWs [41, 70] (see Section 5.1.2). They recently extended their calculations to heterostructures of materials with wurtzite bulk structure [11].

5.4.2.2 Results

Although different in details, these calculations confirm the main features of elastic relaxation expected for an axial NW heterostructure (Fig. 5.2). Figure 5.3 illustrates the strain fields in the NW for

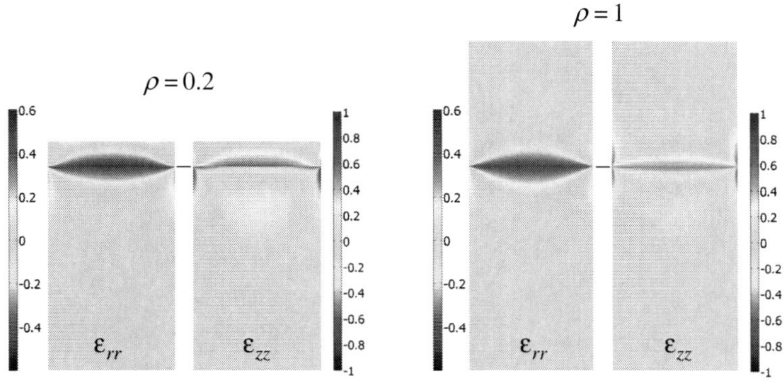

Figure 5.3. Finite element calculation of the radial (ϵ_{rr}) and axial (ϵ_{zz}) dilatations in axial heterostructures for epilayer aspect ratios $\rho = 0.2$ and $\rho = 1$. The scale gives the strains normalized to the intrinsic misfit ϵ_0 between the semi-infinite foot of the NW and the epilayer, which interface is marked by a short segment. Note that even for $\rho = 0.2$, the strain is partitioned between the two materials. See also Color Insert.

two different aspect ratios. In particular, if $h \ll R$, elastic relaxation affects mainly an outer shell of the epilayer with a thickness of the order of h, whereas if $h \gg R$, strain is appreciable and varies strongly over a thickness of the order of R around the interface, the rest of the epilayer being essentially stress-free.

This effect is conveniently quantified by function $f(\rho) = W_e^{NW}/W_e^{2D}$, where W_e^{NW} is the total elastic energy stored in the whole system with a coherent interface and $W_e^{2D} = \pi R^2 h \omega_e^{2D}$ the energy stored in the same volume cut in a planar A/B heterostructure with an epilayer of the same thickness h, $\omega_e^{2D}(\epsilon_0)$ being the energy per unit volume in the epilayer[9] (Fig. 5.4). Function f must satisfy two constraints. First, as noted above, $f(\rho) \to 1$ when $h \to 0$. Second, when $h \gg R$, the system is only strained over a height of order R (Figs. 5.2 and 5.3), hence $W_e^{NW} \sim R^3$ whereas W_e^{2D} scales withe the volume $\pi R^2 h$ of the uniformly strained epilayer, so that $f(\rho) \sim \rho^{-1}$ at large ρ. If A and B are elastically isotropic with the same E and ν, a good approximation is $f(\rho) \simeq \frac{1}{1+P\rho}$, where constant P depends only weakly on Poisson's ratio ($P \simeq 27.3$ for

[9]All the elastic energy is then stored and uniformly distributed in the epilayer B.

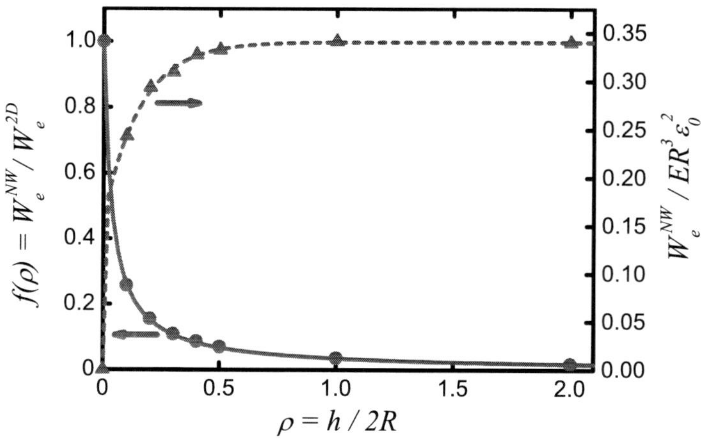

Figure 5.4. Variation with epilayer aspect ratio of the total energy stored in the NW system (triangles, right scale) and of the ratio of this energy to the energy stored in the same volume cut in a 2D system with identical epilayer thickness (disks, left scale). The full red line is a fit with the function f given in the text. Adapted from [37]. See also Color Insert.

$\nu = 1/3$) [37]. Since $P \gg 1$, the elastic energy is considerably reduced with respect to the 2D case even for modest aspect ratios, for instance to about a quarter of its 2D value for $\rho = 0.1$. The variations of f are illustrated in Fig. 5.4.

5.4.3 *Critical Dimensions*

The notion of critical thickness for the plastic relaxation of 2D layers can be extended into that of critical dimensions (epilayer thickness and NW radius) for the introduction of dislocations in NWs. We start by recalling the calculation of the critical thickness in the 2D case before turning to axial heterostructures.

5.4.3.1 The 2D case

We recalled above that, in 2D geometry, for a thin A layer growing on a semi-infinite B substrate, there exists a critical thickness h_c of the layer below which purely elastic relaxation is favored, whereas above h_c dislocations tend to appear to accommodate the misfit ϵ_0 between A and B. Assuming that the two materials are elastically isotropic, with the same Young moduli E and Poisson's ratios ν, the energy per unit area in the coherent case is:

$$w_e^{2D}(\epsilon_0, h) = h\omega_e^{2D}(\epsilon_0) = h\frac{E}{1-\nu}\epsilon_0^2 \tag{5.1}$$

The energy per unit length of a dislocation located in the interface at distance h below the free surface is:

$$\tau_d = \frac{E\left(1-\nu\cos^2\theta\right)b^2}{8\pi\left(1-\nu^2\right)}\left(C + \ln\frac{h}{b}\right) \tag{5.2}$$

with θ the angle between the dislocation line and its Burgers vector (BV) $\vec{\mathbf{b}}$ and C a constant of the order of 1 [33, 56]. In Eq. 5.2, the terms in the last bracket correspond respectively to the core energy and the energy of the strain field of the dislocation. Misfit ϵ_0 may be fully accommodated by a square array of dislocations spaced by distance $D_0(\epsilon_0) = b_{\text{eff}}/\epsilon_0$, where b_{eff} is the edge component of the BV in the interface [56]. The areal energy of this array is:

$$w_d^{2D}(\epsilon_0, h) = \frac{2\epsilon_0}{b_{\text{eff}}}\tau_d \tag{5.3}$$

From Eqs. 5.2 and 5.3, w_d^{2D} includes a thickness-independent term (the core energy) whereas $w_e^{2D} \to 0$ as $h \to 0$ (Eq. 5.1). On the other hand, w_d^{2D} and w_e^{2D} increase respectively logarithmically and linearly with h for large h. Hence, $w_d^{2D} > w_e^{2D}$ at low h and $w_e^{2D} > w_d^{2D}$ at high h. The crossover point is the equilibrium critical thickness h_c.

More generally, the total energy in presence of an array of misfit dislocations spaced by $D > D_0(\epsilon_0)$, and therefore accommodating misfit $\epsilon_a = b_{\text{eff}}/D < \epsilon_0$, may be calculated by splitting it into four terms. The first two are the elastic self-energy[10] and the core energy of the dislocation array, which sum is $w_d^{2D}(\epsilon_a, h)$. To these must be added the elastic energy $w_e^{2D}(\epsilon_0 - \epsilon_a, h)$ stored in the layer due to its residual misfit $(\epsilon_0 - \epsilon_a)$ and the interaction energy between the elastic field of the dislocations and the uniform elastic field in the layer corresponding to the residual misfit.

[10] i.e. the sum of the energies of each individual dislocation isolated in an homogeneous medium.

5.4.3.2 Radius-dependent critical thickness for an axial heterostructure

For an A/B axial heterostructure in a NW of given radius R (Fig. 5.1a), we may adopt a criterion inspired from the 2D case, namely that the critical layer thickness is the thickness at which the energy of the coherent system becomes larger than that of the system with interfacial misfit dislocations. These energies have to be recalculated in NW geometry. Moreover, since strains and stresses vary over distances comparable to the layer dimensions, the appropriate quantities are the total energies of the system rather than the non-uniform local densities of energy. The calculations of the purely elastic energy W_e^{NW} in case of coherent elastic relaxation were reviewed in Section 5.4.2.

The calculation of the energy W_d^{NW} in presence of dislocations is more tricky. Actually, the dislocation configuration itself is somewhat arbitrary. To accommodate the misfit in a planar interface, it is necessary to introduce dislocations along two directions and natural to take them as orthogonal for cubic or elastically isotropic crystals. It is likely (and proved below) that introducing even a single pair of dislocations in the interface will not be energetically favorable if $2R \ll D_0(\epsilon_0)$. Conversely, if $2R \gg D_0(\epsilon_0)$, misfit accommodation should require several such pairs. The few models so far developed have considered a single pair of orthogonal dislocations of maximal length (i.e., passing through the center of the interfacial disk) [31, 32, 37, 126, 127]. As noted by Kästner and Gösele, this is somehow justified by the fact that, since the misfit-induced forces that act on the dislocation are radial, this is an equilibrium position for an interfacial dislocation [66]. Finding the critical thickness by equating the energies in the coherent state and in the state with one such pair is thus an equilibrium criterion considering the introduction of the first pair of dislocations.

Once this dislocation configuration has been chosen, a possible calculation method is to decompose W_d^{NW} in four terms, as outlined for the 2D case (end of Section 5.4.3.1). However, even with such a simple dislocation configuration, the calculation poses some problems. First, the strain field of a dislocation normal to the axis of a cylinder has not been calculated analytically so far. Hence, neither the self-energy of the dislocation nor the interaction energy can be calculated exactly. Second, which part of the misfit is accommodated by the dislocation pair is not clear. Notwithstanding these difficulties, Glas used this method to calculate a radius-dependent critical thickness [37]. The accommodated misfit ϵ_a needed to evaluate $w_e^{2D}(\epsilon_0 - \epsilon_a, h)$ was determined as that misfit that would be accommodated in the 2D case by a dislocation array with the same length of dislocation per unit area. Building on an idea proposed previously for wires parallel to the substrate [89], the dislocation self-energy was evaluated by using the 2D formula Eq. (5.2) with the layer thickness h replaced by the distance $\bar{h}$ between the dislocation and the closest free surface. Since the distance to the NW sidewall obviously varies along the dislocation, Glas took its average $\bar{d} = 2R/\pi$. For h small, $\bar{h} = h$, whereas $\bar{h} = \bar{d}$ for h large. Finally, the interaction energy was neglected.

Using this approach and expression $f(\rho) = \frac{1}{1+P\rho}$ (see Section 5.4.2.2), Glas derived an implicit equation from which the radius-dependent critical thickness $h_c(\epsilon_0, R)$ may be calculated once the material parameters are known [37].

The variations of the critical thickness with R for 60° dislocations are shown in Fig. 5.5(a) for several values of the misfit, taking $b = 0.4$ nm, a value pertaining to GaAs and a good approximation for many cubic semiconductors. For each misfit, the $h_c(\epsilon_0, R)$ curve separates, in the (R, h) plane of layer dimensions, the coherency domain (below, left) from the domain of plastic relaxation (above, right). The figure shows that the critical thickness increases rapidly as the NW radius decreases. Moreover, for each misfit, $h_c(\epsilon_0, R) \rightarrow \infty$ when R tends to some finite value. This value is a critical radius $R_c(\epsilon_0)$, which will be discussed in Section 5.4.3.3.

This approach was refined by Ye and coworkers, who performed a FE calculation of W_d^{NW}. The effect of the edge component of the misfit dislocation was simulated by introducing an extra atomic plane in the epilayer [126, 127]. In this way, the elastic energy of the system is calculated in only one step (without distinguishing self, residual, and interaction energies), to which it suffices to add the core energy of the dislocations. The counterpart is that this can only be done for particular values of the elastic parameters. The

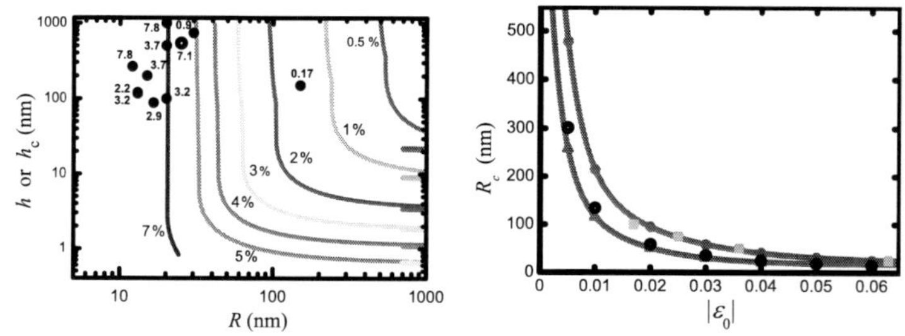

(a) (b)

Figure 5.5. (a) Variations of the critical thickness h_c of a misfitting layer growing on top of a nanowire as a function of the nanowire radius, for various values of the misfit (given near each curve) and $\nu = 1/3$. In the plastically relaxed state, misfit is accommodated by 60° dislocations ($b = 0.4$ nm). Each segment at the right gives the asymptote of the curve immediately above for $R \to \infty$, which is the 2D critical thickness. Symbols give the radii and thicknesses of layers grown without (disks) or with (circles) dislocations, for various values of the misfit (indicated in percent near each point): 0.17% [125], 0.9% [109], 2.2% and 2.9% [37], 3.2% [6, 72], 3.7% [119, 128], 7.8% [59], 7.1% [48, 63]. (b) Critical NW radius above which arbitrarily thick misfitting layers should grow coherently, as a function of misfit. Blue triangles and red disks correspond to plastic relaxation by, respectively, edge and 60° dislocations. b and ν as in (a). The green squares are the critical radii measured by Chuang *et al.* [13] for NWs on misfitting substrates (Fig. 5.1c). Adapted from [37]. See also Color Insert.

critical thicknesses determined in this way are comparable to those determined by Glas, albeit slightly larger.

As yet, there is no systematic experimental determination of critical thicknesses in NWs with axial heterostructures and only few mentions of the state (coherent or not) of the interface in such systems. In most strained axial heterostructures fabricated so far, the layers have been safely kept below their 2D critical thicknesses [73, 99]. However, there are reports of layers grown beyond this limit, all of which, apart from the early work of Hiruma [59], claim the absence of misfit dislocations [5, 16, 37, 48, 63, 72, 109, 119, 125, 128]. In most cases, the calculations of Glas [37] and Ye *et al.* [127] appear to predict correctly the state (coherent or with dislocations) experimentally observed (see the symbols in Fig. 5.5(a)).

5.4.3.3 Critical radius for an axial heterostructure

The calculations reported in Section 5.4.3.2 show that, for each misfit, the critical thickness h_c (ϵ_0, R) tends to infinity for a certain value of the NW radius. This value appears as a critical radius R_c (ϵ_0), below which an epilayer of arbitrary thickness can be grown without interface dislocations being generated.

Since they treat infinite NWs, Ertekin et. al. cannot calculate a critical thickness. These authors were however the first to determine a critical radius, by comparing the energies of the infinite NW without or with a single pair of misfit dislocations [31, 32]. In the latter case, they evaluated the dislocation self-energy by replacing the layer thickness in the 2D formula Eq. (5.2) by the NW radius R and the residual elastic energy by setting $\epsilon_a = b/(2R)$, and they neglected the interaction energy. Ye *et al.* used their FE simulations to determine the critical radii for a small set of heterostructures [127]. In this case as well, the approach of Glas has the advantage of providing an analytical solution. Glas indeed derived a second implicit equation which gives the critical radius for given misfit and material parameters [37]. The critical radii corresponding to two kinds of dislocations are shown in Fig. 5.5(b).

Once again, the critical radii determined by these various authors are comparable. This might stem partly from the fact that they all consider the same dislocation configuration. However, the critical diameters so determined are larger than the spacing D_0 (ϵ_0) expected for an array of misfit dislocations that would relieve the same misfit in the 2D case. This suggests that the energy of the plastically relaxed state might be further reduced by introducing more than one pair of interfacial dislocations. The calculated critical radii would then be overestimated for large misfits.

Nevertheless, the very existence of a critical radius can be demonstrated independently of the particular expressions chosen for the energies involved. We saw earlier that for any misfit, the elastic energy without dislocation W_e^{NW} scales with R^3 if $\rho \gtrsim 1$. This is true in particular for its limit when $\rho \to \infty$ (Fig. 5.4). On the other hand, the total dislocation energy W_d is larger than the energy of the dislocation cores, because of the additional elastic energy of the dislocation strain field, *whatever its form*. Since the core energy depends on the atomic structure of the dislocation [56], it is proportional to the dislocation length and hence to R (neglecting further relaxation at the intersections with the lateral surface, which involves a few atoms at most) and independent of layer height; this remains true irrespective of the precise location of the defects in the interface and even of their number. From these different power dependencies upon R it follows that, for infinitely thick layers, the dislocation energy always dominates, provided R is small enough. In other words, there exists a critical radius below which plastic relaxation is forbidden and coherent layers of arbitrary thickness should be obtainable. The general features of the $h_c\,(\epsilon_0, R)$ curves (Fig. 5.5(a)), in particular the horizontal (2D critical thickness) and vertical (critical radius) asymptotes, are thus independent of the expressions used in the calculations.

For sake of completeness, we mention that Kästner and Gösele adopted another approach, by considering the kinetics of dislocation formation and glide, albeit in an infinitely long wire, from which they derived values of critical radii for axial heterostructures [66].

Unfortunately, to our knowledge, there is no experimental determination of the critical radii of axial heterostructures in NWs.

5.5 Nanowires on a Misfitting Substrate

Nanowires offer great flexibility not only for growing A/B axial heterostructures in the body of the NW, but also for growing a NW of material B on an A substrate (Fig. 5.1c). A and B may be both elementary or compound semiconductors, but the growth of compound semiconductor NWs on Si substrates opens an alternative and promising route to the integration of III–V compounds on Si [13, 14].

The case of a cylindrical NW B grown directly on a misfitting bulk substrate A is closely related to that of an axial A/B heterostructure in a NW (Section 5.4). We may guess that in this case, lateral relaxation will not be as efficient because the substrate has no lateral free surface, so that the region around the interface has less latitude to relax and strain is less evenly partitioned between NW and substrate.

This is true, but only to a limited extent. Assuming again that A and B are elastically isotropic materials with the same elastic constants, we have performed FE calculations for a coherent interface in such a structure [39], which show that elastic strain relaxation is also very efficient in this case. This is now measured by ratio $f'(h) = W_e'^{NW}/W_e^{2D}$, where $W_e'^{NW}$ is the elastic energy stored in the whole system and W_e^{2D} is the energy per area πR^2, calculated in the 2D case for the same epilayer height h, as in Section 5.4.2.2. When $h \to 0$, the epilayer tends to its 2D state in both cases, hence $f' \to 1$ and $f'/f \to 1$. When h increases, f'/f rapidly tends to an asymptotic value close to 1.5. Using the same estimate as before [37] for the energy in the configuration with a pair of interfacial dislocations yields critical radii that are about 2/3 of the critical radii for the axial heterostructure. The misfit does not affect much the NW properties, since a quasi-complete relaxation occurs again over a distance of order R from the substrate.

At variance with the case of axial heterostructures, there are a few experiments that purport to have determined critical radii in such a configuration [13, 15]. Unfortunately, this has been done not by actually observing the structure of the interface but by recording a drastic change of growth mode for some critical NW radius, possibly accompanied by a dramatic decrease of the photoluminescence efficiency of the heterostructure. Chuang et al. and Cirlin *et al.* grew various compound semiconductors on mismatched Si or GaAs substrates. The results of these two groups for the same A/B couples are very close. Their experimental critical radii decrease with increasing mismatch and are approximately twice as large as those calculated by Ertekin *et al.* [13, 31], but very close to the values of Glas for axial heterostructures and 60° dislocations with $b = 0.4$ nm [37] (Fig. 5.5(b)). However, the values that can be calculated for the same defects following the method outlined above for the actual case of a NW on a mismatched substrate are about 40% too small [39].

The present case is reminiscent of that of QDs on a substrate. Although actual QDs assume non-cylindrical shapes with inclined facets, some calculations of strain relaxation in QDs encompass NWs as a limiting case of vertical facets. Such is the case of the work of Johnson and Freund [65], which however predicts critical radii several orders of magnitude larger than the works devoted specifically to cylindrical NWs [31, 37]. This highlights the necessity of performing NW-specific relaxation calculations.

5.6 Radial Heterostructures

5.6.1 *Coherent Elastic Relaxation in Radial Heterostructures*

We now consider a cylindrical core of material A with radius R wrapped by a shell of material B with thickness H (Fig. 5.1b). Here again, we ignore the interface between the composite NW and its bulk substrate and assume that the NW is infinite. The calculation of the elastic state of the NW with a coherent A/B interface was performed analytically in the framework of linear elasticity by several authors under various hypotheses: materials with transverse isotropy about the system axis [123], elastically isotropic materials with identical elastic constants [49, 74], and cubic [116], wurtzite [102], and zinc blende [101] crystals. In addition, Søndergaard *et al.* applied the FE method to NWs of cubic structure where core and shell both have an hexagonal cross section (as encountered in real NWs). This was extended by Grönqvist *et al.* who performed both FE and VFF calculations for zinc blende crystals [46, 106] in the same geometry and also for standard circular cylinders and by Hestroffer *et al.* who developed a VFF model for wurzite NWs with hexagonal cross sections [55].

Let us summarize the general features of the coherent relaxation strain field. For sake of definiteness, we assume that the shell has larger lattice parameters than the core ($\epsilon_0 > 0$). If $H \ll R$, the situation is akin to the 2D case, with the core in its stress-free state and all strain localized in the shell, which is compressed in the tangential (θ, z) plane and thus expands radially.[11] As H increases, the strain gets partitioned between core and shell. In the tangential plane, the core is dilated by the shell and the shell compressed by the core in the vicinity of the A/B interface: $\epsilon_{\theta\theta}$ and ϵ_{zz} are both positive

[11]We use standard cylindrical coordinates, with z along the NW axis.

in the core and negative in the shell. The Poisson effect imposes the reverse signs to ϵ_{rr}. Whereas, for infinite NWs, ϵ_{zz} is uniform, and in particular independent of the radial position, in the plane strain in the shell decrease rapidly away from the interface over a distance of the order of R. Here again, the outer free surface of the NW insures efficient surface relaxation. As a consequence, the elastic energy stored in the NW per unit axial length, at fixed core radius, increases with H but rapidly saturates.

5.6.2 *Critical Dimensions for Nanowires with Radial Heterostructures*

In radial heterostructures, two types of dislocations have been considered as candidates to relieve the misfit between core and shell, namely straight edge dislocations along the NW axis and edge dislocation loops normal to the axis, in both cases located at the core/shell interface [49, 74, 90, 91, 101, 102]. Each type accommodates the misfit in one of the two perpendicular tangential directions. The strain fields of such dislocations have been calculated analytically only for an elastically isotropic circular cylinder where core and shell have the same elastic constants [49, 90]. In this case, the strains are uniform in the core. For anisotropic wurtzite or zinc blende core and shell with different elastic constants, Raychaudhuri and Yu simply used the formula for a 2D layer of thickness H on a semi-infinite substrate [102]. Ovid'ko and coworkers [49, 90, 91] and Raychaudhuri and Yu [102] computed the critical dimensions of the system by the same equilibrium method as used in the cases of 2D and axial heterostructures, namely by comparing the energies of the system without and with one dislocation (of either type).[12]

The conclusions of these analyses are the following. If plastic relaxation occurs, on an energetic basis, the first dislocation to form

[12] In presence of a dislocation, the total energy comprises the four terms mentioned in Sections 5.4.3.1 and 5.4.3.2, namely self-energy and core energy of the dislocation, residual misfit energy and interaction energy. Raychaudhuri and Yu evaluate the residual misfit energy by considering that core and shell optimally share the misfit accommodated by the dislocation. Ovid'ko and coworkers seem to consider that the residual misfit energy is equal to the misfit energy before the introduction of the dislocation ($\epsilon_a = 0$). This might lead to overestimating the critical dimensions.

will always be a straight dislocation rather than a loop. Critical dimensions can thus be assessed by considering only the former.[13] For each misfit ϵ_0 and core radius R, there exists a critical shell thickness $H_c(\epsilon_0, R)$ under which no dislocation is generated. For each misfit ϵ_0, there exists a critical core radius $R_c(\epsilon_0)$ such that an arbitrarily thick shell can be grown around a core of radius $R \leq R_c$ without generating any misfit dislocation; in other words, $H_c(\epsilon_0, R) \to \infty$ if $R \to R_c(\epsilon_0)$.[14]

Core-shell NWs are advantageous for misfit accommodation, since the critical shell thickness is always larger than the 2D critical thickness for the same misfit, approaching it only when $R \to \infty$. However, the calculated critical radii are rather small [67], around 10 nm for $\epsilon_0 = 1\%$ [102] and of only a few nm for $\epsilon_0 = 2\%$ [91, 102], much smaller indeed than the axial critical radii for the same misfit.

There are reports of misfitting core/shell heterostructures free of interfacial dislocations in systems such as Ge/Si [77], GaAs/(Ga,In)P [105], or GaN/(Ga,In)N [100]. On the other hand, there are very few reports of defects at the interface between a core and a shell of uniform thickness. However, Lin *et al.* report dislocation loops in their GaP core ($R \simeq 10$ nm)/GaN shell ($H \simeq 4$ nm) NWs [75]. The core radius is indeed well beyond the critical radius for the misfit of about 21% between zinc blende GaP and GaN (however, GaN here adopts its standard wurtzite structure).

Misfitting shells can also relax by developing a surface corrugation. This is the analogous of the well-known Asaro–Tiller–Grinfeld effect that affects misfitting 2D layers. Similarly to the 2D case [19], dislocations then tend to appear at the loci of maximum stress concentration [94]. Hence, forbidding or limiting surface diffusion may prevent ripple and dislocation formation [45].

---

[13] This conclusion is to be taken cautiously, since the loop has a finite length and relieves the misfit only locally, whereas the straight dislocation is infinite.

[14] Somewhat surprisingly, Ovid'ko and coworkers predict that, at given ϵ_0, it is also possible to grow shells *above* another critical value without dislocation [49, 91]. From the discussion by Raychaudhuri *et al.* of their own results [102], it appears that this might be an artefact induced by the consideration of a single dislocation only.

5.7 Using Strain to Engineer the Physical Properties of Heterostructres in Nanowires

In this section, we give a few examples of the effect of strain relaxation on the electronic and optical properties of NWs with axial or radial heterostructures. Section 5.4 has shown that, thanks to the efficient strain relaxation at their sidewalls, NWs offer the possibility of growing more or less strongly mismatched axial heterostructures consisting of a B insertion in an A NW. In this way, QDs of low bandgap may be inserted between potential barriers of high bandgap, or else a barrier of high bandgap in a material of lower bandgap.

Fabricating QDs in a NW offers many advantages. At variance with the QDs grown in the Stranski-Krastanov mode, which form spontaneously with uncontrolled geometry at undetermined positions, QDs in NWs can be controlled in both shape and position. The NW radius, which defines the QD lateral size, can be tailored by using calibrated seed particles or holes in a mask, and the thickness of the insertion via its growth time. Coupled QDs can also be realized. Moreover, a shell can be grown around the axial heterostructure to prevent radiative recombinations at the free surface. The shell can be thickened to transform the NW into a waveguide, or *photonic nanowire*.

Niquet and coworkers calculated the effect of strain on the electronic and optical properties of QDs inserted in (111)-oriented zinc blende NWs. In this orientation, strain generates a piezoelectric field, which is however weaker in NWs than in 2D quantum wells, thanks to strain relaxation (Section 5.4). For the highly mismatched InAs/GaAs system, due to the the very inhomogeneous strain in the QD, if $h < R$, there is a potential well trapping the electrons at the outer surface, which disappears for thicker QDs. When h increases, the strain and the piezoelectric field separate electrons and holes, which strongly reduces the oscillator strength of the interband transitions [84]. However, an InP shell around the axial heterostructure reduces strain relaxation, which restores a significant overlap between the wavefunctions of electrons and heavy holes. Even without a shell, the less strained InAs/InP system is more promising [85]. Bright excitonic transitions are predicted, together with a polarization of the light emitted perpendicularly and parallely to the NW axis for low and high QD thickness h, respectively, which signs the transition from a disk-like to a rod-like behavior. The transition energy between the first electron and hole levels increases when h diminishes, due to confinement but also due to strain.

Strong room-temperature photoluminescence was indeed observed in such structures when a shell was present [112]. Björk et al. fabricated 1D NW resonant tunneling diodes by inserting an InAs QD between thin InP barriers in an InAs NW [7]. The NW geometry offers great flexibility for realizing stacked QD structures: Gudiksen *et al.* observed a strong photoluminescence signal from superlattices of GaAs QDs in GaP NWs and reported that light emission is strongly polarized along the axis [47]. The photonic properties may now match those of Stranski-Krsatanov QDs. Borgström *et al.* obtained single photon emission from As-rich Ga(P,As) QDs surrounded by P-rich Ga(P,As) barriers [8]. Tribu *et al.* observed photon antibunching from CdSe QDs in a ZnSe NW [117].

NW heterostructurs are also interesting for their transport properties. InP barriers in InAs NWs were studied theoretically by Niquet and Camacho Mojica, who verified that, due to the change in strain relaxation, the barrier height is close to the bulk value in thick InP layers ($h \geq 1.5R$), but tends to the 2D limit in thin ones [85]. Due to strain relaxation, the piezoelectric field is much lower than in the corresponding 2D structure. Björk *et al.* studied experimentally a wide range of thicknesses and indeed observed a discontinuity very close to that of the bulk stress-free materials for thick barriers [7].

Charge transport has also been studied in strained core-shell heterostructures, which might constitute ideal 1D channels. In agreement with calculations that predict a type-II band alignment in Ge-Si core-shell NW, Lu *et al.* observed a 1D hole gas in such undoped structures. They also observed Coulomb blockade and ballistic transport up to room temperature [77].

5.8 Summary and Conclusions

By selecting the appropriate material system, growth method, and growth conditions, it is possible to fabricate axial or core-shell heterostructures (and combinations thereof) in semiconductor NWs. The NW geometry offers great flexibility for combining in such heterostructures materials differing markedly in terms of lattice constant, without introducing extended defects. This stems from the ease with which the internal misfit strain may relax elastically at the NW sidewalls, which constitute a free boundary. This allows the grower to associate materials that cannot be coherently joined not only in 2D planar geometry but even as QDs on a substrate. The elastic state of the system and the critical dimensions (thicknesses and radii) under which no interfacial dislocation is formed can be calculated analytically or numerically for these various types of heterostructures. These calculated critical dimensions are broadly consistent with the experimental observations. Since strain affects the electronic and optical properties of semiconductor NWs, these calculations are useful guides to the choice of the geometry and material parameters likely to produce heterostructures with specific strain-related physical properties.

Acknowledgments

I thank my colleagues Jean-Christophe Harmand, Gilles Patriarche and Ludovic Largeau (CNRS - Laboratoire de Photonique et de Nanostructures), and Vladimir Dubrovskii and George Cirlin (Saint Petersburg Academic University) for many fruitful discussions. Part of this work was realized in the framework of the projects FILEMON35 (ANR-05-NANO-016) and BONAFO (ANR-08-NANO-031) of Agence Nationale de la Recherche.

References

1. Akopian, N., Patriarche, G., Liu, L., Harmand, J.-C., and Zwiller, V. (2010). Crystal phase quantum dots, *Nano Lett.* **10**, pp. 1198–1201.
2. Barton, M. V. (1941). The circular cylinder with a band of uniform pressure on a finite length of the surface, *J. Appl. Mech.* **8**, pp. A97–A104.

3. Bauer, J., Gottschalch, V., Paetzelt, H., and Wagner, G. (2008). VLS growth of GaAs/(InGa)As/GaAs axial double-heterostructure nanowires by MOVPE, *J. Cryst. Growth* **310**, pp. 5106–5110.
4. Bellet-Amalric, E., Elouneg-Jamroz, M., Bougerol, C., Den Hertog, M., Genuist, Y., Bounouar, S., *et al.* (2010). Epitaxial growth of ZnSe and ZnSe/CdSe nanowires on ZnSe, *Phys. Stat. Sol. C* **7**, pp. 1526–1529.
5. Björk, M. T., Ohlsson, B. J., Sass, T., Persson, A. I., Thelander, C., Magnusson, M. H., *et al.* (2002). One-dimensional heterostructures in semiconductor nanowhiskers, *Appl. Phys. Lett.* **80**, pp. 1058–1060.
6. Björk, M. T., Ohlsson, B. J., Sass, T., Persson, A. I., Thelander, C., Magnusson, M. H., *et al.* (2002). One-dimensional steeplechase for electrons realized, *Nano Lett.* **2**, pp. 87–89.
7. Björk, M. T., Ohlsson, B. J., Thelander, C., Persson, A. I., Deppert, K., Wallenberg, L. R., *et al.* (2005). Nanowire resonant tunneling diodes, *Appl. Phys. Lett.* **81**, pp. 4458–4460.
8. Borgström, M. T., Verheijen, M. A., Immink, G., de Smet, T., and Bakkers, E. P. A. M. (2006). Interface study on heterostructured GaP-GaAs nanowires, *Nanotechnology.* **17**, pp. 4010–4013.
9. Bougerol, C., Songmuang, R., Camacho, D., Niquet, Y. M., and Daudin, B. (2010). Structural properties of GaN nanowires and GaN/AlN insertions grown by molecular beam epitaxy, *J. Phys. Conf. Ser.* **209**, p. 012010.
10. Burke, R. A., Weng, X., Kuo, M.-W., Song, Y.-W., Itsuno, A. M., Mayer, T. S., *et al.* (2010). Growth and characterization of unintentionally doped GaSb nanowires, *J. Electron. Mater.* **39**, pp. 355–364.
11. Camacho Mojica, D., and Niquet, Y.-M. (2010). Stark effect in GaN/AlN nanowire heterostructures: Influence of strain relaxation and surface states, *Phys. Rev. B* **81**, p. 195313.
12. Caroff, P., Messing, M. E., Borg, B. M., Dick, K. A., Deppert, K., and Wernersson, L.-E. (2009). InSb heterostructure nanowires: MOVPE growth under extreme lattice mismatch, *Nanotechnology.* **20**, p. 495606.
13. Chuang, L. C., Moewe, M., Chase, C., Kobayashi, N. P., Chang-Hasnain, C., and Crankshaw, S. (2007). Critical diameter for III-V nanowires grown on lattice-mismatched substrates, *Appl. Phys Lett.* **90**, p. 043115.
14. Cirlin, G. E., Dubrovskii, V. G., Samsonenko, Y. B., Bouravleuv, A. D., Durose, K., Proskuryakov, Y. Y., *et al.* (2010). Self-catalyzed, pure zincblende GaAs nanowires grown on Si(111) by molecular beam epitaxy, *Phys. Rev. B* **82**, p. 035302.

15. Cirlin, G. E., Dubrovskii, V. G., Soshnikov, I. P., Sibirev, N. V., Samsonenko, Y. B., Bouravleuv, A. D., *et al.* (2009). Critical diameters and temperature domains for MBE growth of III-V nanowires on lattice mismatched substrates, *Phys. Stat. Sol. Rapid Res. Lett.* **3**, pp. 112–114.
16. Clark, T. E., Nimmatoor, P., Lew, K.-K., Pan, L., Redwing, J. M., and Dickey, E. C. (2008). Diameter dependent growth rate and interfacial abruptness in vapor-liquid-solid $Si/Si_{1-x}Ge_x$ heterostructure nanowires, *Nano Lett.* **8**, pp. 1246–1252.
17. Cornet, D. M., and LaPierre, R. R. (2007). InGaAs/InP core-shell and axial heterostructure nanowires, *Nanotechnology.* **18**, p. 385305.
18. Cui, Y., Wei, Q., Park, H., and Lieber, C. M. (2001). Nanowire nanosensors for highly sensitive and selective detection of biological and chemical species, *Science* **293**, pp. 1289–1292.
19. Cullis, A. G., Pidduck, A. J., and Emeny, M. T. (1995). Misfit dislocation sources at surface ripple troughs in continuous heteroepitaxial layers, *Phys Rev. Lett.* **75**, pp. 2368–2371.
20. Dheeraj, D. L., Patriarche, G., Zhou, H., Harmand, J. C., Weman, H., and Fimland, B. O. (2009). Growth and structural characterization of GaAs/GaAsSb axial heterostructured nanowires, *J. Cryst. Growth* **311**, pp. 1847–1850.
21. Dheeraj, D. L., Patriarche, G., Zhou, H., Hoang, T. B., Moses, A. F., Grønsberg, S., *et al.* (2008). Growth and characterization of wurtzite GaAs nanowires with defect-free zinc blende GaAsSb inserts, *Nano Lett.* **8**, pp. 4459–4463.
22. Dick, K. A. (2008). A review of nanowire growth promoted by alloys and non-alloying elements with emphasis on Au-assisted III-V nanowires, *Prog. Cryst. Growth Charac. Mater.* **54**, pp. 138–173.
23. Dick, K. A., Deppert, K., Karlsson, L. S., Wallenberg, L. R., Samuelson, L., and Seifert, W. (2005). A new understanding of Au-assisted growth of III-V semiconductor nanowires, *Adv. Funct. Mater.* **15**, pp. 1603–1610.
24. Dick, K. A., Deppert, K., Mårtensson, T., Mandl, B., Samuelson, L., and Seifert, W. (2005). Failure of the vapor-liquid-solid mechanism in Au-assisted MOVPE growth of InAs nanowires, *Nano Lett.* **5**, pp. 761–764.
25. Dick, K. A., Kodambaka, S., Reuter, M. C., Deppert, K., Samuelson, L., Seifert, W., *et al.* (2007). The morphology of axial and branched nanowire heterostructures, *Nano Lett.* **7**, pp. 1817–1822.
26. Duan, X., and Lieber, C. M. (2000). General synthesis of compound semiconductor nanowires, *Adv. Mater.* **12**, pp. 298–302.

27. Dubrovskii, V. G., Cirlin, G. E., Soshnikov, I. P., Tonkikh, A. A., Sibirev, N. V., Samsonenko, Y. B., *et al.* (2005). Diffusion-induced growth of GaAs nanowhiskers during molecular beam epitaxy: theory and experiment, *Phys. Rev. B* **71**, p. 205325.
28. Dubrovskii, V. G., Sibirev, N. V., Cirlin, G. E., Bouravleuv, A. D., Samsonenko, Y. B., Dheeraj, D. L., *et al.* (2009). Role of nonlinear effects in nanowire growth and crystal phase, *Phys. Rev. B* **80**, p. 205305.
29. Eaglesham, D. J., and Cerullo, M. (1990). Dislocation-free Stranski-Krastanow growth of Ge on Si(100), *Phys. Rev. Lett.* **64**, pp. 1943–1946.
30. Ercolani, D., Rossi, F., Li, A., Roddaro, S., Grillo, V., Salviati, G., et al. (2009). InAs/InSb nanowire heterostructures grown by chemical beam epitaxy, *Nanotechnology.* **20**, p. 505605.
31. Ertekin, E., Greaney, P. A., and Chrzan, D. C. (2005). Equilibrium limits of coherency in strained nanowire heterostructures, *J. Appl. Phys.* **97**, p. 114325.
32. Ertekin, E., Greaney, P. A., Sands, T. D., and Chrzan, D. C. (2002). Equilibrium analysis of lattice-mismatched nanowire heterostructures, in V. Klimov, J. M. Buriak, D. D. M. Wayner, F. Priolo, B. White, and L. Tsybeskov (eds.), *Quantum confined semiconductor nanostructures, Mat. Res. Soc. Proc.*, Vol. 737 (Mat. Res. Soc., Pittsburgh), p. F10.4.
33. Fitzgerald, E. A. (1991). Dislocations in strained-layer epitaxy: theory, experiment, and applications, *Mater. Sci. Rep.* **7**, pp. 87–142.
34. Fröberg, L. E., Wacaser, B. A., Wagner, J. B., Jeppesen, S., Ohlsson, B. J., Deppert, K., *et al.* (2008). Transients in the formation of nanowire heterostructures, *Nano Lett.* **8**, pp. 3815–3818.
35. Givargizov, E. I. (1975). Fundamental aspects of VLS growth, *J. Cryst. Growth* **31**, pp. 20–30.
36. Glas, F. (1995). Correlated static atomic displacements and transmission electron microscopy contrast in compositionally homogeneous disordered alloys, *Phys. Rev. B* **51**, pp. 825–839.
37. Glas, F. (2006). Critical dimensions for the plastic relaxation of strained axial heterostructures in free-standing nanowires, *Phys. Rev. B* **74**, p. 121302.
38. Glas, F. (2010). Chemical potentials for Au-assisted vapor-liquid-solid growth of III-V nanowires, *J. Appl. Phys.* **108**, p. 073506.
39. Glas, F. (2011). Unpublished.
40. Glas, F., Guille, C., Hénoc, P., and Houzay, F. (1987). TEM study of the molecular beam epitaxy island growth of InAs on GaAs, in A. G. Cullis

and P. D. Augustus (eds.), *Microscopy of Semiconducting Materials 1987, Inst. Phys. Conf. Ser.*, Vol. 87 (The Institute of Physics, Bristol), pp. 71–76.

41. Glas, F., Harmand, J. C., and Patriarche, G. (2007). Why does wurtzite form in nanowires of III-V zinc blende semiconductors? *Phys. Rev. Lett.* **99**, p. 146101.
42. Glas, F., Harmand, J. C., and Patriarche, G. (2010). Growth, structure and phase transitions of epitaxial nanowires of III-V semiconductors, *J. Physics: Conf. Ser.* **209**, p. 012002.
43. Glas, F., Harmand, J. C., and Patriarche, G. (2010). Nucleation antibunching in catalyst-assisted nanowire growth, *Phys. Rev. Lett.* **104**, p. 135501.
44. Goldstein, L., Glas, F., Marzin, J. Y., Charasse, M. N., and Le Roux, G. (1985). Growth by molecular beam epitaxy and characterization of InAs/GaAs strained-layer superlattices, *Appl. Phys. Lett.* **47**, p. 1099.
45. Goldthorpe, I. A., Marshall, A. F., and McIntyre, P. C. (2009). Inhibiting strain-induced surface roughening: dislocation-free Ge/Si and Ge/SiGe core-shell nanowires, *Nano Lett.* **9**, pp. 3715–3719.
46. Grönqvist, J., Søndergaard, N., Boxberg, F., Guhr, T., Åberg, S., and Xu, H. Q. (2009). Strain in semiconductor core-shell nanowires, *J. Appl. Phys.* **106**, p. 053508.
47. Gudiksen, M. S., Lauhon, L. J., Wang, J., Smith, D. C., and Lieber, C. M. (2002). Growth of nanowire superlattice structures for nanoscale photonics and electronics, *Nature* **415**, pp. 617–620.
48. Guo, Y. N., Zou, J., Paladugu, M., Wang, H., Gao, Q., Tan, H. H., *et al.* (2006). Structural characteristics of GaSb/GaAs nanowire heterostructures grown by metal-organic chemical vapor deposition, *Appl. Phys. Lett.* **89**, p. 231917.
49. Gutkin, M. Y., Ovid'ko, I. A., and Sheinerman, A. G. (2000). Misfit dislocations in wire composite solids, *J. Phys.: Condens. Matter* **12**, pp. 5391–5401.
50. Hanke, M., Eisenschmidt, C., Werner, P., Zakharov, N. D., Syrowatka, F., Heyroth, F., *et al.* (2007). Elastic strain relaxation in axial Si/Ge whisker heterostructures, *Phys. Rev. B* **75**, p. 161303.
51. Harmand, J.-C., Glas, F., and Patriarche, G. (2010). Growth kinetics of a single $InP_{1-x}As_x$ nanowire, *Phys. Rev. B* **81**, p. 235436.
52. Harmand, J. C., Glas, F., Patriarche, G., Tchernycheva, M., Sartel, C., Liu, L., *et al.* (2011). Growth of III-arsenide/phosphide nanowires by molecular beam epitaxy, in J. Li, D. Wang and R. R. LaPierre (eds.),

Advances in III–V semiconductor nanowires and nanodevices (Bentham Science Publishers, Oak Park, IL), pp. 68–88.

53. Harmand, J. C., Patriarche, G., Péré-Laperne, N., Mérat-Combes, M.-N., Travers, L., and Glas, F. (2005). Analysis of vapor-liquid-solid mechanism in Au-assisted GaAs nanowire growth, *Appl. Phys. Lett.* **87**, p. 203101.
54. Heiß, M., Gustafsson, A., Conesa-Boj, S., Peiró, F., Morante, J. R., Abstreiter, G., *et al.* (2009). Catalyst-free nanowires with axial $InGa_{1-x}As_x$/GaAs heterostructures, *Nanotechnology.* **20**, p. 075603.
55. Hestroffer, K., Mata, R., Camacho, D., Leclere, C., Tourbot, G., Niquet, Y. M., *et al.* (2010). The structural properties of GaN/AlN core-shell nanocolumn heterostructures, *Nanotechnology.* **21**, 41, p. 415702.
56. Hirth, J. P., and Lothe, J. (1982). *Theory of dislocations* (Wiley, New York).
57. Hiruma, K., Murakoshi, H., Yasawa, M., and Katsuyama, T. (1996). Self-organized growth of GaAs/InAs heterostructure nanocylinders by organometallic vapor phase epitaxy, *J. Cryst. Growth* **163**, pp. 226–231.
58. Hiruma, K., Yasawa, M., Haraguchi, K., Ogawa, K., Katsuyama, T., Koguchi, M., *et al.* (1993). GaAs free-standing quantum-size wires, *J. Appl. Phys.* **74**, pp. 3162–3171.
59. Hiruma, K., Yasawa, M., Katsuyama, T., Ogawa, K., Haraguchi, K., Koguchi, M., *et al.* and Kakibayashi, H. (1995). Growth and optical properties of nanometer-scale GaAs and InAs whiskers, *J. Appl. Phys.* **77**, pp. 447–462.
60. Holonyak Jr., N., Wolfe, C. M., and Moore, J. S. (1965). Vapor-liquid-solid growth of gallium phosphide, *Appl. Phys. Lett.* **6**, pp. 64–65.
61. Hsiao, C. H., Chang, S. J., Hung, S. C., Cheng, Y. C., Huang, B. R., Wang, S. B., *et al.* (2010). ZnSe/ZnCdSe heterostructure nanowires, *J. Cryst. Growth* **312**, pp. 1670–1675.
62. Huang, Y., Duan, X. F., and Lieber, C. M. (2005). Nanowires for integrated multicolor nanophotonics, *Small* **1**, pp. 142–147.
63. Jeppsson, M., Dick, K. A., Wagner, J. B., Caroff, P., Deppert, K., Samuelson, L., *et al.* (2008). GaAs/GaSb nanowire heterostructures grown by MOVPE, *J. Cryst. Growth* **310**, pp. 4115–4121.
64. Johansson, J., Karlsson, L. S., Dick, K. A., Bolinsson, J., Wacaser, B. A., Deppert, K., *et al.* (2009). Effects of supersaturation on the crystal structure of gold seeded III-V nanowires, *Cryst. Growth Des.* **9**, pp. 766–773.

65. Johnson, H. T., and Freund, L. B. (1997). Mechanics of coherent and dislocated island morphologies in strained epitaxial material systems, *J. Appl. Phys.* **81**, pp. 6081–6090.

66. Kästner, G., and Gösele, U. (2004). Stress and dislocations at cross-sectional heterojunctions in a cylindrical nanowire, *Philos Mag.* **84**, pp. 3803–3824.

67. Kavanagh, K. L. (2010). Misfit dislocations in nanowire heterostructures, *Semicond. Sci. Technol.* **25**, p. 024006.

68. Keating, P. N. (1966). Effect of invariance requirements on the elastic strain energy of crystals with application to the diamond structure, *Phys. Rev.* **145**, pp. 637–645.

69. Kelzenberg, M. D., Turner-Evans, D. B., Kayes, B. M., and Filler, M. A. (2008). Photovoltaic measurements in single-nanowire silicon solar cells, *Nano Lett.* **8**, pp. 710–714.

70. Koguchi, M., Kakibayashi, H., Yasawa, M., Hiruma, K., and Katsuyama, T. (1992). Crystal structure change of GaAs and InAs whiskers from zinc-blende to wurtzite type, *Jpn. J. Appl. Phys.* **31**, pp. 2061–2065.

71. Krogstrup, P., Yamasaki, J., Sørensen, C. B., Johnson, E., Wagner, J. B., Pennington, R., *et al.* (2009). Junctions in axial III-V heterostructure nanowires obtained via an interchange of group III elements, *Nano Lett.* **9**, pp. 3689–3693.

72. Larsson, M. W., Wagner, J. B., Wallin, M., Håkansson, P., Fröberg, L. E., and Samuelson, L., *et al.* (2007). Strain mapping in free-standing heterostructured wurtzite InAs/InP nanowires, *Nanotechnology.* **18**, p. 015504.

73. Lauhon, L. J., Gudiksen, M. S., and Lieber, C. M. (2004). Semiconductor nanowire heterostructures, *Phil. Trans. R. Soc. London A* **362**, pp. 1247–1260.

74. Liang, Y., Nix, W. D., Griffin, P. B., and Plummer, J. D. (2005). Critical thickness enhancement of epitaxial SiGe films grown on small structures, *J. Appl. Phys.* **97**, p. 043519.

75. Lin, H.-M., Chen, Y.-L., Yang, J., Liu, Y.-C., Yin, K.-M., Kai, J.-J., *et al.* (2003). Synthesis and characterization of core-shell GaP@GaN and GaN@GaP nanowires, *Nano Lett.* **3**, pp. 537–541.

76. Lu, W., and Lieber, C. M. (2006). Semiconductor nanowires, *J. Phys. D: Appl. Phys.* **39**, pp. R387–R406.

77. Lu, W., Xiang, J., Timko, B. P., Wu, Y., and Lieber, C. M. (2005). One-dimensional hole gas in germanium/silicon nanowire heterostructures, *Proc. Nat. Acad. Sci.* **102**, pp. 10046–10051.

78. Martin, R. M. (1970). Elastic properties of ZnS structure semiconductors, *Phys. Rev. B* **1**, pp. 4005–4011.
79. Matthews, J. W., and Blakeslee, A. E. (1974). Defects in epitaxial multilayers. 1. Misfit dislocations, *J. Cryst. Growth* **27**, pp. 118–125.
80. Mohan, P., Motohisa, J., and Fukui, T. (2005). Controlled growth of highly uniform, axial/radial direction-defined, individually addressable InP nanowire arrays, *Nanotechnology.* **16**, pp. 2903–2907.
81. Mohseni, P. K., Maunders, C., Botton, G. A., and LaPierre, R. R. (2007). GaP/GaAsP/GaP core-multishell nanowire heterostructures on (111) silicon, *Nanotechnology.* **18**, p. 445304.
82. Motohisa, J., Noborisaka, J., Takeda, J., Inari, M., and Fukui, T. (2004). Catalyst-free selective-area MOVPE of semiconductor nanowires on (111)B oriented substrates, *J. Cryst. Growth* **272**, p. 180–185.
83. Mura, T. (1991). *Micromechanics of defects in solids* (Kluwer, Dordrecht).
84. Niquet, Y. M. (2006). Electronic and optical properties of InAs/GaAs nanowire superlattices, *Phys. Rev. B* **74**, p. 155304.
85. Niquet, Y.-M., and Camacho Mojica, D. (2008). Quantum dots and tunnel barriers in InAs/InP nanowire heterostructures: Electronic and optical properties, *Phys. Rev. B* **77**, p. 115316.
86. Noborisaka, J., Motohisa, J., Hara, S., and Fukui, T. (2005). Fabrication and characterization of freestanding GaAs/AlGaAs core-shell nanowires and AlGaAs nanotubes by using selective-area metalorganic vapor phase epitaxy, *Appl. Phys. Lett.* **87**, p. 093109.
87. Ohlsson, B. J., Björk, M. T., Persson, A. I., Thelander, C., Wallenberg, L. R., Magnusson, M. H., *et al.* (2002). Growth and characterization of GaAs and InAs nano-whiskers and InAs-GaAs heterostructures, *Physica E* **13**, pp. 1126–1130.
88. Ouattara, L., Mikkelsen, A., Sköld, N., Eriksson, J., Knaapen, T., Ćavar, E., *et al.* (2007). GaAs/AlGaAs nanowire heterostructures studied by scanning tunneling microscopy, *Nano Lett.* **7**, pp. 2859–2864.
89. Ovid'ko, I. A. (2002). Relaxation mechanisms in strained nanoislands, *Phys. Rev. Lett.* **88**, p. 046103.
90. Ovid'ko, I. A., and Sheinerman, A. G. (2004). Misfit dislocation loops in composite nanowires, *Philos. Mag.* **84**, pp. 2103–2118.
91. Ovid'ko, I. A., and Sheinerman, A. G. (2006). Misfit dislocation in nanocomposites with quantum dots, nanowires and their ensembles, *Adv. Phys.* **55**, pp. 627–689.

92. Paladugu, M., Zou, J., Guo, Y. N., Zhang, X., Joyce, H. J., Gao, Q., et al. (2009). Evolution of wurtzite structured GaAs shells around InAs nanowire cores, *Nanoscale Res. Lett.* **4**, pp. 846–849.

93. Paladugu, M., Zou, J., Guo, Y.-N., Zhang, X., Kim, Y., Joyce, H. J., *et al.* (2008). Nature of heterointerfaces in GaAs/InAs and InAs/GaAs axial nanowire heterostructures, *Appl. Phys Lett.* **93**, p. 101911.

94. Pan, L., Lew, K.-K., Redwing, J. M., and Dickey, E. C. (2005). Stranski-Krastanow growth of germanium on silicon nanowires, *Nano Lett.* **5**, pp. 1081–1085.

95. People, R., and Bean, J. C. (1985). Calculation of critical layer thickness versus lattice mismatch for Ge_xSi_{1-x}/Si strained-layer heterostructures, *Appl. Phys. Lett.* **47**, pp. 322–324.

96. Persson, A. I., Larsson, M. W., Stenstrom, S., Ohlsson, B. J., Samuelson, L., and Wallenberg, L. R. (2004). Solid-phase diffusion mechanism for GaAs nanowire growth, *Nature Mater.* **3**, pp. 678–681.

97. Plante, M. C., and LaPierre, R. R. (2006). Au-assisted growth of GaAs nanowires by gas source molecular beam epitaxy: tapering, sidewall faceting and crystal structure, *J. Cryst. Growth* **310**, pp. 356–363.

98. Plante, M. C., and LaPierre, R. R. (2009). Analytical description of the metal-assisted growth of III-V nanowires: axial and radial growths, *J. Appl. Phys.* **105**, p. 114304.

99. Poole, P. J., Lefebvre, J., and Fraser, J. (2003). Spatially controlled nanoparticle-free growth of InP nanowires, *Appl. Phys. Lett.* **83**, pp. 2055–2057.

100. Qian, F., Li, Y., Park, H.-G., Dong, Y., Ding, Y., Wang, Z. L., *et al.* (2008). Multi-quantum-well nanowire heterostructures for wavelength-controlled lasers, *Nature Mater.* **7**, pp. 701–706.

101. Raychaudhuri, S., and Yu, E. T. (2006). Calculation of critical dimensions for wurtzite and cubic zinc blende coaxial nanowire heterostructures, *J. Vac. Sci. Technol. B* **24**, pp. 2053–2059.

102. Raychaudhuri, S., and Yu, E. T. (2006). Critical dimensions in coherently strained coaxial nanowire heterostructures, *J. Appl. Phys.* **99**, p. 114308.

103. Regolin, I., Sudfeld, D., S. Lüttjohann, V. K., Prost, W., Kästner, J., Dumpich, G., *et al.* (2007). Growth and characterisation of GaAs/InGaAs/GaAs nanowhiskers on (111) GaAs, *J. Cryst. Growth* **298**, pp. 607–611.

104. Renard, J., Songmuang, R., Bougerol, C., Daudin, B., and Gayral, B. (2008). Exciton and biexciton luminescence from single GaN/AlN quantum dots in nanowires, *Nano Lett.* **8**, pp. 2092–2096.

105. Sköld, N., Karlsson, L. S., Larsson, M. W., Pistol, M.-E., Seifert, W., Trägårdh, J., *et al.* (2005). Growth and optical properties of strained GaAs-$Ga_xIn_{1-x}P$ core-shell nanowires, *Nano Lett.* **5**, pp. 1943–1947.

106. Søndergaard, N., He, Y., Fan, C., Han, R., Guhr, T., and Xu, H. Q. (2009). Strain distributions in lattice-mismatched semiconductor core-shell nanowires, *J. Vac. Sci. Technol. B* **27**, pp. 827–830.

107. Soshnikov, I. P., Cirlin, G. E., Tonkikh, A. A., Samsonenko, Y. B., Dubrovskii, V. G., Ustinov, V. M., *et al.* (2005). Atomic structure of MBE-grown GaAs nanowhiskers, *Phys. Sol. State* **47**, pp. 2121–2126.

108. Spirkoska, D., Arbiol, J., Gustafsson, A., Conesa-Boj, S., Glas, F., Zardo, I., *et al.* (2009). Structural and optical properties of high quality zinc-blende/wurtzite GaAs nanowire heterostructures, *Phys. Rev. B* **80**, p. 245325.

109. Svensson, C. P. T., Seifert, W., Larsson, M. W., Wallenberg, L. R., Stangl, J., Bauer, G., *et al.* (2005). Epitaxially grown GaP/$GaAs_{1-x}P_x$/GaP double heterostructure nanowires for optical applications, *Nanotechnology.* **16**, pp. 936–939.

110. Tambe, M. J., Lim, S. K., Smith, M. J., Allard, L. F., and Gradečak, S. (2008). Realization of defect-free epitaxial core-shell GaAs/AlGaAs nanowire heterostructures, *Appl. Phys. Lett.* **93**, p. 151917.

111. Tateno, K., Zhang, G., and Nakano, H. (2008). Growth of GaInAs/AlInAs heterostructure nanowires for long-wavelength photon emission, *Nano Lett.* **8**, pp. 3645–3650.

112. Tchernycheva, M., Cirlin, G. E., Patriarche, G., Travers, L., Zwiller, V., Perinetti, U., *et al.* (2007). Growth and characterization of InP nanowires with InAsP insertions, *Nano Lett.* **7**, pp. 1500–1504.

113. Tchernycheva, M., Travers, L., Patriarche, G., Glas, F., Harmand, J.-C., Cirlin, G. E., *et al.* (2007). Au-assisted molecular beam epitaxy of InAs nanowires: Growth and theoretical analysis, *J. Appl. Phys.* **102**, p. 094313.

114. Thelander, C., Agarwal, P., Brongersma, S., Eymery, J., Feiner, L. F., Forchel, A., *et al.* (2006). Nanowire-based one-dimensional electronics, *Materials Today* **9**, pp. 28–35.

115. Timoshenko, S. P., and Goodier, J. N. (1951). *Theory of elasticity* (McGraw-Hill, New York).

116. Trammell, T. E., Zhang, X., Li, Y., Chen, L.-Q., and Dickey, E. C. (2008). Equilibrium strain-energy analysis of coherently strained core-shell nanowires, *J. Cryst. Growth* **310**, pp. 3084–3092.

117. Tribu, A., Sallen, G., Aichele, T., André, R., Poizat, J.-P., Bougerol, C., et al. (2008). A high-temperature single-photon source from nanowire quantum dots, *Nano Lett.* **8**, pp. 4326–4329.

118. Varahramyan, K. M., Ferrer, D., Tutuc, E., and Banerjee, S. K. (2009). Band engineered epitaxial Ge-Si_xGe_{1-x}core-shell nanowire heterostructures, *Appl. Phys. Lett.* **95**, p. 033101.

119. Verheijen, M. A., Immink, G., de Smet, T., Borgström, M. T., and Bakkers, E. P. A. M. (2006). Growth kinetics of heterostructured GaP-GaAs nanowires, *J. Am. Chem. Soc.* **128**, pp. 1353–1359.

120. Wagner, J. B., Sköld, N., Wallenberg, L. R., and Samuelson, L. (2010). Growth and segregation of GaAs-$Al_xIn_{1-x}P$ core-shell nanowires, *J. Cryst. Growth* **312**, pp. 1755–1760.

121. Wagner, R. S., and Ellis, W. C. (1964). Vapor-liquid-solid mechanism of single crystal growth, *Appl. Phys. Lett.* **4**, pp. 89–90.

122. Wang, N., Cai, Y., and Zhang, R. (2008). Growth of nanowires, *Mater. Sci. Engin. R* **60**, pp. 1–51.

123. Warwick, C. M., and Clyne, T. W. (1991). Development of composite coaxial cylinder stress analysis model and its application to SiC monofilament systems, *J. Mater. Sci.* **26**, pp. 3817–3827.

124. Wen, C.-Y., Reuter, M. C., Bruley, J., Tersoff, J., Kodambaka, S., Stach, E. A., *et al.* (2009). Formation of compositionally abrupt axial heterojunctions in silicon-germanium nanowires, *Science* **326**, pp. 1247–1250.

125. Wu, Y., Fan, R., and Yang, P. (2002). Block-by-block growth of single-crystalline Si/SiGe superlattice nanowires, *Nano Lett.* **2**, pp. 83–86.

126. Ye, H., Lu, P., Yu, Z., and Han, L. (2009). Critical lateral dimension for a nanoscale-patterned heterostructure using the finite element method, *Semicond. Sci. Technol.* **24**, p. 025029.

127. Ye, H., Lu, P., Yu, Z., Song, Y., Wang, D., and Wang, S. (2009). Critical thickness and radius for axial heterostructure nanowires using finite-element method, *Nano Lett.* **9**, pp. 1921–1925.

128. Zhang, G., Tateno, K., Gotoh, H., Sogawa, T., and Nakano, H. (2010). Structural, compositional, and optical characterizations of vertically aligned AlAs/GaAs/GaP heterostructure nanowires epitaxially grown on Si substrate, *Jpn. J. Appl. Phys.* **49**, p. 015001.

129. Zhong, Z., and Sun, Q. P. (2002). Analysis of a transversely isotropic rod containing a single cylindrical inclusion with axisymmetric eigenstrains, *Int. J. Sol. Struct.* **39**, pp. 5763–5765.

130. Zhou, H. L., Hoang, T. B., Dheeraj, D. L., van Helvoort, A. T. J., Liu, L., Harmand, J. C., *et al.* (2009). Wurtzite GaAs/AlGaAs core-shell nanowires grown by molecular beam epitaxy, *Nanotechnology.* **20**, p. 415701.

Chapter 6

Metamorphic HEMT Technology Exemplified by InAlAs/InGaAs/GaAs HEMTs

William E. Hoke and Colin S. Whelan

Raytheon Company, 362 Lowell St., Andover, MA 01810 USA

william_e_hoke@raytheon.com

The metamorphic growth process has enabled the development of high quality epitaxial films on lattice mismatched substrates beyond the limitations of pseudomorphic growth. This capability has permitted the fabrication of new device structures as well as less expensive existing structures. The metamorphic approach has been applied to several material systems and device structures and is illustrated here for the InAlAs/InGaAs/GaAs high electron mobility transistor (HEMT). The constraints of lattice matching and pseudomorphic growth are first examined to develop the motivations for metamorphic growth. Various material properties of metamorphic films and metamorphic HEMT (MHEMT) structures on GaAs are characterized. Examples of microwave devices and circuits incorporating GaAs metamorphic HEMTs are given. A new application of integrating GaAs metamorphic circuits with silicon CMOS on a common silicon substrate is presented. Finally, reliability results of metamorphic HEMTs are reviewed.

Lattice Engineering: Technology and Applications

Edited by Shumin Wang

ISBN 978-981-4316-29-3 (Hardcover), 978-981-4364-25-6 (eBook)

www.panstanford.com

6.1 Introduction

The metamorphic growth process is general and has been applied to numerous material systems. Attempting to survey these various developments in proper detail is well beyond the scope of one chapter. Instead the case of metamorphic growth of InAlAs/InGaAs materials on GaAs with application to HEMT structures is examined. This material system has been extensively explored and should serve as a useful example for other material systems.

Two key characteristics for achieving low defect density epitaxial growth of one material on top of another crystalline material are compatible crystal structures and lattice constants. Figure 6.1 displays the lattice constants and bandgaps for zincblende binary III-V semiconductors as well as silicon and germanium. The solid line between GaAs and InAs represents ternary InGaAs lattice constants and bandgaps. At first sight, the epitaxial growth of one material on top of another should require a very high degree of lattice match so that the atoms at the heterojunction precisely line up to avoid a high density of dislocations. There are few closely lattice matched binary material systems in nature. One pair, as shown in Fig. 6.1, is AlAs and GaAs. Defining the lattice mismatch as the

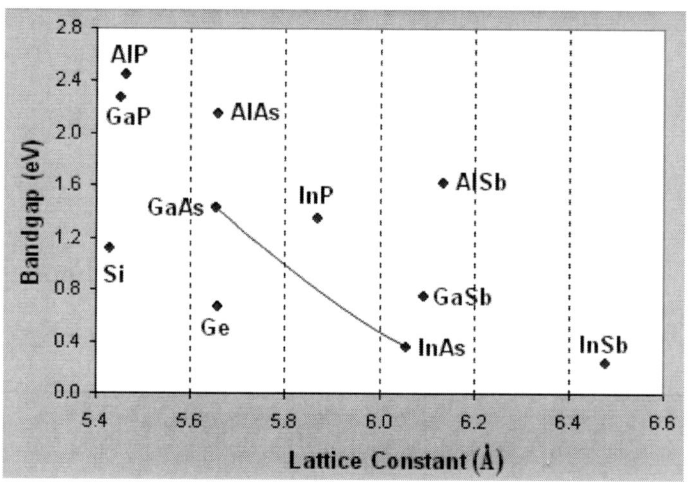

Figure 6.1. Lattice constants and bandgaps for binary zincblende III-V semiconductors as well as germanium and silicon.

difference between the two material lattice constants divided by the substrate lattice constant, AlAs and GaAs have a mismatch of only 0.0014. This characteristic has been utilized in a wide range of heterojunction structures, including the first HEMTs [1, 2]. The application of ternary materials to lattice match binary materials has also been extensively developed such as the growth of $In_{0.53}Ga_{0.47}As$ and $In_{0.52}Al_{0.48}As$ on InP.

An important step in expanding the realm of possible epitaxial heterojunction systems without creating dislocations was the realization that near perfect lattice match was not strictly required, since crystal bonds can withstand a limited amount of strain before rupturing. If the thickness of the mismatched epitaxial layer did not exceed the critical thickness for strain-driven dislocation formation, the film would elastically distort so that the in-plane lattice constant matched the underlying crystal. The growth of a strained but dislocation-free or pseudomorphic InGaAs/GaAs HEMT was demonstrated on GaAs [3], which exhibited an increased current density compared to the near lattice matched AlGaAs/GaAs HEMT. This effort was soon followed by an AlGaAs/InGaAs pseudomorphic HEMT (PHEMT), which further increased the channel charge [4]. The PHEMT structure evolved for improved power performance to an AlGaAs/InGaAs/AlGaAs heterojunction [5, 6] in which both AlGaAs barriers were selectively doped. The GaAs PHEMT structure has demonstrated excellent reliability [7] and has been the dominant microwave power transistor structure for many years. Recently, GaN HEMTs, with a pseudomorphic AlGaN barrier layer, are beginning to replace the GaAs PHEMT in applications where maximum microwave power performance is required.

Addressing lattice mismatch by pseudomorphic growth is still limited by the critical thickness for dislocation formation, which has been modeled by various approaches, for example Refs. [8–10]. For the PHEMT, the indium concentration in the InGaAs channel layer can be increased by reducing its thickness. For thicknesses much less than 100 Å, however, the quantum size effect increases the energy of the quantum well levels, compensating for the effect of increased indium content. Also the increased strain from higher indium content can drive three-dimensional growth. Figures 6.2 and 6.3 [11] display cross-sectional and plan-view transmission

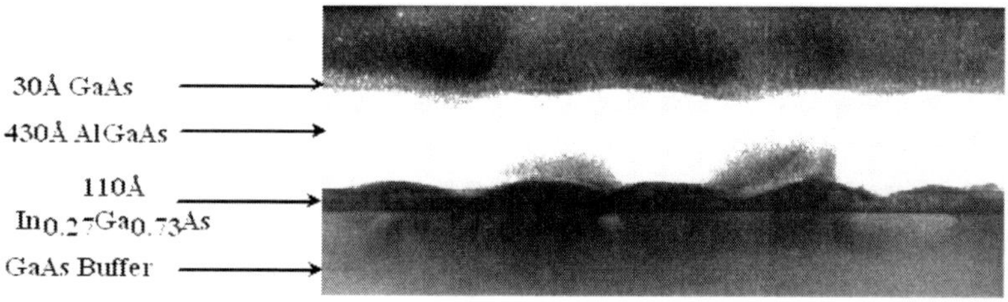

Figure 6.2. Cross-sectional TEM of a GaAs PHEMT with a 110 Å $In_{0.27}Ga_{0.73}As$ channel layer.

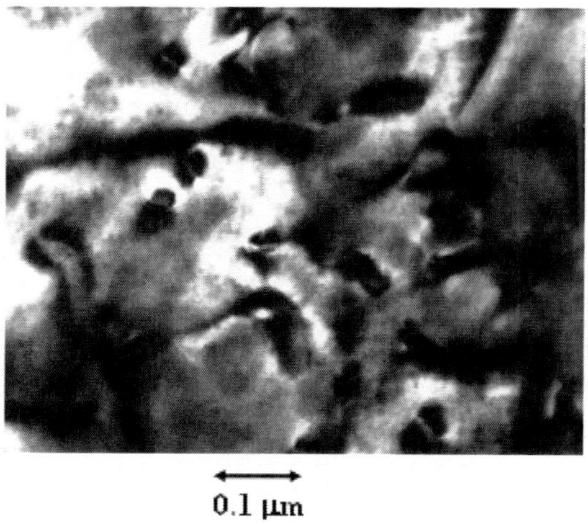

Figure 6.3. Plan-view TEM of the sample in Fig. 6.2.

electron micrographs (TEM) of a PHEMT with a 110 Å $In_{0.27}Ga_{0.73}As$ channel layer, respectively. The compressive strain has caused undulations in the cross-sectional view of the InGaAs channel thickness with resultant dislocation formation in the plan view. The resultant material properties were seriously degraded and unacceptable for device fabrication [11]. Compared to a PHEMT with a 110 Å $In_{0.20}Ga_{0.80}As$ layer, the electron mobility dropped from over 6000 to 400 cm^2/Vs and the InGaAs channel photoluminescence (PL) intensity dropped by a factor of 100. Strain-induced layer undulations have also been detailed in compressively strained SiGe layers on Si [12].

Different approaches have been investigated to increase the critical layer thickness of pseudomorphic films. Lower molecular beam epitaxy (MBE) growth temperatures have been shown to increase the critical layer thickness [13, 14] of InGaAs films on GaAs substrates due to kinetic barriers to dislocation formation. The use of strain compensation by incorporating pseudomorphic layers with compressive and tensile strain has been successfully demonstrated [15, 16]. These approaches are not amenable to thick, mismatched layers well beyond the critical thickness. Also the use of InGaAs substrates has been explored. The structure in Fig. 6.2 has been grown on an $In_{0.065}Ga_{0.935}As$ substrate resulting in planar interfaces and a mobility of 6340 cm^2/Vs [17]. Due to differences in liquidus and solidus compositions of melt grown boules, however, the growth of crack-free ternary substrate boules is quite difficult to achieve and large area InGaAs substrates are not available commercially.

6.2 MHEMT Opportunities

Metamorphic film growth and properties will be illustrated using the InGaAs and InAlAs ternary systems. In Fig. 6.1, the InGaAs (InAlAs) lattice constants vary from that of GaAs (AlAs) through a lattice match with InP, $In_{0.53}Ga_{0.47}As$, and $In_{0.52}Al_{0.48}As$, to InAs. Advantages of increasing the indium content in the HEMT InGaAs channel layer are the increase in mobility as well as the increase in conduction band discontinuity with InAlAs for higher channel charge density. For InGaAs and InAlAs films with thicknesses greater than approximately 100 Å, there are compositions between GaAs and InP and between InP and InAs that are inaccessible by the pseudomorphic approach. Additionally InAs substrates are conductive and not acceptable for HEMT devices. The metamorphic growth approach enables utilizing these inaccessible compositions in device structures. Figure 6.4 illustrates MHEMT structures with indium channel concentrations of 35%, 53% that lattice matches InP, and 75%. The graded or metamorphic buffer layer transforms the lattice constant to permit growth of the desired channel and barrier layers. The last structure is an example of metamorphic growth that transforms the lattice constant to 64% indium content to maintain

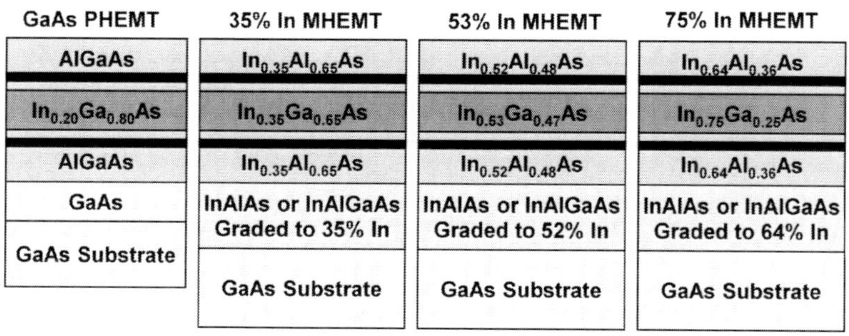

Figure 6.4. Schematic cross sections of a GaAs PHEMT and three GaAs MHEMTs with respective channel compositions of $In_{0.20}Ga_{0.80}As$, $In_{0.35}Ga_{0.65}As$, $In_{0.53}Ga_{0.47}As$, and $In_{0.75}Ga_{0.25}As$. The 75% In MHEMT has a pseudomorphic $In_{0.75}Ga_{0.25}As$ channel layer on top of the $In_{0.64}Al_{0.36}As$ barrier layer. The black lines represent pulse doping on both sides of the InGaAs channel layer with thin spacer layers between the doping pulses and the channel layer.

an acceptable bandgap for the $In_{0.64}Al_{0.36}As$ barrier layers followed by pseudomorphic growth of the $In_{0.75}Ga_{0.25}As$ channel layer.

Historically, metamorphic growth on GaAs substrates as opposed to InP substrates has also been driven by cost and availability. There was considerable activity in metamorphic development around the year 2000. At that time 100 mm substrates cost approximately \$200 for GaAs and over \$1000 for InP. Additional concerns are that InP substrates cleave easier than GaAs, increasing the risk of wafer breakage, and InP via hole formation for circuit fabrication is more difficult. Consequently, as illustrated in Fig. 6.4, MHEMTs on GaAs with $In_{0.53}Ga_{0.47}As$ channel layers were also developed as an alternative to the InP HEMT. Currently, substrate cost and availability considerations have moved to larger diameter substrates. A 150-mm GaAs substrates cost approximately \$175 whereas 150 mm InP substrates are very expensive (> \$5000) due to lack of market interest and 100 mm substrates cost approximately \$800.

6.3 MHEMT Material Results

The metamorphic buffer layer has two important functions. First, it transforms the in-plane lattice constant from that of the underlying layer or substrate whereas pseudomorphic growth does not. Altering the lattice constant requires the formation of a high density of dislocations and other crystalline defects. Initial dislocation densities near the bottom of the metamorphic buffer layer can be 10^{10} cm^{-2} or greater. The second function of the metamorphic buffer layer is to contain most of the dislocations so that the dislocation density in the device layers is considerably less, preferably 10^6 cm^{-2} or less.

For the InGaAs MHEMT on a GaAs substrate, the lattice constant must be expanded. One approach would be to grow a thick buffer layer with the desired lattice constant. For example, for the 53% InGaAs MHEMT, a thick $In_{0.52}Al_{0.48}As$ layer or an InP layer could be used. This approach has not worked well, probably due to the high initial strain causing three-dimensional growth. The most common approach is to grade the indium concentration over approximately a micron thickness. With this grading rate, a two-dimensional growth front is maintained. Another consideration is that the HEMT buffer layer must be resistive, making graded InGaAs layers not acceptable for HEMT applications. The most commonly graded HEMT buffer layers utilize indium graded ternary InAlAs layers [18, 19] or quaternary InAlGaAs layers [20, 21]. Compositional overshoot [18] is also used in which the indium concentration is graded 5–10% beyond the desired composition. As with pseudomorphic growth, the top layers will tolerate a certain amount of strain and are only approximately 80% relaxed, [22] making compositional overshoot necessary. The thermal conductivity of these graded buffer layers is low, however, and can be increased by using graded InGaP [23].

Cross-sectional TEM of a 53% In MHEMT [24] are displayed in Fig. 6.5. The metamorphic buffer layer contains a high density of dislocations, but the subsequently grown $Al_{0.5}In_{0.5}As$ buffer layer does not. In contrast to Fig. 6.2, the top device layers exhibit planar interfaces despite the InGaAs and InAlAs compositions being twice that in Fig. 6.2 and several times thicker. A plan view [24] of the 200 Å 53% In MHEMT is shown in Fig. 6.6. No dislocations are observed at this magnification, which is in stark contrast to the plan view of the 110 Å $In_{0.27}Ga_{0.73}As$ layer shown in Fig. 6.3. After examining several plan views, the dislocation density is estimated at 1×10^6 cm^{-2}.

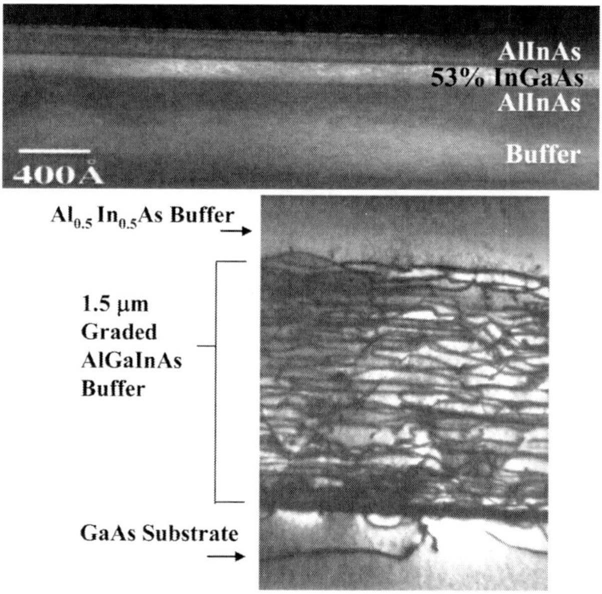

Figure 6.5. Cross-sectional TEM of a 53% InGaAs MHEMT. Top: device layers. Bottom: metamorphic buffer layer.

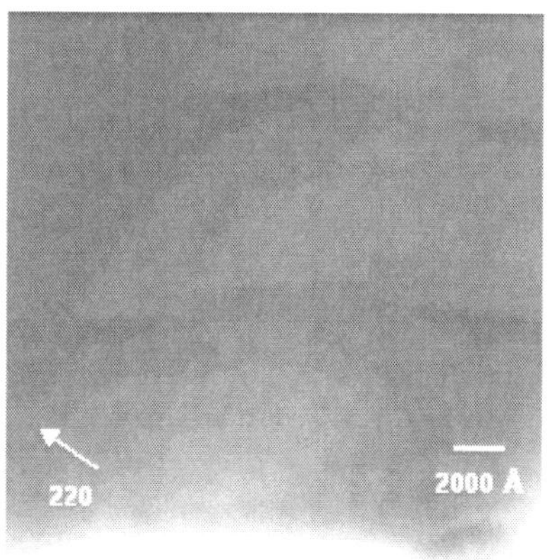

Figure 6.6. Plan-view TEM of a 53% In MHEMT exhibiting no dislocations in this sample view. The dislocation density determined from multiple TEM samples was 1×10^6 cm^{-2}.

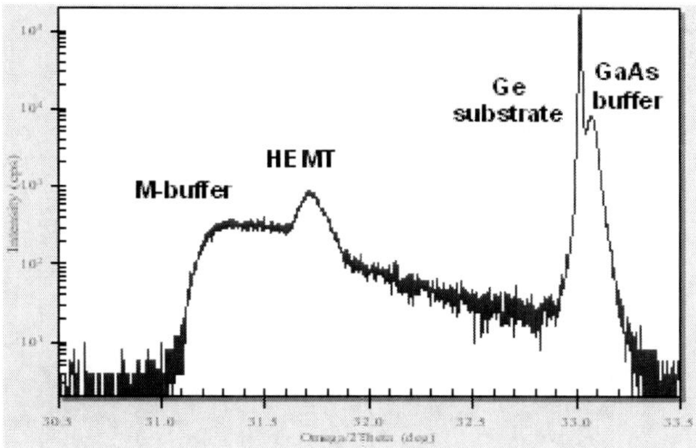

Figure 6.7. (004) X-ray spectrum of an MHEMT with indium-graded InAlAs metamorphic buffer layer with compositional overshoot and $In_{0.53}Ga_{0.47}As$ and $In_{0.52}Al_{0.48}As$ device layers. A 1 micron-thick GaAs buffer layer was first grown on a germanium substrate.

The metamorphic buffer layer transforms the lattice constant. Figure 6.7 displays the (004) x-ray diffraction spectrum of a 53% In MHEMT grown on a thick GaAs buffer layer on a germanium substrate [25]. The broad signal in moving from the GaAs and germanium peaks to lower angles is the indium-graded InAlAs buffer layer. A peak is observed for the $In_{0.52}Al_{0.48}As$ and $In_{0.53}Ga_{0.47}As$ HEMT device layers. The diffraction signal at angles less than the device layers is from the compositional overshoot due to InAlAs material with indium concentrations greater than 52%.

The surface morphology of graded metamorphic structures exhibits a characteristic cross hatch pattern when viewed under the microscope, which is indicative of two-dimensional growth. The cross hatch will be weak or absent if three-dimensional growth occurs due to excessive strain at the growth surface. A 25 μm × 25 μm atomic force microscopy (AFM) surface map of an MHEMT is shown in Fig. 6.8 [26]. A linear graded AlGaInAs buffer layer was used to transform the lattice constant to 64% indium content. Despite the high indium content, the AFM root-mean-square (RMS) roughness was only 11 Å. This value is considerably less than the thinnest layer in the structure, which is the 40 Å spacer layer between the channel and dopant pulse.

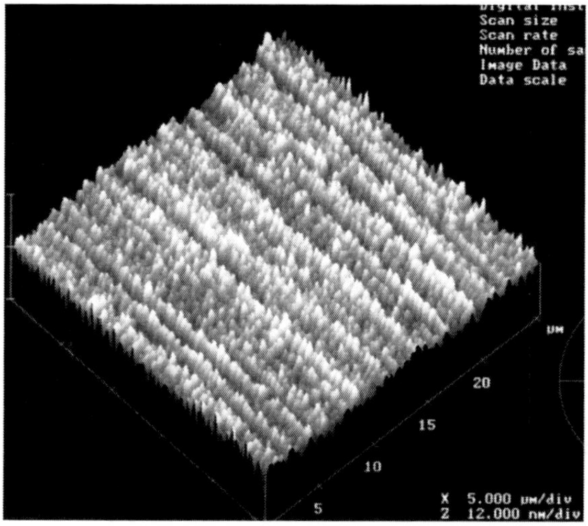

Figure 6.8. 25 μm × 25 μm AFM surface map of a MHEMT graded to $In_{0.64}Al_{0.36}As$. The RMS roughness is 11 Å.

N-type MHEMTs typically use silicon doping in the barrier layers to provide channel electrons. A concern is whether MHEMT-related defects enhance diffusion of dopant atoms. Figure 6.9 is a secondary ion mass spectrometry (SIMS) depth profile of a silicon double pulse doped MHEMT containing an $In_{0.65}Ga_{0.35}As$ channel layer [11]. The uncalibrated gallium signature indicates the position of the channel layer. The measured full-width-half-maximum (FWHM) of the doping pulses, 43 Å, is limited by the depth resolution of the SIMS technique, indicating that the doping pulses are quite sharp. Supporting evidence for abrupt doping pulses is that the MHEMT structure had a room temperature mobility of 10,520 cm^2/Vs for a sheet density of 4.2×10^{12} cm^{-2}. In this structure, the top spacer layer between channel and top pulse was 50 Å and 40 Å for the bottom spacer layer. Significant silicon migration into the spacer layers or channel layer would have degraded the mobility.

The low field transport quality of MHEMTs is evaluated by the mobility. InAs has a significantly higher electron mobility than GaAs. Consequently, incorporating increasing amounts of indium into the MHEMT InGaAs channel layer should result in higher mobilities if additional MHEMT buffer grading does not degrade channel

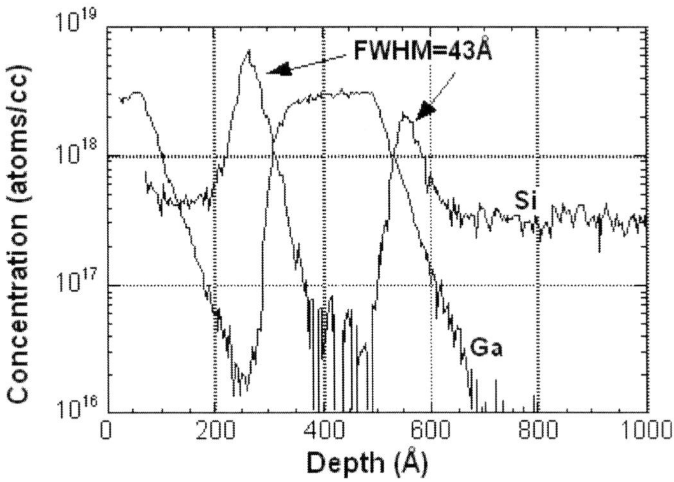

Figure 6.9. SIMS depth profile of a silicon double pulse doped MHEMT with an $In_{0.65}Ga_{0.35}As$ channel layer. The uncalibrated gallium signature indicates the InGaAs channel position. The measured doping pulse widths are limited by the SIMS depth resolution.

properties. Figure 6.10 displays the room temperature mobilities as a function of InGaAs channel composition for a double pulse doped GaAs PHEMT, two single pulse doped InP HEMTs, and several MHEMT structures [26]. All samples had sheet channel densities within 10% of 3.0×10^{12} cm^{-2}. The spacer thicknesses between the doping pulses and InGaAs channel layer were 30–40 Å except for the GaAs PHEMT, which had a top spacer of 30 Å and a bottom spacer of 50 Å. Figure 6.10 indicates that the MHEMT mobility increases nearly linearly with increasing indium content of the InGaAs channel layer. Also the mobilities of MHEMTs with indium contents similar to the lattice matched or pseudomorphic InP HEMTs have nearly the same mobilities, indicating little mobility degradation from the metamorphic grading process.

The data point with the highest indium content had a composite 100 Å $In_{0.65}Ga_{0.35}As$/100 Å $In_{0.80}Ga_{0.20}As$ channel layer grown on top of a metamorphic buffer layer graded to $In_{0.64}Al_{0.36}As$. The two InGaAs channel layers are clearly seen in the TEM of Fig. 6.11. The MHEMT had excellent electrical properties despite grading to 64%

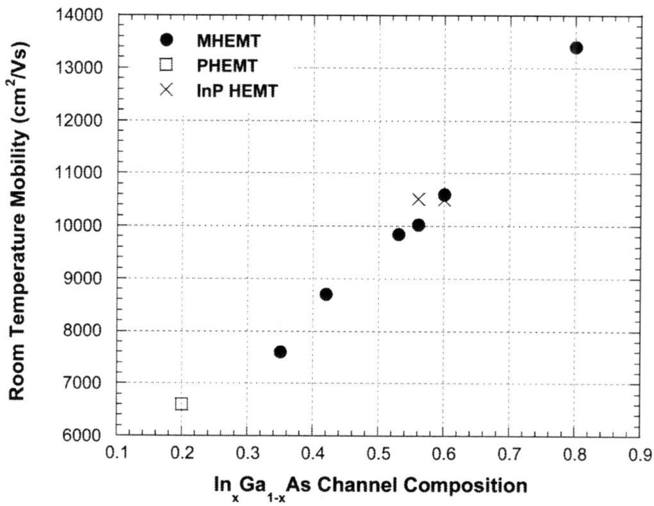

Figure 6.10. Room temperature mobility as a function of InGaAs channel composition for a GaAs PHEMT, two InP HEMTs, and several MHEMTs. All samples had a sheet carrier concentration of $3.0 \pm 0.3 \times 10^{12}$ cm^{-2}.

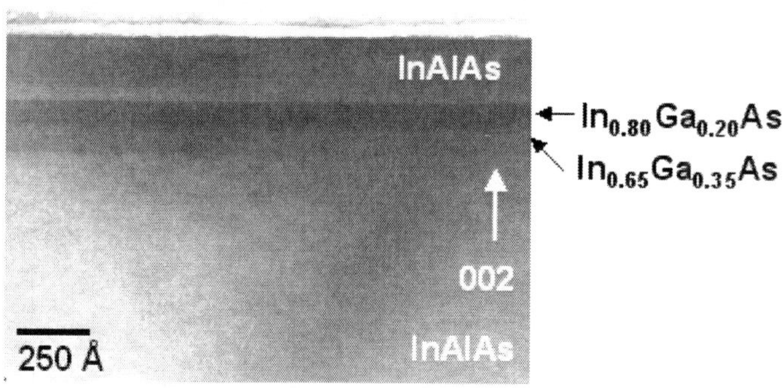

Figure 6.11. Cross-sectional TEM of a MHEMT containing a composite InGaAs channel layer of 100Å $In_{0.65}Ga_{0.35}As$/100Å $In_{0.80}Ga_{0.20}As$.

indium content. For a sheet density of 2.8×10^{12} cm^{-2}, the mobility (cm^2/Vs) was 13,500 at room temperature and 45,400 at 77 K.

The optical quality of metamorphic HEMTs is often evaluated by PL. The MHEMT channel layer is a quantum well bordered by larger bandgap barrier layers. Optical excitation creates electron and

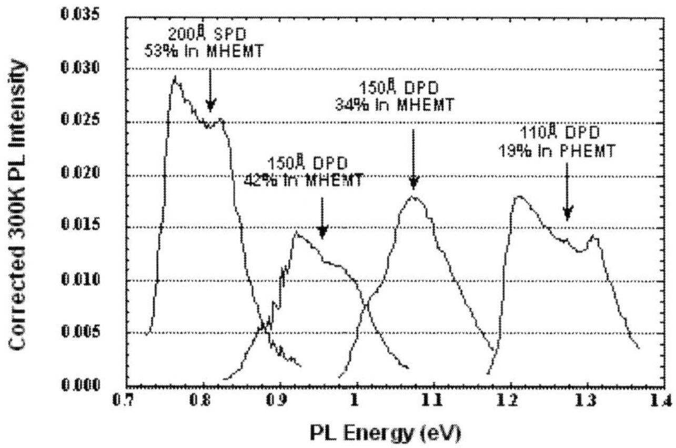

Figure 6.12. Room temperature PL spectra of a GaAs PHEMT and MHEMT structures with different InGaAs channels. The intensities were corrected for the wavelength spectral response of the PL system. DPD is double pulse doped, and SPD is single pulse doped.

hole pairs in the material. Nonradiative recombination centers in the barrier and channel layers reduce the channel PL intensity. Figure 6.12 displays room temperature PL spectra from MHEMT structures with different indium compositions as well as a reference double pulse doped PHEMT structure [27]. In order that the PL intensities could be compared, the spectra were obtained with the same laser power, spot size, and optics. Also the wavelength spectral response of the system was corrected. The highest energy spectrum is from a GaAs PHEMT, which has no defects from metamorphic grading.

Several results are illustrated in Fig. 6.12. Within the typical variations in intensity measurements, there are no significant differences in PL intensities between the PHEMT sample and the metamorphic samples, indicating that crystalline defects from metamorphic grading are not causing a significant increase in nonradiative recombination. With increasing indium channel composition and consequently increased metamorphic grading, there is no trend of reduced PL intensity. Figure 6.12 further illustrates that the metamorphic approach permits growth of GaAs MHEMTs with arbitrary indium concentrations. As just discussed, MHEMTs with indium concentrations greater than 53% have been grown with excellent electrical properties. PL measurements are not shown in Fig. 6.12 due to detector limitations. The results in Fig. 6.12 are also pertinent to optical applications of metamorphic technology.

6.4 MHEMT Device Results and Reliability

Metamorphic growth enables the fabrication of device structures containing InGaAs and InAlAs layers on GaAs with compositions varying from GaAs and AlAs to InAs. Consequently, the device designer has more freedom in choosing the material properties for a specific device application. There are natural material trade-offs in designing a layer structure. Figure 6.10 shows the attractive increase in mobility with increasing indium content of the InGaAs channel layer. Also the conduction band discontinuity between the InAlAs barrier and InGaAs channel layer increases with indium content, resulting in higher channel charge. For example, pulse doping on both sides of a GaAs PHEMT with an $In_{0.2}Ga_{0.8}As$ channel layer is required to achieve a channel sheet density of 3×10^{12} cm^{-2}. An MHEMT with an $In_{0.5}Ga_{0.5}As$ channel layer requires only one doping pulse to achieve the same charge density. As shown in Fig. 6.13, the on-state breakdown voltage, however, falls off with increasing indium content in the InGaAs channel layer primarily due to the

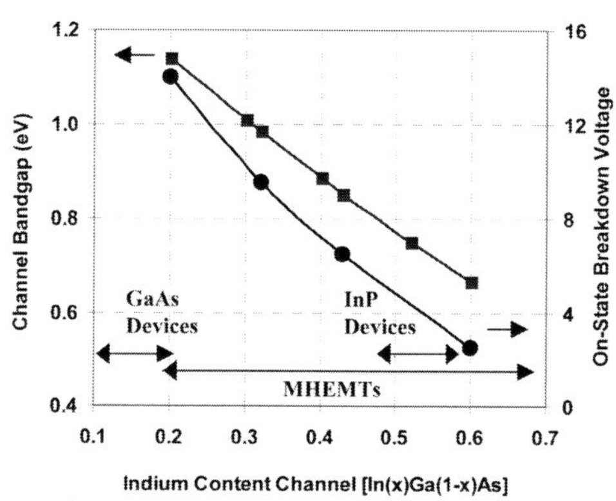

Figure 6.13. Reduction in channel bandgap with concomitant reduction in on-state breakdown with increasing indium content in the MHEMT InGaAs channel layer.

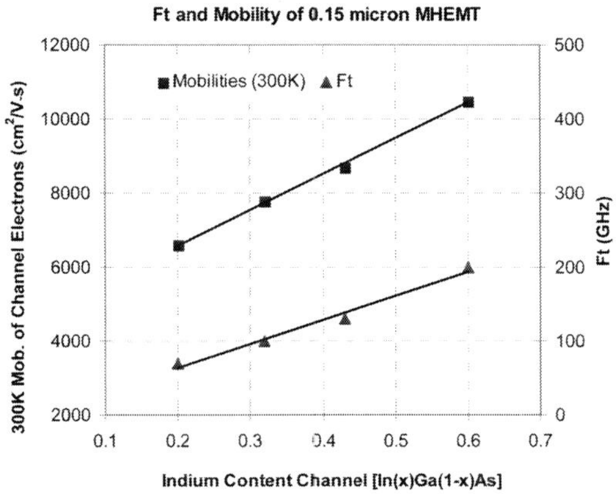

Figure 6.14. MHEMT mobility and F_t as a function of InGaAs channel composition for transistors with 0.15 μm gate lengths.

reduction of the InGaAs bandgap [28]. A similar trend in the off-state breakdown with indium content is also observed [29].

The effect of material properties is also evident in other device characteristics. Due to the improved transport properties of the InGaAs channel layer with increasing indium content, F_t (frequency at which the current gain falls to 0 dB), increases with indium content. Figure 6.14 displays F_t measurements for MHEMTs with 0.15 μm gate lengths as a function of indium content in the InGaAs channel layer [28]. The trend in F_t values parallels the trend in MHEMT mobility.

An important application of metamorphic transistors is low-noise amplifiers (LNA). With increasing indium content of the channel layer, the higher mobility coupled with improved carrier confinement from the deeper quantum well improves the noise properties of the device. Figure 6.15 displays the minimum noise figure, F_{min}, at 25 GHz as a function of the drain current for metamorphic transistors with four 75 micron gates and a gate length of 0.15 μm [30]. Data from one GaAs PHEMT with a 19% InGaAs channel layer are also plotted. The lowest noise figure with the PHEMT is 1.4 dB. The noise figure drops significantly with the metamorphic devices. A minimum value of 0.65 dB was observed with the 60% In MHEMT. The MHEMTs also exhibit a

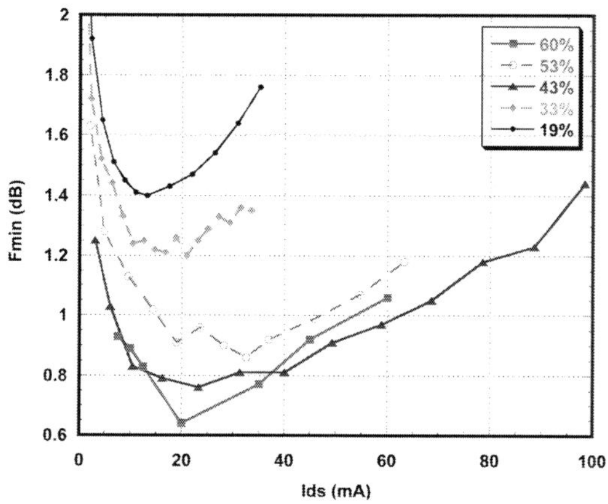

Figure 6.15. F_{min} as a function of drain current at 25 GHz measured on transistors with four 0.15 μm × 75 μm gate fingers. The performances of one GaAs PHEMT (19%) and four MHEMTs with different indium concentrations are shown.

wider low-noise region than the PHEMT due to the improved carrier confinement in the deep InGaAs quantum well.

A greater indium channel content also leads to higher associated gain. The same devices in Fig. 6.15 were measured at the minimum noise match at 25 GHz, and the results are plotted in Fig. 6.16 [30]. The associated gain increases with indium content. At a drain current of 20 mA, the 60% MHEMT has 4 dB more gain than the GaAs PHEMT. The 53% and 60% MHEMTs also achieved high gains at low current drives, which are attractive for low power receiver applications.

A natural application of metamorphic growth technology is the MHEMT with indium compositions in the pseudomorphically forbidden lattice constant region between GaAs and InP. MHEMTs with approximately 30–40% indium content are attractive due to improved mobility over GaAs PHEMTs while providing a larger on-state breakdown voltage [Ref. 28 and Fig. 6.13] and larger off-state breakdown [29] compared to lattice matched InP HEMTs. Furthermore, a large conduction band discontinuity exists between the $In_{0.3}Al_{0.7}As$ barrier and the $In_{0.3}Ga_{0.7}As$ channel layer [30, 31], which increases channel sheet density and carrier confinement.

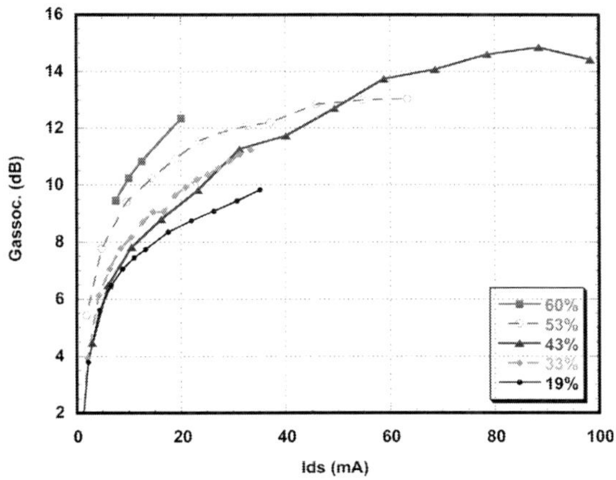

Figure 6.16. Associated gain versus drain current measured at 25 GHz for the transistors in Fig. 6.15.

Excellent microwave gain, noise, and power performance has been demonstrated with nominally 30–40% MHEMTs [32–35]. Figure 6.17 compares the gain and noise performance of 40% MHEMT and GaAs PHEMT devices as a function of drain current at 25 GHz [36]. Both devices used the same gate geometry of two 50 μm long gates with gate length of 0.15 μm. The devices were biased at 1 V between source and drain where the devices yielded the lowest noise figure. The MHEMT shows clear improvements over the PHEMT for noise figure and gain. The MHEMT exhibits a broad minimum in the noise figure of 1.18 dB compared to 1.60 dB for the PHEMT. The associated gain is approximately 3 dB higher with the MHEMT.

MHEMT power devices with nominally 40% indium content were fabricated with a 0.6 mm periphery using twelve 50 μm gate fingers with a gate length of 0.15 μm [36]. The channel charge density was 3.0×10^{12} cm^{-2}. The devices were tested at 35 GHz using a source-drain bias of 5 V, indicating good on-state breakdown characteristics. The gate-drain breakdown at 1 mA/mm was −13 V. Figure 6.18 displays results when tuning for a compromise between gain and power. At 1 dB gain compression, the device produced 26.7 dBm of output power, 7.5 dB of associated gain, and 46.2% power added efficiency (PAE). At 2 dB compression, the output

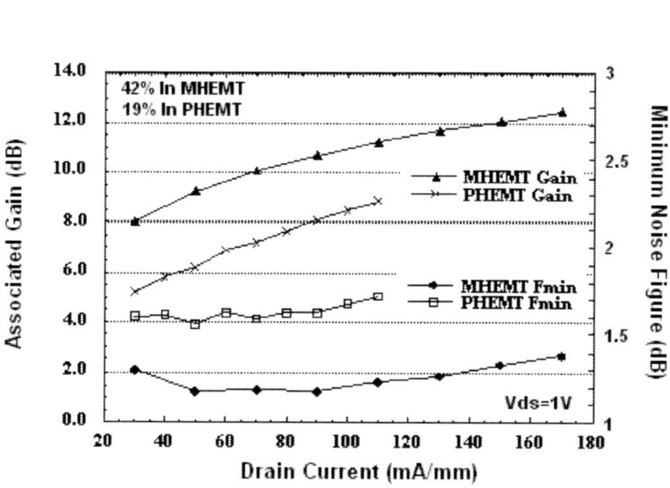

Figure 6.17. Associated gain and minimum noise figure versus drain current at 25 GHz for 42% indium MHEMT and 19% In PHEMT devices with the same gate geometry (2 × 50 μm × 0.15 μm). The devices were biased at 1 V between source and drain.

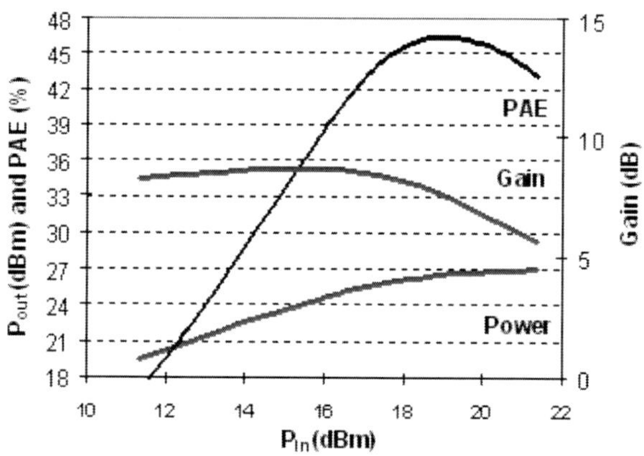

Figure 6.18. Output power, PAE, and associated gain at 35 GHz for a 40% In MHEMT device with 12 × 0.15 μm × 50 μm gates. The source-drain bias was 5 V.

power was 27.0 dBm, which normalizes to 830 mW/mm of gate periphery. Increasing the channel charge density to 3.4×10^{12} cm^{-2} resulted in a power density of 920 mW/mm at 6 V drain biasing.

Another MHEMT application is the analog structure to the InP HEMT [20, 37–40]. Many groups have demonstrated MHEMT

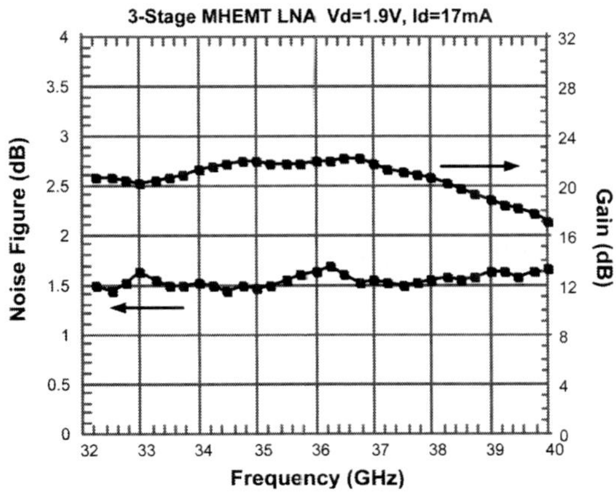

Figure 6.19. Noise figure and gain versus frequency for a Ka-band three-stage 53% In MHEMT LNA. The device was biased at a drain voltage of 1.9 V and drain current of 17 mA.

material, device, and circuit properties very comparable to the InP HEMT. One of the earliest MHEMT circuit demonstrations was a three-stage amplifier fabricated by Rohdin *et al.* [41]. The gain from 0–20 GHz was greater than 30 dB with 8 dB of gain at 50 GHz. The performance of a wideband Ka-band three-stage MHEMT low-noise amplifier (LNA) is shown in Fig. 6.19 [24]. The circuit was biased at a drain voltage of 1.9 V and drain current of 17 mA, thus requiring only 32 mW of DC power. The gain was at least 20 dB with noise figure of 1.5–1.7 dB over the broad band of 32–38 GHz. Additional MHEMT low-noise circuit results are described in Ref. [42].

MHEMTs and metamorphic PIN (MPIN) photodiodes have been integrated on the same wafer to form monolithic optoelectronic integrated circuits (OEICs) and absorb at the long haul fiber standard of 1550 nm [43, 44]. A schematic cross section of the layer structure is shown in Fig. 6.20. The material was grown by MBE in one continuous growth run. Selective etches were used to define the structure. The MHEMT and MPIN photodiode performances were tested separately and found not to be degraded by merging of the two processes [43]. A micrograph of the monolithic OEIC is shown in Fig. 6.21 and consists of a MPIN photodiode, a six-stage dc-45 GHz MHEMT traveling wave amplifier [45], and a buffer

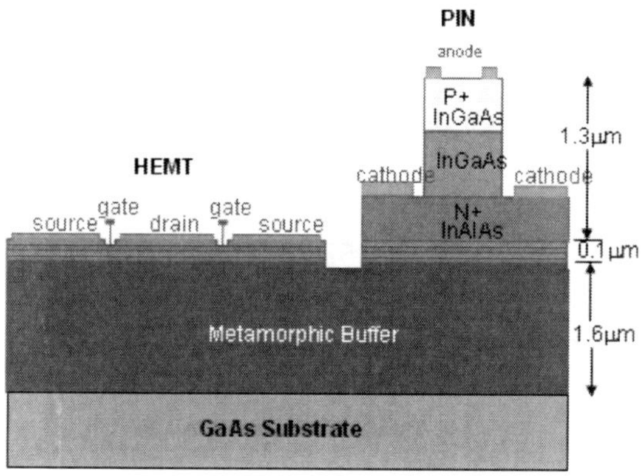

Figure 6.20. Schematic cross section of a monolithic MHEMT and MPIN OEIC structure.

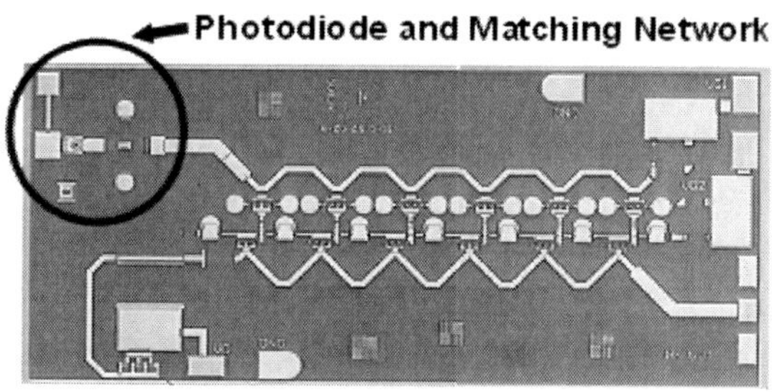

Figure 6.21. 1.7 mm × 3.7 mm OEIC chip consisting of a MPIN photodiode, buffer amplifier, six-stage MHEMT traveling wave amplifier, and matching circuitry.

amplifier. The match between the photodiode and traveling wave amplifier, TWA, was made with a wide bandwidth lossy matching network or a buffer amplifier. Figure 6.22 shows the output power as a function of frequency for a monolithic OEIC with lossy match. For comparison, the performance is plotted of a hybrid OEIC circuit also using a lossy match between the discrete photodiode and TWA. The monolithic OEIC displays a larger bandwidth than the discrete circuit, most likely from reduced parasitics from monolithic integration. Incorporating the buffer amplifier increases the monolithic OEIC output power 5–6 dB without compromising the bandwidth.

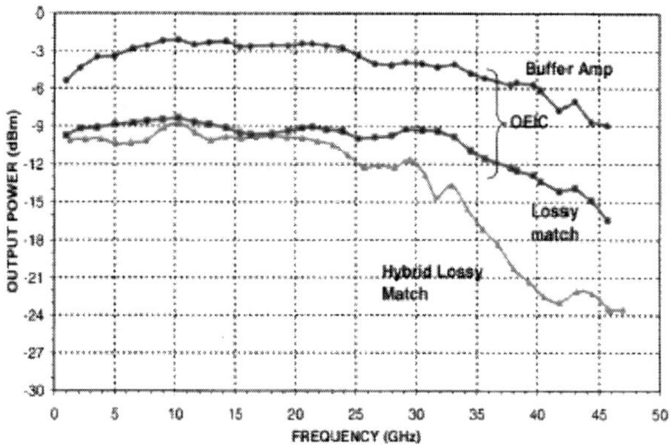

Figure 6.22. Optical to electrical power conversion performance as a function of frequency of a metamorphic monolithic OEIC and hybrid OEIC with a wide bandwidth lossy match. The performance of a metamorphic monolithic OEIC with a buffer amplifier is also shown.

A recent application of metamorphic growth is the monolithic integration of III-V devices with silicon devices in close proximity on the same silicon wafer [46, 47]. The monolithic integration of silicon circuitry with III-V circuitry has been pursued for over 20 years with limited success. With the recent development of wafer transfer technology such as Smart-Cut^{TM} [48], semiconductor layers have been transferred onto silicon substrates with diameters up to 300 mm [49]. In the case of SOLES (semiconductor on lattice engineered substrates) wafers, germanium and silicon layers are transferred onto a handle silicon substrate using isolating SiO_2 layers for wafer bonding. A schematic of the layer structure is shown in Fig. 6.23. SOLES wafers up to 200 mm have been fabricated. In the SOLES wafer, the germanium layer acts as the template for III-V growth as it closely lattice-matches GaAs as shown in Fig. 6.1. The top silicon layer is used for CMOS fabrication with the underlying SiO_2 layer for circuit isolation.

The integration process starts with CMOS fabrication, since its processing temperatures are several hundred degrees higher than the metamorphic growth temperatures. The CMOS is then protected with a SiO_2 cap layer, and windows are etched down to the germanium template layer. GaAs is then blanket grown on the

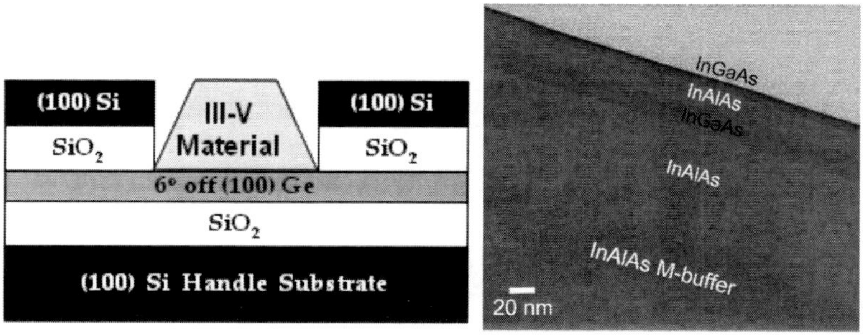

Figure 6.23. Left: Schematic of a SOLES wafer for Si/III-V integration. Right: TEM of MHEMT device layers grown in a window on the germanium layer in a SOLES wafer.

wafer by MBE resulting in epitaxial growth in the windows and polycrystalline material grown on the SiO_2 cap layer [47]. Antiphase boundaries due to GaAs growth on germanium are minimized by using misoriented germanium layers [50] as well as annealing the germanium prior to growth to form bilayer steps [51]. The GaAs layer acts as the base layer for metamorphic grading. Figure 6.23 displays a cross-sectional TEM of MHEMT device layers grown in a SOLES window on the germanium template layer. The initial Si/III-V integration demonstration circuit was a differential amplifier with unity gain frequency greater than 25 GHz [46, 47]. The amplifier incorporated silicon CMOS and metamorphic heterojunction bipolar transistor (HBT) circuitry. Figure 6.24 shows fabricated CMOS

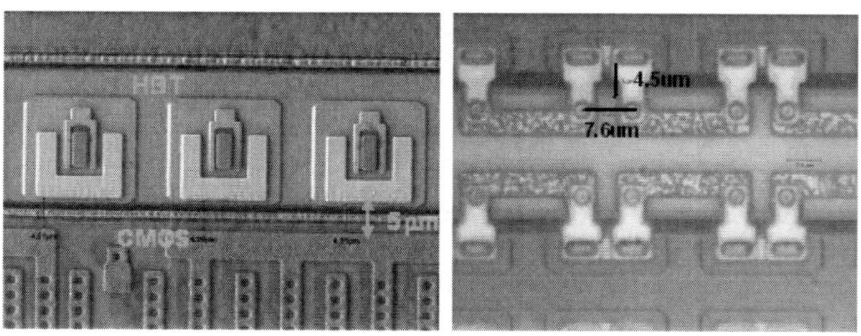

Figure 6.24. Left: Three metamorphic HBTs within 5 microns of CMOS circuitry on a SOLES wafer. Right: Electrical contacts between CMOS and HBT elements.

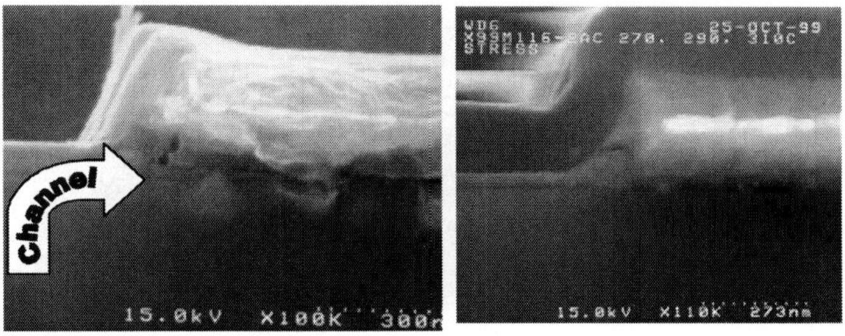

Figure 6.25. Left: Cross section of a Pd, Au ohmic on MHEMT device layers after 72 hours at 250°C with excess metal diffusion penetrating deep beneath the InGaAs channel layer. Right: Cross section of a MHEMT sample having an improved ohmic metal stack after a series of stress bakes at 270, 290, 310°C for 24 hours each. The ohmic stack increases the optimum alloy temperature and eliminates excessive diffusion.

circuitry in close proximity to three metamorphic HBTs. Electrical contacts between the HBTs and silicon CMOS are also displayed in Fig 6.24. In the near future, MHEMTs will also be integrated with silicon CMOS. It is expected that metamorphic growth will play an important future role in monolithically integrating other III-V device structures with silicon circuitry.

Finally, MHEMT devices have demonstrated excellent reliability [52, 53]. Similar to GaAs and InP HEMT technologies, the reliability of MHEMTs is governed by the metal-to-semiconductor contacts. Ohmic contacts serve to dope the semiconductor during the alloy process and then must remain stable for the lifetime of the device. Contact resistance will continue to rise if ohmic diffusion is excessive. Mean time to failures (MTTFs) of only 1000 hours at 125°C were found when using conventional metal systems containing Ni, Ge, and Au or Pd, Ge due to excessive metal diffusion (Fig. 6.25 Left). Non-alloyed ohmics using Mo/Au and other refractory metals were attempted but exhibited severe Au diffusion at the contact edges due to thinning of the refractory metal there. Improved metal layers were designed to reduce diffusion and enhance stability (Fig. 6.25 Right). Figure 6.26 displays DC Arrhenius reliability data of these MHEMT devices with gate length of 0.15 μm and gate width of 100 μm measured at five temperatures [53]. The devices were stressed with a drain bias of 1 V and current density

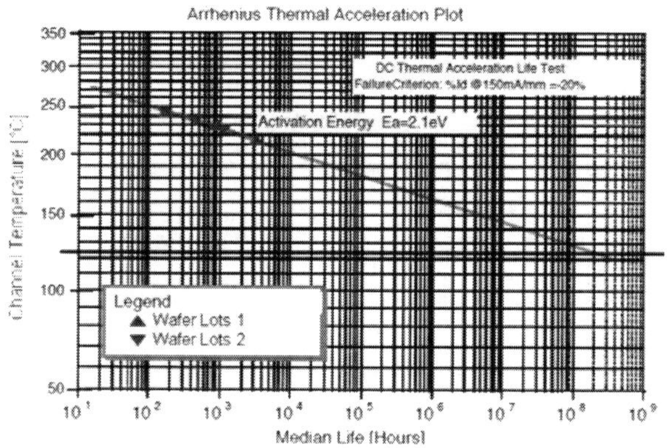

Figure 6.26. MTTF for a 20% drop in current for 2 × 0.15 μm × 50 μm MHEMT devices. Each data point represents approximately 12–15 devices.

of 150 mA/mm at temperatures of 215, 225, 230, 235, and 245°C. Each data point represents approximately 12–15 devices. Defining a failure as a 20% drop in current measured at $V_{gs} = 0$ V, the extrapolated MTTF at 125°C is ~1 × 10^8 hours.

6.5 Summary

Metamorphic growth enables epitaxial deposition of layer structures with moderate dislocation densities that are not possible by the pseudomorphic approach. The metamorphic buffer layer transforms the lattice constant of the substrate or underlying template layer to that of the desired material while confining most of the dislocations that are necessarily formed due to the alteration of the lattice constant. Device structures such as the 35% InGaAs HEMT are now possible on GaAs substrates. Consequently, the compositions of the critical device layers can be tailored to optimize performance. The material, device, and circuit properties of MHEMTs grown on GaAs match or closely approach the properties of analogous lattice matched or pseudomorphic structures on InP, offering potential cost savings and increases in substrate size. Furthermore, MHEMTs grown on GaAs have exhibited excellent reliability.

Acknowledgments

The authors would like to thank P. Balas, J. Bettencourt, K. Herrick, T. Kazior, T. Kennedy, C. Laighton, S. Lardizabal, J. LaRoche, C.-D. Lee, R. Leoni, S. Lichwala, P. Lyman, P. Marsh, J. Mosca, N. Pisello, J. Sanctuary, A. Torabi, and Y. Zhang for their contributions to this work.

References

1. Dingle, R., Stormer, H. L., Gossard, A. C., and Wiegman, W. (1978). Electron mobilities in modulation-doped semiconductor heterojunction superlattices, *Appl. Phys. Lett.*, **33**, pp. 665–667.
2. Mimura, T., Hiyamizu, S., Fujii, T., and Nanbu, K. (1980). A new field-effect transistor with selectively doped GaAs/n-$Al_xGa_{1-x}As$ heterojunctions, *Jpn. J. Appl. Phys.*, **19**, pp. L225–L227.
3. Rosenberg, J. J., Benlamri, M., Kirchner, P. D., Woodall, J. M., and Pettit, J. P. (1985). An $In_{0.15}Ga_{0.85}As$/GaAs pseudomorphic single quantum well HEMT, *IEEE Electron Device Lett.*, **6**, pp. 491–493.
4. Ketterson, A., Maloney, M., Masselink, W. T., Peng, C. K., Klem, J., Fischer, R., *et al.* (1985). High transconductance InGaAs/AlGaAs pseudomorphic modulation-doped field-effect transistors, *IEEE Electron Device Lett.*, **6**, pp. 628–630.
5. Inoue, K., Nishii, K., Matsuno, T., and Onuma, T. (1987). Strained N-AlGaAs/InGaAs/N-AlGaAs selectively-doped double-heterojunction FET, *International Electron Devices Meeting, Washington, D.C., Technical Digest*, **33**, pp. 422–425.
6. Smith, P., Lester, L., Chao, P., Lee, B., Smith, R., Ballingall, J. *et al.* (1987). Millimeter wave double heterojunction pseudomorphic power HEMTs, *International Electron Devices Meeting, Washington, D.C., Technical Digest*, **33**, pp. 854–856.
7. Huang, J. C., Saledas, P., Wendler, J., Platzker, A., Boulais, W., Shanfield, S., *et al.* (1993). A double-recessed $Al_{0.24}Ga_{0.76}As/In_{0.16}Ga_{0.84}As$ pseudomorphic HEMT for Ka- and Q-band power applications, *IEEE Electron Device Lett.*, **14** pp. 456–458.
8. Matthews, J. W., and Blakeslee, A. E. (1974). Defects in epitaxial layers, *J. Cryst. Growth*, **27**, pp. 118–125.

9. People, R., and Bean, J. C. (1985). Calculation of critical layer thickness versus lattice mismatch for Ge_xSi_{1-x}/Si strained-layer heterostructures, *Appl. Phys. Lett.*, **47**, pp. 322–324.
10. Dodson, B. W., and Tsao, J. Y. (1989). Scaling relations for strained-layer relaxation, *Appl. Phys. Lett.*, **55**, pp. 1345–1347.
11. Hoke, W. E., Lemonias, P. J., Mosca, J. J., Lyman, P. S., Torabi, A., Marsh, P. F., *et al.* (1999). Molecular beam epitaxial growth and device performance of metamorphic high electron mobility transistor structures fabricated on GaAs substrates, *J. Vac. Sci. Technol. B*, **17**, pp. 1131–1135.
12. Cullis, A. G. (1996). Strain-induced modulations in the surface morphology of heteroepitaxial layers, *MRS Bulletin*, **21**, pp. 21–26.
13. Whaley, G. J., and Cohen, P. I. (1988). The growth of strained InGaAs on GaAs: kinetics versus energetics, *J. Vac. Sci. Technol. B*, **6**, pp. 625–626.
14. Ekenstedt, M. J., Wang, S. M., and Andersson, T. G. (1991). Temperature-dependent critical layer thickness for $In_{0.36}Ga_{0.64}As/GaAs$ single quantum wells, *Appl. Phys. Lett.*, **58**, pp. 854–855.
15. Thiis, P. J., Tiemeijer, L. F., Binsma, J. J., and Van Dongen, T. (1994). Progress in long-wavelength strained-layer InGaAs(P) quantum well semiconductor lasers and amplifiers, *IEEE J. Quantum Elect.*, **30**, 477–499.
16. Chiu, T. H., Cunningham, J. E., Woodward, T. K., and Sizer, T. (1993). Growth of strain-balanced InAsP/InGaP superlattices for 1.06 μm optical modulators, *Appl. Phys. Lett.*, **62**, pp. 340–342.
17. Hoke, W. E., Lyman, P. S., Mosca, J. J., Hendriks, H. T., Torabi, A., Bonner, W.A., *et al.* (1997). Improved pseudomorphic high electron mobility transistor structures on InGaAs substrates, *J. Appl. Phys.*, **81**, pp. 968–973.
18. Cordier, Y., and Ferre, D. (1999). InAlAs buffer layers grown lattice mismatched on GaAs with inverse steps, *J. Cryst. Growth*, **201/202**, pp. 263–266.
19. Lubyshev, D., Liu, W. K., Stewart, T., Cornfeld, A. B., Patton, J., Mirecki Millunchick, J., *et al.* (2000). MBE growth of high quality metamorphic HEMT structures on GaAs, *Proceedings of the Conference on InP and Related Materials Williamsburg, Va*, pp. 392–395.
20. Haupt, M., Kohler, K., Ganser, P., Emminger, S., Muller, S., and Rothemund, W. (1996). Growth of high quality $Al_{0.48}In_{0.52}As/Ga_{0.47}In_{0.43}As$ heterostructures using strain relaxed $Al_xGa_yIn_{1-x-y}As$ buffer layers on GaAs, *Appl. Phys. Lett.*, **69**, pp. 412–414.

21. Wakita, A., Rohdin, H., Robbins, V., Moll, N., Su, C.-Y., Nagy, A. *et al.* (1998). Low–noise bias reliability of AlInAs/GaInAs MODFETs with linearly graded low-temperature buffer layers grown on GaAs substrates, *Proceedings of the Conference on InP and Related Materials Tsukuba, Japan*, pp. 223–226.

22. Olsen, J. A., Hu, E. L., Lee, S. R., Fritz, I. J., Howard, A. J., Hammons, B. E. *et al.* (1996). X-ray reciprocal-space mapping of strain relaxation and tilting in linearly graded InAlAs layers, *J. Appl. Phys.*, **79**, pp. 3578–3584.

23. Yuan, K., Radhakrishnan, K., Zheng, H. Q., and Yoon, S. F. (2001). Novel $In_{0.52}Al_{0.48}As/In_{0.53}Ga_{0.47}As$ metamorphic high electron mobility transistors on GaAs substrate with $In_xGa_{1-x}P$ graded buffer layers, *Mater. Sci. Semicond. Process.*, **4**, pp. 641–645.

24. Hoke, W. E., Kennedy, T. D., Torabi, A., Whelan, C. S., Marsh, P. F., Leoni, R. E., *et al.* (2003). Properties of metamorphic materials and device structures on GaAs substrates, *J. Cryst. Growth*, **251**, pp. 804–810.

25. Lubyshev, D., Fastenau, J. M., Wu, Y., Liu, W. K., Bulsara, M. T., Fitzgerald, E. A. *et al.* (2008). Molecular beam epitaxial growth of metamorphic high electron mobility transistors and metamorphic heterojunction bipolar transistors on Ge and Ge-on-insulator/Si substrates, *J. Vac. Sci. Technol. B.*, **26**, pp. 1115–1119.

26. Hoke, W. E., Kennedy, T. D., Torabi, A., Whelan, C. S., Marsh, P. F., Leoni, R. E., *et al.* (2003). High indium metamorphic HEMT on a GaAs substrate, *J. Cryst. Growth*, **251**, pp. 827–831.

27. Hoke, W. E., Lyman, P. S., Whelan, C. S., Mosca, J. J., Torabi, A., Chang, K. L., *et al.* (2000). Growth and characterization of metamorphic $In_x(AlGa)_{1-x}As/In_xGa_{1-x}As$ high electron mobility transistor material and devices with $X = 0.3$–0.4, *J. Vac. Sci. Technol. B*, **18**, pp. 1638–1641.

28. Whelan, C. S., Marsh, P. F., Hoke, W. E., McTaggart, R. A., McCarroll, C. P., and Kazior, T. E. (2000). GaAs metamorphic HEMT (MHEMT): An attractive alternative to InP HEMTs for high performance low noise and power applications, *Proceedings of the Conference on InP and Related Materials Williamsburg, Va*, pp. 337–340.

29. Cordier, Y., Bollaert, S., Zaknoune, M., Dipersio, J., and Ferre, D. (1999). InAlAs/InGaAs metamorphic high electron mobility transistors on GaAs substrate: Influence of indium content on material properties and device performance, *Jpn. J. Appl. Phys.*, **38**, pp. 1164–1168.

30. Whelan, C. S., Lardizabal, S. M., Buhles, P. M., Hoke, W. E., Marsh, P. F., McTaggart, R. A., *et al.* (2000). A comparison of channel indium content in low noise metamorphic HEMTs with $In_xGa_{1-x}As$ $(0.3 < X$

$\leq$ 0.6), *Proceedings of the Conference on InP and Related Materials, Williamsburg, Va*, pp. 349–352.

31. Win, P., Druelle, Y., Cappy, A., Cordier, Y., Favre, J., and Bouillet, C. (1992). Metamorphic $In_{0.3}Ga_{0.7}As/In_{0.29}Al_{0.71}As$ layer on GaAs: A new structure for high performance high electron mobility transistor realization, *Appl. Phys. Lett.*, **61**, pp. 922–924.
32. Chertouk, M., Heiss, H., Xu, D., Kraus, S., Klein, W., Bohm, G., *et al.* (1996). Metamorphic InAlAs/InGaAs HEMTs on GaAs substrates with a novel composite channel design, *IEEE Electron Device Lett.*, **17**, pp. 273–275.
33. Zaknoune, M., Bonte, B., Gaquiere, C., Cordier, Y., Druelle, Y., Theron, D., *et al.* (1998). InAlAs/InGaAs metamorphic HEMT with high current density and high breakdown voltage, *IEEE Electron Device Lett.*, **19**, pp. 345–347.
34. Contrata, W., and Iwata, N. (1999). Double-doped $In_{0.35}Al_{0.65}As/In_{0.35}Ga_{0.65}As$ power heterojunction FET on GaAs substrate with 1 W output power, *IEEE Electron Device Lett.*, **20**, pp. 369–371.
35. Whelan, C. S., Marsh, P. F., Lardizabal, S. M., Hoke, W. E., McTaggart, R. A., and Kazior, T. E. (2000). Low noise and power metamorphic HEMT devices and circuits with $X = 30\%$ to 60% In_xGaAs channels on GaAs substrates, *Proceedings of the Gallium Arsenide and other Semiconductor Applications Symposium, Paris, France*, pp. 189–192.
36. Whelan, C. S., Hoke, W. E., McTaggart, R. A., Lardizabal, S. M., Lyman, P. S., Marsh, P. F., *et al.* (2000). Low noise $In_{0.32}(AlGa)_{0.68}As/In_{0.43}Ga_{0.57}As$ metamorphic HEMT on GaAs substrate with 850 mW/mm output power density, *IEEE Electron Device Lett.*, **21**, pp. 5–8.
37. Wang, G.-W., Chen, Y.-K., Schaff, W. J., and Eastman, L. F. (1988). A 0.1 μm gate $Al_{0..5}In_{0.5}As/Ga_{0.5}In_{0.5}As$ MODFET fabricated on GaAs substrates, *IEEE Trans. Electron Devices*, **35**, pp. 818–822.
38. Harmand, J. C., Matsuno, T., and Inoue, K. (1989). Lattice-mismatched growth and transport properties of InAlAs/InGaAs heterostructures on GaAs substrates, *Jpn. J. Appl. Phys.*, **28**, pp. L1101–L1103.
39. Gill, D. M., Kane, B. C., Svensson, S. P., Tu, D.-W., Uppal, P. N., and Byer, N. E. (1996). High-performance, 0.1 μm InAlAs/InGaAs high electron mobility transistors on GaAs, *IEEE Electron Device Lett.*, **17**, pp. 328–330.
40. Kawano, M., Kuzuhara, T., Kawasaki, H., Sasaki, F., and Tokuda, H. (1997). InAlAs/InGaAs metamorphic low-noise HEMT, *IEEE Microwave Guided Wave Lett.*, **7**, pp. 6–8.

41. Rohdin, H., Nagy, A., Robbins, V., Su, C-Y, Madden, C., Wakita, A., *et al.* (1995). Low-noise, high-speed $Ga_{0.47}In_{0.53}As/Al_{0.48}In_{0.52}As$ 0.1-μm MODFETs and high-gain/bandwidth three-stage amplifier fabricated on GaAs substrate, *Proceedings of the International Conference on InP and Related Materials Hokkaido, Japan*, pp. 73–76.

42. Whelan, C. S., Marsh, P. F., Leoni, R. E., Hoke, W. E., Lardizabal, S. M., Lichwala, S. J., *et al.* (2003). Metamorphic low noise amplifiers and optical components, *Int. J. High Speed Electron. and Syst.*, **13**, pp. 65–89.

43. Leoni, R. E., Whelan, C. S., Marsh, P. F., Zhang, Y., Hunt, J. G., Laighton, C. S., *et al.* (2003). Metamorphic optoelectronic integrated circuits, *IEEE Radio Frequency Integrated Circuits Symposium, Philadelphia, IEEE MTT-S Digest, Philadelphia, Pa,* pp. 147–150.

44. Hoke, W. E., Leoni, R. E., Whelan, C. S., Kennedy, T. D., Torabi, A., Marsh, P. F., *et al.* (2004). Material properties and performance of metamorphic optoelectronic integrated circuits grown by molecular beam epitaxy on GaAs substrates, *J. Vac. Sci. Technol. B*, **22**, pp. 1554–1557.

45. Leoni, R. E., Lichwala, S. J., Hunt, J. G., Whelan, C. S., Marsh, P. F., Hoke, W. E., *et al.* (2001). A DC – 45 GHz metamorphic HEMT traveling wave amplifier, *GaAs Integrated Circuits Symposium Digest, Baltimore, Md.,* pp. 133–136.

46. Kazior, T. E., LaRoche, J. R., Lubyshev, D., Fastenau, J. M., Liu, W. K., Urtega, M., *et al.* (2009). A high performance differential amplifier through the direct monolithic integration of InP HBTs and Si CMOS on silicon substrates, *IEEE Microwave Symp Dig*, pp. 1113–1116.

47. Liu, W. K., Lubyshev, D., Fastenau, J. M., Wu, Y., Bulsara, M. T., Fitzgerald, E. A., *et al.* (2009). Monolithic integration of InP-based transistors on Si substrates using MBE, *J. Cryst Growth.*, **311**, pp. 1979–1983.

48. Smart-Cut™ is a registered trademark of Soitec.

49. Maleville, C., and Mazure, C. (2004). Smart-Cut™ technology: from 300-mm ultrathin SOI production to advanced engineered substrates, *Solid-State Electron.*, **48**, pp. 1055–1063.

50. Hudait, M. K., and Krupanidhi, S. B. (2000). Atomic force microscopic study of surface morphology in Si-doped epi-GaAs on Ge substrates: effect of off-orientation, *Mater. Res. Bull.*, **35**, pp. 909–919.

51. Sieg, R. M., Ringel, S. A., Ting, S. M., Samavedam, S. B., Currie, M., Langdo, T., *et al.* (1998). Toward device-quality GaAs growth by molecular beam epitaxy on offcut $Ge/Si_{1-x}Ge_x/Si$ substrates, *J. Vac. Sci. Technol. B*, **16**, pp. 1471–1474.

52. Marsh, P. F., Whelan, C. S., Hoke, W. E., Leoni, R. E., and Kazior, T. E. (2002). Reliability of metamorphic HEMTs on GaAs substrates, *Microelectron Reliab.*, **42**, pp. 997–1002.

53. Whelan, C. S., Herrick, K., Leoni, R. E., Marsh, P. F., Zhang, Y., Lardizabal, S., *et al.* (2003). W-band metamorphic low noise and power amplifiers, *Proceeding of 11th European GaAs and Other Compound Semiconductors Application Symposium, Munich, Germany,* pp. 365–367.

Chapter 7

Metamorphic Heterojunction Bipolar Transistors

Hong Wang
School of Electrical and Electronic Engineering,
Nanyang Technological University,
Nanyang Avenue, 639798 Singapore
ewanghong@ntu.edu.sg

InP-based heterojunction bipolar transistors (HBTs) have demonstrated superior high-frequency performance over their SiGe and GaAs counterparts. The rapid development of metamorphic buffer growth technology allows the fabrication of InP-based HBTs on GaAs or Si substrates with lower substrate and fabrication costs. Metamorphic HBTs (MHBTs) — growing InP-based HBTs on GaAs substrates have made substantial progresses in the last decade. In this chapter, the developments of the MHBT technologies, including device epitaxial structures, processing technologies, and the state-of-the-art microwave and millimeter-wave performance, will be reviewed. Some of the critical issues such as thermal resistance and device reliability will be discussed.

Lattice Engineering: Technology and Applications
Edited by Shumin Wang

ISBN 978-981-4316-29-3 (Hardcover), 978-981-4364-25-6 (eBook)
www.panstanford.com

7.1 Introduction

In late 1990s, the advances in the metamorphic growth technology — growing InP-based devices on GaAs substrates have resulted in metamorphic high-electron mobility transistors (MHEMTs) with excellent DC and microwave characteristics. The use of metamorphic technology has eliminated the InP substrate related issues such as the limited substrate size, high cost, and brittle nature of the InP material. Major emphasis has been made on MHEMTs with various aspects from material growth and device fabrication to circuit applications. Circuits based on MHEMTs are now commercially available. The great success in metamorphic HEMTs has been the motivation to investigate the fabrication and characterization of InP-based HBTs on GaAs, i.e., metamorphic heterojunction bipolar transistors (MHBTs). However, the initial progress made in realizing high-performance MHBTs was limited. This is due to a number of factors that include (i) the bipolar transistor structure is much more sensitive to defects than the majority carrier device such as HEMTs; and (ii) material growth and device fabrication for InP-based HBTs are more complicate and less mature. The major breakthrough of InP-based MHBTs on GaAs started in 2000. A MHBT with indium content as high as 53% in the InGaAs base was reported by Wang *et al.* [1, 2]. The devices grown on $In_xGa_{1-x}P$ linear graded metamorphic buffer presented reasonable DC and high frequency performance. In the past 10 years, many research teams have been working the development of MHBT technology. MHBTs on GaAs substrates using different metamorphic buffers were investigated. Significant progresses have been made on the improving device DC and high frequency performance. MHBTs on GaAs substrates have demonstrated f_T and f_{max} higher than 268 GHz [3]. The results are consistent with InP-based HBTs, which has the same layer structure grown on a lattice-matched InP substrates. In addition, other types of MHBTs include InP-based HBTs on Ge and Ge-on-insulator/Si substrates [4], and 6 Å InAs HBTs on GaAs [5] or InP [6] substrates have also been studied.

From the circuit application point of view, MHBTs have many advantages over existing production technologies (e.g., lattice-matched InP or GaAs HBTs). The metamorphic technology not only allows the fabrication of high performance HBTs on larger substrates (e.g., GaAs) with more robust backside process but also provides more freedom to tune the material composition to improve the device performance.

In this chapter, we will discuss the key components of MHBT technology. The metamorphic buffer and high performance MHBT structures will be reviewed in Section 7.2. In Section 7.3, device engineering and processing technologies that are being developed to realize high performance MHBTs will be covered. The DC and high frequency characteristics of MHBTs will also be presented. In Section 7.4, the key issues associated with MHBTs for circuit applications such as thermal resistance of metamorphic buffer and reliability will be discussed. Section 7.5 summarizes the device performance attained on different MHBTs.

7.2 MHBT Layer Structures

Growth of the metamorphic buffer layer is key technology to make the high-speed performance of InP-based devices available on the more mature GaAs substrates. For high-frequency electronics devices, InP-based high-electron mobility transistors (HEMTs) and HBTs have demonstrated excellent performance due to the superior electronic transport properties of the InP and InGaAs materials. However, one of major issues for cost-effective volume production of InP-based devices and circuits is the InP substrates, which are more expensive and brittle. The InP wafer size for commercial production is still below 6 inch diameter.

A variety of M-buffers have been used to grow InP-based devices on GaAs substrates. The basic requirement of the metamorphic buffer is to accommodate the lattice mismatch between the GaAs substrate and the InP-based alloys of the device structure. The M-buffer must absorb this strain while minimizing the propagation of dislocations into the device layers. The resulting surface should have low roughness and minimal warp for device reliability and processing requirements. Different M-buffer schemes have been explored for better device performance such as graded InGaAs, InAlAs, InGaAlAs, InGaP, and InAlP. Bulk layers of InAlAs and InP were also studied. In general, the InGaAs and InGaP alloys have lower electrical resistance. Graded In(Ga)AlAs alloys are popular for production of MHEMTs. As to MHBTs, different buffers schemes were reported by different research groups. Both As- and P-based M-buffers were used for device demonstration.

Since the invention of the HBTs, many types of devices have been developed based on different material systems such as Si/SiGe on Si, AlGaAs/GaAs or InGaP/GaAs on GaAs, and InAlAs/InGaAs or InP/InGaAs on InP, etc. Each innovation attempts to alleviate some of the technical problems associated with the device performance, manufacture cost, or device reliability. The principle of operation of a MHBT is similar to that of a conventional HBT. The main concern between the MHBTs and HBTs is that in MHBTs the HBT structure is formed on a thick metamorphic buffer that has high threading dislocation density and surface roughness. The III-V HBT is, in general, a device structure that has a large bandgap emitter and a narrow bandgap heavily doped base layer. The device performance is largely affected by the material quality of the base layer. So far, for the MHBTs, InP/InGaAs single HBTs on InP M-buffer demonstrated impressive high-frequency performance. On the other hand, double HBTs have shown better breakdown voltage, which is important for power applications. Other structures such as InP/InGaAs SHBTs or DHBTs on InAlAs or InGaP M-buffer were also reported in literatures.

The epitaxial growths for MHBTs can be performed by MOCVD or molecular beam epitaxy (MBE). For ternary M-buffers, graded buffer schemes were widely used. For example, InAlAs M-buffers with a linear grade to a high indium contents, followed by a thinner inverse grade back to the target device composition were used for some of the device demonstrations. The M-buffer thickness was normally in the range of 0.9 to 1.5 μm. In general, the graded M-buffer with the inverse grade results in nearly complete relaxation and a very low density of threading dislocations at the surface. The morphology of the M-buffers normally shows a crosshatched pattern, which is the result of continual introduction of strain and consequent relaxation through a network of misfit dislocations. The anisotropic nature of the crosshatching is explained by unequal nucleation and glide properties of the orthogonal dislocations. The magnitude of

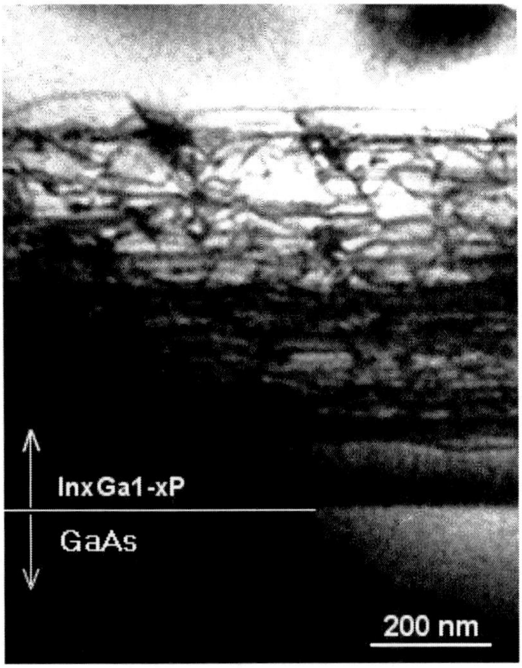

Figure 7.1. TEM image of $In_xGa_{1-x}P$ linear graded buffer grown on GaAs substrate for MHBT.

the surface roughness is dependent upon the growth conditions, epilayer thickness, and total dislocation density. It has been reported that thicker active layer structures may partially mask the underlying morphology of the M-buffer, especially the crosshatching of the InAlAs graded layer. The typical AFM root-mean-square (rms) roughness value of an MHBT structures is in the range of 20–100 Å.

The dislocation evolution through the M-buffers is normally studied by TEM. Figure 7.1 shows a typical TEM overview image of a graded InGaP M-buffer in an MHBT structure. It can be clearly seen that the dislocations are well confined in a small region at bottom part of the M-buffer. In this case, three regions with distinct transitions appear within the 1.1 μm thick M-buffer: An area of high dislocation density within the first 3000 Å next to the substrate, a 2000 Å middle region with reduced dislocations, and the top section with minimal dislocations.

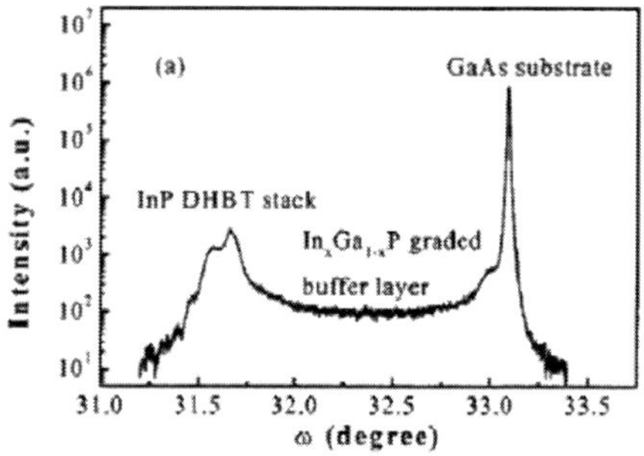

Figure 7.2. (004) XRD rocking curve of a metamorphic DHBT structure grown on GaAs substrate by solid-source MBE. Linearly graded (x: 0.48 to 1) $In_xGa_{1-x}P$ is served as the buffer layer.

The material quality of the MHBT layers could also be evaluated by high-resolution x-ray diffraction (XRD) rocking curves. Figure 7.2 shows the (004) XRD rocking curve measured for a metamorphic double HBT (DHBT) sample with graded InGaP M-buffer grown by solid-source MBE. The peak located at 33.1° corresponds to the GaAs substrate, and the peak located around 31.6° corresponds to the DHBT stack. The XRD measurement confirms the gradual change in lattice constant from the GaAs to the InP-based HBT materials by the steady intensity between these two distinct peaks, which is associated to the $In_xGa_{1-x}P$ linearly graded (x: 0.48 to 1) buffer layer. Unlike the MHEMT structures, the electrical performance of the epi-wafers can be effective assessed by using Hall measurements, the electrical performance of MHBT structures can only be verified based on the fabricated devices. To ensure a quick assessment of MHBT structures, large area HBTs are normally fabricated for DC test to finally confirm the material quality. The DC gain, junction leakage, breakdown voltages are the typical parameters to monitor.

7.3 Fabrication of MHBTs and Device Characteristics

MHBTs are generally fabricated on relaxed M-buffers on semi-insulating GaAs substrates. The relaxed M-buffer acts as a "virtual substrate", which is suitable for the growth of different HBT heterostructures with required material composition. Both single

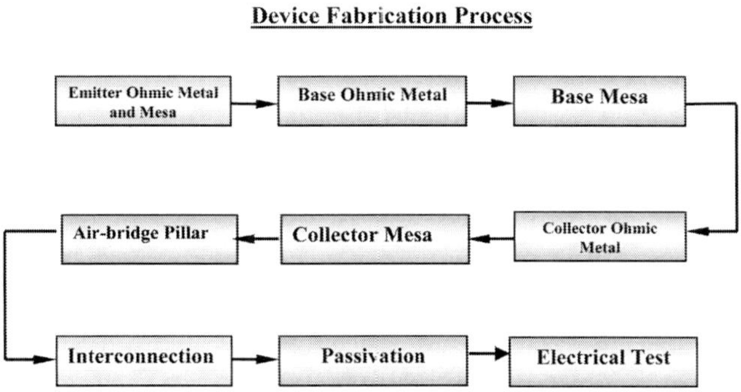

Figure 7.3. Typical process flow for the fabrication of InP/InGaAs MHBTs.

and double HBT structures using InGaAs as the base with indium content higher than 53% can be realized. The fabrication process for the InP/InGaAs MHBTs is essentially the same as the one for InP/InGaAs lattice-matched HBTs (LHBTs). A relatively large mesa height could be necessary for certain MHBTs structures to achieve reasonable device isolation. Figures 7.3 and 7.4 shows a typical process flow for the InP/InGaAs MHBTs fabrication and a SEM picture of a MHBT on GaAs substrate with emitter size of 1.6 × 10 μm^2, respectively. Developing high-yield fabrication

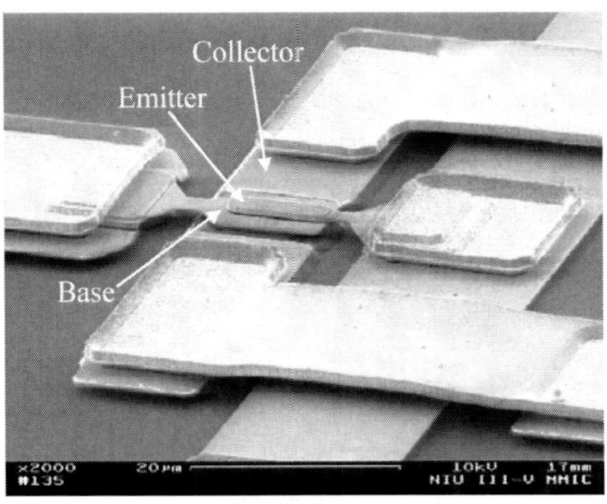

Figure 7.4. SM picture of a fabricated InP MHBT with 1.6 × 10 μm^2 emitter size.

process is one of the major challenges for MHBT due to high roughness and dislocation density of the substrates.

H. Wang *et al.* were the first to report on InP/InGaAs MHBT with indium composition of 53% in the InGaAs base grown by solid-source MBE in 2000 [1]. The device with 5 × 5 μm^2 emitter and 47 nm p$^+$-InGaAs based doped at 2 × 10^{19} cm^{-3} demonstrated a peak current gain of 40 and a common-emitter breakdown voltage (BV_{CEO}) higher than 9 V, a current gain cut-off frequency (f_T) of 48 GHz and a maximum oscillation frequency (f_{max}) of 42 GHz. A minimum noise figure of 2.9 dB and associated gain of 19.5 dB were measured at a collector current level of 2.6 mA at 2 GHz. The devices were made on an InGaP M-buffer. Later, Y. M. Kim *et al.* [7] and Z. Griffith *et al.* [3] reported high performance InP/InGaAs/InP DHBTs on GaAs with InP M-buffer. f_T and f_{max} of 268 and 339 GHz were measured, respectively, from the devices with 0.5 μm emitter. Values of f_T, f_{max}, I_{cbo}, and β were found consistent with InP-based DHBTs of the same layer structure grown on a lattice-matched InP substrate. Other MHBTs with reasonable device performance were also demonstrated. For example, M. D. Lange *et al.* [6] reported metamorphic 6.00 Å HBTs on InP by MBE in 2004. The devices were designed for low power consumption by taking advantage of the small bandgap of high-indium composition (86%) InGaAs in the base layer. The 6.00 Å lattice of the device structures was accommodated by linearly graded InAlAs buffers on semi-insulating InP substrates. The 6.00 Å MHBT demonstrated a V_{BE} of 0.4 V for operation at 1 mA. f_T higher than 100 GHz was achieved. Metamorphic InAs bipolar junction transistors (BJTs) with high common emitter current gain, grown on semi-insulating GaAs substrates, were also reported [5]. Recently, the MBE growth of InP-based MHBTs on Ge and GeOI/Si substrates was demonstrated. The DC and high frequency parameters of MHBTs grown on GeOI/Si substrates were comparable with reported InP/InGaAs MHBTs on GaAs substrates [4]. This opened a new opportunity to integrate the InP-based HBTs on Si platform.

Figure 7.5 shows the common-emitter I–V characteristics of an InP/InGaAs MHBT with a linearly graded $In_xAl_{1-x}As$ metamorphic buffer. The emitter size of the device is 1.6 × 10 μm^2. The Gummel plot for the device is shown in Fig. 7.6. Figure 7.7 displays the high

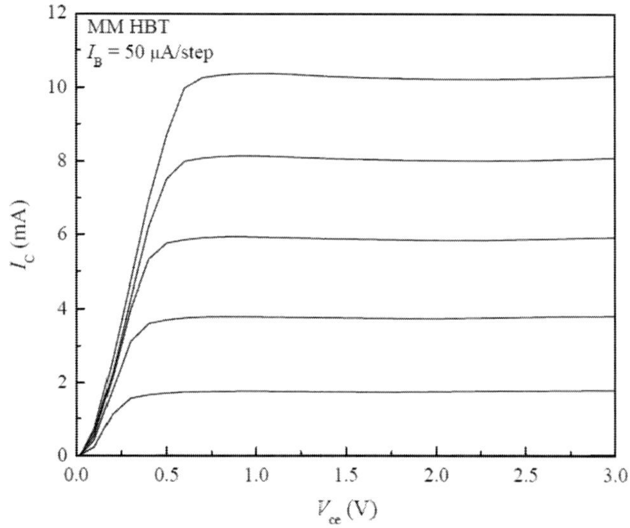

Figure 7.5. Common-emitter I–V characteristics of InP/InGaAs MHBT with a linearly graded $In_xAl_{1-x}As$ metamorphic buffer.

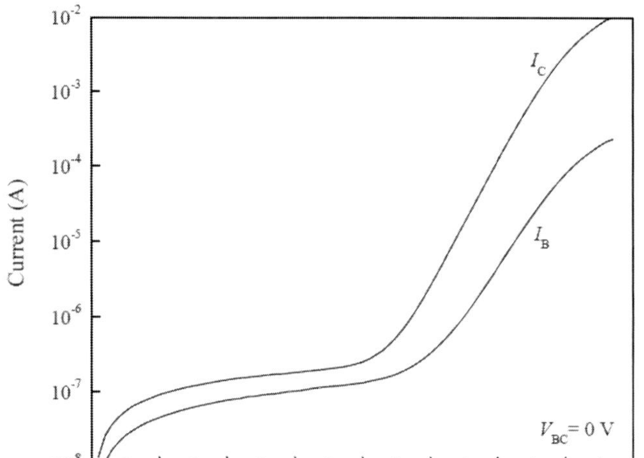

Figure 7.6. Gummel plots of InP/InGaAs MHBT with a linearly graded $In_xAl_{1-x}As$ metamorphic buffer.

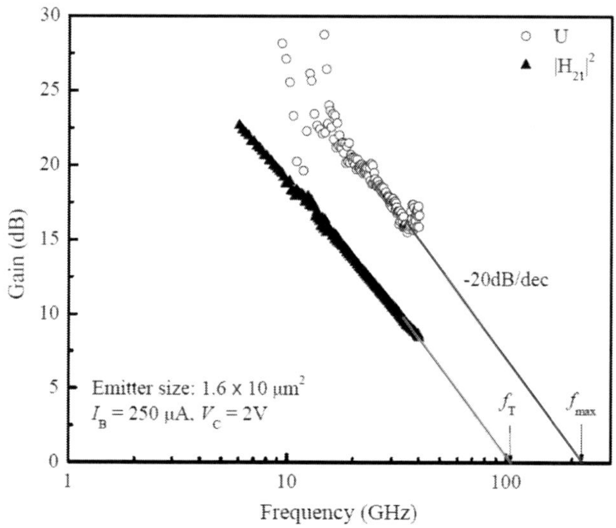

Figure 7.7. High frequency current gain and unilateral gain as a function of frequencies from 6 to 40 GHz.

frequency current gain and unilateral gain in the frequency range from 6 to 40 GHz. By extrapolating these two plots at −20 dB/decade and from the intersection with the 0 dB reference, one determines the unity current gain cutoff frequency and the maximum frequency of oscillation, respectively. At a V_{CE} of 2V and an I_C of 10 mA, an f_T of 100 GHz and an f_{max} of 220 GHz are obtained.

7.4 Issues of MHBTs

7.4.1 *Thermal Characteristics of MHBTs*

One of the important issues of the metamorphic technology is the relatively high thermal resistance of the buffer layer, which may potentially limit the device power performance. The thermal resistance of an HBT with a typical emitter-up mesa structure is largely determined by the thermal resistance of the layer underneath the collector such as the sub-collector and the buffer layers as well as the substrate. For the MHBTs, the low substrate thermal conductivity of GaAs (0.44 W/cm K) compared to InP (0.68

W/cm K) could result in a high device thermal resistance. More importantly, ternary or quaternary III-V compounds are widely used as the metamorphic buffers. The significant decrease in thermal conductivity caused by the alloy-disorder scattering due to the random distribution of group III (and/or V) atoms in the cationic (and/or anionic) sublattices in the ternary or quaternary buffer may play an important role to further degrade the device thermal characteristics. Defects and dislocations in a layer may cause the Rayleigh scattering of phonons and thereby greatly reduce the thermal conductivity. This was observed in GaAs [8] and GaN [9]. For metamorphic buffer, it is unclear whether the presence of a large amount of dislocations significantly worsens the material thermal conductivity.

Different III-V compounds have been considered as strain-relief buffers for MHBTs. Kim *et al.* [10] have compared the thermal properties of MHBT using $In_xAl_{1-x}P$ and InP as the buffer layer. A much lower thermal resistance was observed in MHBTs using a uniform InP buffer. This could be attributed to the high thermal conductivity of the InP material. The thermal conductivities of other materials such as AlGaAsSb and InAlAs [11], which may potentially serve as a metamorphic buffer for HBTs, were found to have an effective thermal conductivity of at least three times lower than that of InP. Proper design of the device layer structure to achieve better thermal characteristics is critical for MHBTs.

The device thermal resistance (R_{th}) of an HBT can be determined by measuring the base-emitter junction voltages at different temperatures which was stated in reference [12]. Figure 7.8 shows the extracted values of R_{th} for the MHBTs using different buffer schemes and the control LHBT in the temperature range of 30° to 150°C. The detailed device layer structures are list in Table 7.1. Both the MHBT and LHBT show an increase of R_{th} with an increase in temperature. This could be explained by the decrease of material thermal conductivity with increasing temperatures [13]. The values of R_{th} for LHBT fabricated by author's research team are in the range of 1285 to 1500 K/W. For the MHBTs with different M-buffers, the device using linearly graded InGaP buffer shows R_{th} in the range of 1600–1800 K/W, which is significantly lower than the one using InAlAs as the strain-relief buffer (2320 to 2615 K/W). Considering

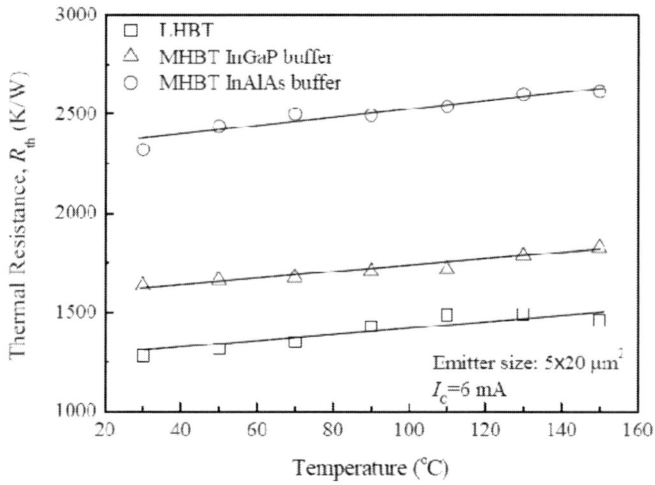

Figure 7.8. R_{th} measured from the MHBTs using InGaP and InAlAs metamorphic buffers and referenced LHBT in the temperature range of 30° to 150°C.

the similar device layer structure for the MHBTs and the referenced LHBT used for testing, the difference in the metamorphic buffer scheme could be the major cause for the different values of R_{th} measured from the MHBTs. Using InGaP as the strain-relief buffer would provide benefit to the device thermal properties as compared to the widely used InAlAs.

The heat removal from a planar device on a homogenous substrate has been studied by Joy and Schlig [14]. They solved the thermal Laplace equation with appropriate boundary conditions leading to a widely accepted compact analytic expression. Although Joy and Schlig's method has been widely applied to different device such as Si bipolar transistors and even GaAs HBTs [15], the presence of an inhomogeneous multilayer structure in the substrate region for the MHBTs makes the use of Joy and Schlig's analytic method difficult. Another approach can be used to deal with the HBT structures consisting of multiple layers. The total thermal resistance of the device can be expressed as the sum of a number of partial thermal resistances from a set of thin slabs, which is closer to the nature of HBT structures. The method was initially used by Licari and Enlow [16] for calculation of heat transfer in hybrid

Table 7.1. Layer structure of MHBTs on GaAs substrate and referenced LHBTs

Emitter Cap	InGaAs	200 nm	n-type	2×19 cm^{-3}
	InAlAs	50 nm	n-type	5×18 cm^{-3}
Emitter	InAlAs	100 nm	n-type	4×17 cm^{-3}
Base	InGaAs	50 nm	p-type	4×19 cm^{-3}
Collector	InGaAs	40 nm	n-type	1×16 cm^{-3}
	InGaAs	10 nm	p-type	1×18 cm^{-3}
	InP	10 nm	n-type	1×18 cm^{-3}
	InP	390 nm	n-type	1×16 cm^{-3}
Subcollector	InGaAs	30 nm	n-type	8×18 cm^{-3}
	InP	250 nm	n-type	8×18 cm^{-3}
InAlAs Buffer	InAlAs	250 nm	Undoped	
Metamorphic Buffer	$In_{0.52}Ga_{0.48}$	$AlAs \Rightarrow In_{0.52}$	Nil	
	$P \Rightarrow InP$ 1.5 μm	$Al_{0.48}As$ 1.1 μm		
Substrate	650 μm S. I. – GaAs		650 μm S. I. – InP	

package systems and recently being successfully applied to study the structural dependence of the thermal resistance of trench-isolated SiGe HBTs [17]. As shown in Fig. 7.9, for an HBT made on a multiple-layer substrate such as the metamorphic structure, the downward transfer of heat flux in the substrate region is considered to be

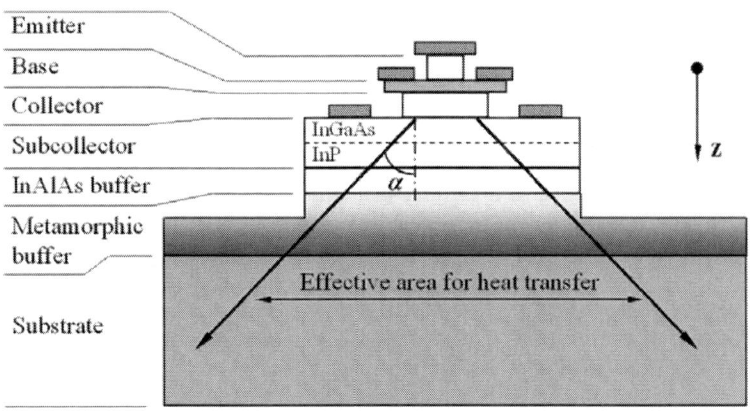

Figure 7.9. Schematic of an HBT made on a metamorphic buffer. The downward transfer of heat flux in sub-collector and substrate regions is illustrated to be confined within a virtual boundary with an angle of α.

uniformly confined within a virtual boundary with an angle of α. Assuming that the temperature in the base-emitter junction is uniform, and ignoring the interface effect in thermal transport, the thermal resistance for downward heat transfer through the collector, sub-collector, metamorphic buffer, and semi-insulating substrate can be calculated by:

$$R_{\text{th}} = \sum_i R_{th,i} = \sum_i \frac{\theta_i d_i}{A_i} \tag{7.1}$$

where θ_i, d_i, and A_i are the thermal resistivity, thickness, and area of a specified slab, respectively. If the substrate has a much large planar dimension as compared to its thickness, utilizing the constant angle approach yields a value for the R_{th}

$$R_{\text{th}} = \int_0^w \frac{\theta(z)}{(L_{emit} + 2z \cdot \tan\alpha)(W_{emit} + 2z \cdot \tan\alpha)} dz \tag{7.2}$$

where w is considered to be the total thickness, including the collector, sub-collector, buffer layer, and substrate. L_{emit} and W_{emit} are the emitter length and width, respectively. The variation of the thermal resistivity in a graded ternary metamorphic buffer can be determined by using the conventional quadratic relationship between material composition and thermal resistivity as:

$$\theta_{A_x B_{1-x} C}(z) = x(z)\theta_{AC} + [1 - x(z)]\theta_{BC} + k_{(AB)C}\, x(z)[1 - x(z)] \tag{7.3}$$

where $k_{(AB)C}$ is the bowing factor of the thermal resistivity. The room temperature thermal resistivities of the binary compounds as well as the bowing factor of ternary compounds used for calculation are listed in Table 7.2. The $k_{(\text{InAl})\text{As}}$ of 60 K cm/W given in the table are extracted based on the experimental value of the effective thermal conductivity (~10 W/m K) of a 1.5 μm $\text{In}_x\text{Al}_{1-x}\text{As}$ linearly graded metamorphic buffer on GaAs substrate reported in [11]. The other data are obtained from previous publications [18–20]. For $k_{(\text{InGa})\text{P}}$, there is a larger discrepancy between the one used by Adachi for the theoretical calculation of the thermal resistivity of InGaAsP and the lately reported experimental data in [18].

Figure 7.10 compares the theoretical calculation with the experimental data for the thermal resistance of LHBT and MHBTs. The angle of spreading heat, α, is assumed to be 45° in our calculation. This assumption has been validated by three-dimensional

Table 7.2. Room Temperature thermal resistivities of III-V binary compounds and bowing parameters for the ternary materials

Material	Thermal resistivity, θ (K cm/W)	Material	Alloy-disorder Bowing, k (K cm/W)
GaAs	2.27 Ref. [20]	InGaAs	72 Ref. [19]
InP	1.47 Ref. [20]	InGaP	72 Ref. [19]
AlAs	1.10 Ref. [20]		19.9 Ref. [18]
InAs	3.70 Ref. [20]	InAlAs	60* Ref. [11]
GaP	1.30 Ref. [20]		

* Estimated.

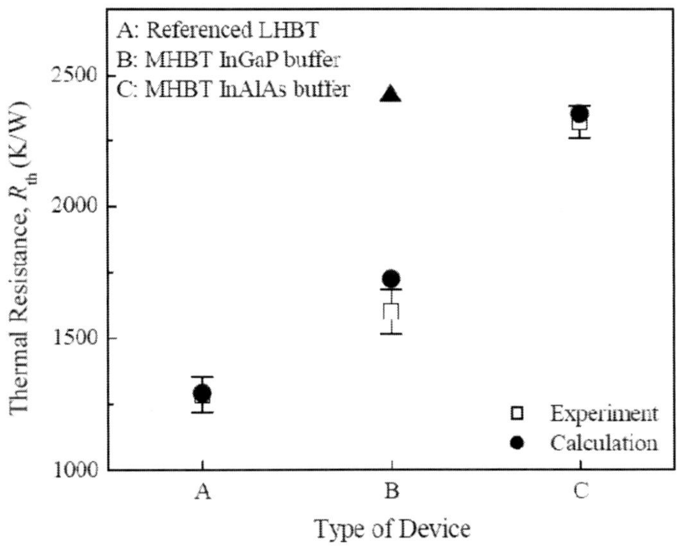

Figure 7.10. Comparison of theoretical calculations with the experimental data for the thermal resistance of MHBTs and the referenced LHBT. The angle of spreading heat, α, is set 45° during calculation. The open symbols are experimental data, and the solid symbols are the results from calculation. For device B using InGaP metamorphic buffer, the result represented by the solid circle was calculated based on a $k_{(\mathrm{InGa})\mathrm{P}}$ of 19.9 K cm/W measured by Both and Herrmann [18], while the value of the triangle symbol was calculated by using the $k_{(\mathrm{InGa})\mathrm{P}}$ of 72 K cm/W given by Adachi [19].

thermal simulations [15] and agrees with the value recommended in literature [21]. The results show that for the device with InGaP metamorphic buffer, using the bowing factor, $k_{(\mathrm{InGa})\mathrm{P}}$, given by Adachi yields a significant overestimation of device thermal resistance, while the calculation of thermal resistance based on $k_{(\mathrm{InGa})\mathrm{P}}$ of 19.9 K cm/W reported by W. Both and F. P. Herrmann [18] shows a better prediction of the experimental data. The simple physical model yields an acceptable accuracy for both the lattice-matched and metamorphic HBTs. Dislocations in the metamorphic buffer region seems to play a negligible effect on the overall thermal resistance.

For the MHBTs, the contributions from each individual layer (from the sub-collector to the substrate) to the total thermal resistances for different devices can be calculated, which is shown in Fig. 7.11. It is seen that for the MHBT using an InGaP metamorphic buffer, the metamorphic buffer and substrate contribute about

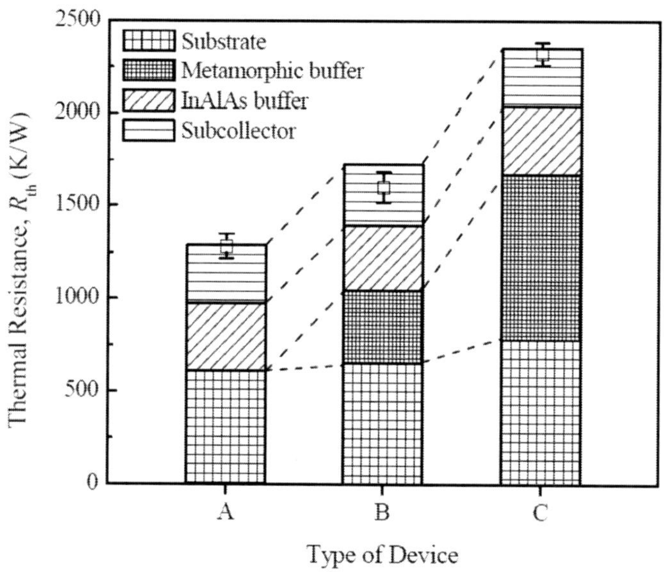

Figure 7.11. Calculations of contributions from the different layers in MHBTs and LHBT (from sub-collector to substrate) to the total thermal resistances. The open symbols with error bars are the experimental data. (Device A: LHBT, Device B: MHBT with InGaP metamorphic buffer, and Device C: MHBT with InAlAs metamorphic buffer).

50% to the total thermal resistance, which is slightly larger than the referenced InP HBT. However, using a linearly graded InAlAs as a strain-relief buffer results in a much larger contribution. The contributions from an InAlAs metamorphic buffer and GaAs substrate to the total thermal resistance could be more than 70%. It seems that the 250 nm InAlAs buffer placed between the substrate and sub-collector provides a substantial contribution to the device thermal resistance. Replacing this layer by a higher thermal conductivity InP or thinning the InAlAs could improve the device thermal properties. The use of InAlAs and InGaAs in the structural design should be avoided if the device thermal property is an important concern.

The total thermal resistance and the contribution of each layer in the substrate region could be largely affected by the layer structure. For example, the calculation results presented in Fig. 7.11 show that even though the same GaAs substrates are used for device B (1.5 μm InGaP metamorphic buffer) and C (1.1 μm InAlAs metamorphic buffer), the thermal resistance contributed from the GaAs substrate are different. This could be clearly illustrated by varying the InP sub-collector thickness. Figure 7.12 plots the calculated R_{th} of device using InGaP buffer and the contributions of thermal resistances from different layers as a function of the InP sub-collector thickness. If the InP sub-collector thickness is increased from 0.2 to 1.5 μm, the R_{th} of MHBT could be reduced from the current ~1700 to 1500 K/W, which is closer to the value measured from the referenced lattice-matched InP HBTs (1285 K/W). Considering the layer structure shown in Fig. 7.12, although the increase in InP sub-collector thickness slightly increases the thermal resistance contributed for sub-collector region, the increase in the effective area for heat transfer in the subsequent low thermal conductivity buffer layers results in a reduction of the total device thermal resistance. Notice that, as the thickness of InP sub-collector is increased, contributions of thermal resistance from GaAs substrate and other layers become smaller. This also explains the experimental observations reported by Kim *et al.* that the MHBTs on GaAs with InP metamorphic buffers exhibit a similar R_{th} to the referenced HBTs fabricated on InP substrate [10]. Of course, a thick InP sub-collector (or buffer) would complicate the device fabrication and lower the

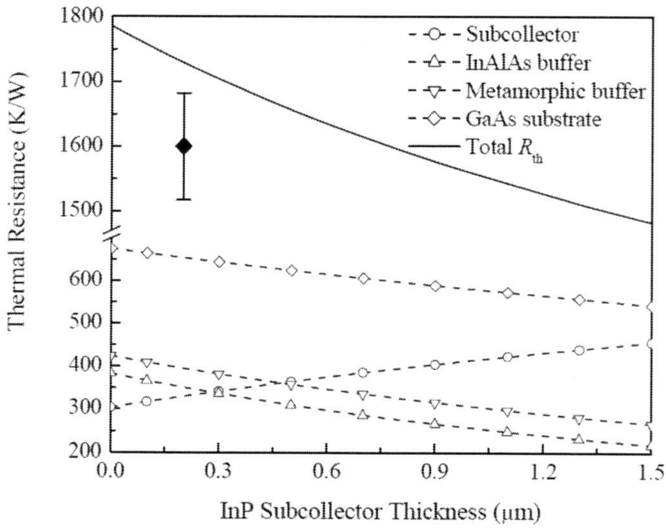

Figure 7.12. Calculated R_{th} of device B (MHBT with InGaP metamorphic buffer) and contributions of thermal resistances from different layers as a function of InP subcollector thickness. The experimental result (solid symbol) for the device with a 250 nm InP sub-collector is plotted for reference.

throughput of epitaxial growth. Proper design of device structures with careful considerations on these different factors is important. The results presented here should be applicable to the other MHBTs with different M-buffer schemes and substrates.

7.4.2 *Stability and Reliability of MHBTs*

MHEMTs biased at low current and voltage for low noise applications have been shown to be reliable, which has demonstrated comparable lifetime to that of lattice-matched HEMTs on InP substrates [22]. However, compared to HEMTs, HBTs could be more sensitive to defects as they are minority carrier devices. For example, the early reliability study on AlGaAs/GaAs lasers on Si substrate (also a minority carrier device) indicated that the rapid formation and propagation of dark line defect could cause significant degradation of the device performance preventing the stable CW lasing operation. The fast degradation has been observed at a current density as low as 1 kA/cm^2 [23]. The residual strain in the device active layer could accelerate the defect generation and propagation [24]. Thus, for the MHBTs with a compositionally graded strain relief buffer layer, the temperature and bias stability are critical for practical system applications. So far, only limited results were reported. However, the results on the bias-temperature stress tests performed on MHBTs with InAlAs M-buffer report by Wang *et al.* are promising [25]. The stress testing was conducted using an HP4156 Semiconductor Parameter Analyzer with a Cascade probe station and thermal chunk. Both MHBTs and the reference LHBTs with 5×10 μm^2 emitter were chosen for the study. Stress current density as high as 100 kA/cm^2 was used, which is much higher than 25 kA/cm^2 required for normal device operation. The use of high current density is to shorten the stress time so that the stress tests could be performed at wafer level without using any package and yield the results in a few hours [26].

Figure 7.13 shows the degradation of collector currents of a MHBT and the reference LHBT over the stress period of 3600 seconds at a temperature of 150°C. It can be seen that both devices show similar degradation behavior where the decrease of collector current gets saturated within the initial tens of seconds. This suggests that some process-induced defects (pre-defects) had been

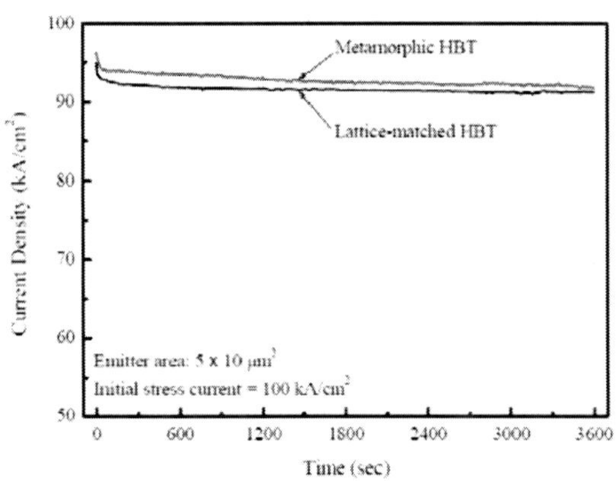

Figure 7.13. Collector current density as a function of current stress time at a constant base current I_B for MHBT and LHBT.

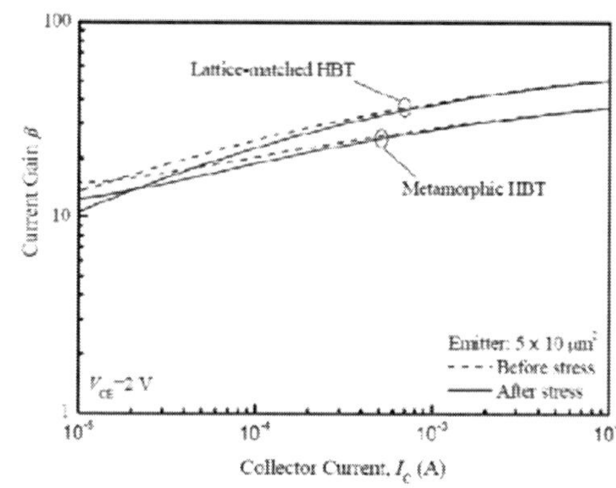

Figure 7.14. Current gain versus collector current for InP/InGaAs MHBT on GaAs and LHBT before and after stress.

incorporated in the devices during the material growth or device fabrication. The severe stress condition may trigger active defect formation. The decrease of collector current for the MHBT and LHBT after 1 hour stress is 4.9% and 3.8%, respectively.

Figure 7.14 compares the effect of stress tests on the current gain as a function of collector current for the MHBT and LHBT. The decrease of β due to the bias-temperature stress occurs only in the low collector current regime. The change of current gain at collector current higher than 1 mA is negligible. The modification of the common-emitter I–V characteristics for a MHBT induced by the stress is shown in Fig. 7.15. Only slight changes on the I–V characteristics can be observed. The experimental results shown here suggest that InP MHBTs grown on GaAs substrate with optimized strain relief buffer layer are as stable as that grown on lattice-matched InP substrate. By proper growth optimization of the strain relief buffer layer, MHBTs on GaAs substrate with electrical stability similar to that of HBTs on InP substrate could be achieved.

7.5 MHBT Performance

After the demonstration of the first MHBT with high indium content (53%) on GaAs in 2000, a steady improvement of high-frequency performance of InP-based MHBTs on GaAs has been

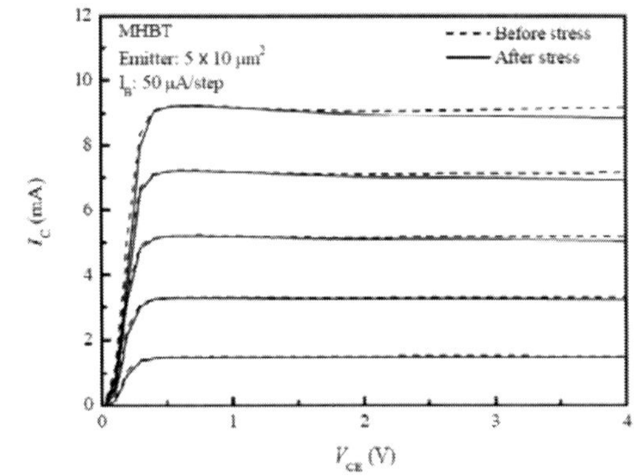

Figure 7.15. Comparison of common-emitter $I-V$ characteristics for the MHBT before and after stress.

made. The devices have demonstrated impressive high-frequency performance, which are comparable to the lattice-match InP-based HBTs with similar emitter size. Similar to InP LHBTs, a highly doped thin base is critical for achieving a high f_T and f_{max}. Data for some of the InP HBTs with similar emitter size are also included for comparison. Table 7.3 summarizes the performance of MHBTs in terms of f_T, f_{max}, and other key device DC parameters. So far, the results on RF noise and power performance of MHBTs are still limited [2].

Despite the technical challenges on the growth of MHBT structures, the advantage of being able to realize different HBTs with large freedom for energy band design has led to increased

Table 7.3. Performance of MHBTs

Substrate	M-Buffer	Base layer	Emitter (μm^2)	Gain	f_T (GHz)	f_{max} (GHz)	Year	Ref.
GaAs	InGaP	$In_{0.53}Ga_{0.47}As$	5 × 5	40	48	42	2000	[1]
GaAs	InAlAs	$In_{0.53}Ga_{0.47}As$	1.6 × 10	35	100	220	2003	[27]
GaAs	InP	$In_{0.53}Ga_{0.47}As$	0.5 × 7	35	268	339	2004	[3]
InP	InAlAs	$In_{0.86}Ga_{0.14}As$	1.6 × 10	40	100	—	2004	[6]
GaAs	InAs	InAs	10 × 25	180	—	—	2003	[5]
GaAs	InP	$GaAs_{0.55}Sb_{0.45}$	1 × 24	20	100	90	2007	[28]
GeOI/Si	InAlAs/GaAs	$In_{0.53}Ga_{0.47}As$	0.5 × 5	40	220	220	2009	[4]

research activities on MHBTs. The potential of MHBTs can be more fully exploited if they can be integrated onto the mature Si platform. This will allow high-performance HBT technology for mass consumer markets. The development of MHBTs was no doubt one of the important topics in the development of high-frequency semiconductor electronics devices since 2000. The MHBT technology developed so far clearly demonstrated the feasibility for the heterogonous integration of different device structures on the substrates with large lattice mismatch. Further advances in the field of MHBTs can be expected.

References

1. Wang, H., Ng, G. I., Zheng, H., Xiong, Y. Z., Chua, L. H., Yuan, K., *et al.* (2000). Demonstration of aluminum-free Metamorphic InP/$In_{0.53}Ga_{0.47}As$/InP double heterojunction bipolar transistors on GaAs substrates, *IEEE Electron Device Lett.*, **21**, pp. 427–429.
2. Wang, H., Ng, G. I., Zheng, H., Yang, H., Xiong, Y. Z., Halder, S., *et al.* (2001). Metamorphic InP/InGaAs heterojunction bipolar transistors on GaAs substrate: DC and microwave performances, *IEEE Trans. Electron Devices*, **48**, pp. 2671–2676.
3. Griffith, Z., Kim, Y. M., Dahlström, M., Gossard, A. C., and Rodwell, M. J. W. (2004). InGaAs–InP metamorphic DHBTs grown on GaAs with lattice-matched device performance and f_T , f_{max} > 268 GHz, *IEEE Electron Device Lett.*, **25**, pp. 675–677.
4. Liu, W. K., Lubyshev, D., Fastenau, J. M., Wu, Y., Bulsara, M. T., Fitzgerald, E. A., *et al.* (2009). Monolithic integration of InP-based transistors on Si substrates using MBE, *J. Cryst. Growth*, **311**, pp. 1979–1983.
5. Averetta, K. L., Wu, X., Kochc, M. W., and Wicksc, G. W. (2003). Low-voltage InAsP/InAs HBT and metamorphic InAs BJT devices grown by molecular beam epitaxy, *J. Cryst. Growth*, **251**, pp. 852–857.
6. Lange, M. D., Cavus, A., Monier, C., Sandhu, R. S., Block, T. R., Gambin, V. F., *et al.* (2004). Metamorphic 6.00 Å heterojunction bipolar transistors on InP by molecular-beam epitaxy, *J. Vac. Sci. Technol.*, **22**, pp. 1570–1574.
7. Kim, Y. M., Dahlström, M., Lee, S., Rodwell, M. J. W., and Gossard, A. C. (2002). High-performance InP/$In_{0.53}Ga_{0.47}As$/InP double HBTs on GaAs substrates, *IEEE Electron Device Lett.*, **23**, pp. 297–299.

8. Jackson, A. W., Ibbetson, J. P., Gossard, A. C., and Mishra, U. K. (1999). Reduced thermal conductivity in low-temperature-grown GaAs, *Appl. Phys. Lett.*, **74**, pp. 2325–2327.

9. Kotchetkov, D., Zou, J., Balandin, A. A., Florescu, D. I., and Pollak, F. H. (2001). Effect of dislocations on thermal conductivity of GaN layers, *Appl. Phys. Lett.*, **79**, pp. 4316–4318.

10. Kim, Y. M., Dahlström, M., Rodwell, M. J. W., and Gossard, A. C. (2003). Thermal properties of metamorphic buffer materials for growth of InP double heterojunction bipolar transistors on GaAs substrate, *IEEE Trans. Electron Devices*, **50**, pp. 1411–1413.

11. Kim, Y. M., Rodwell, M. J. W., and Gossard, A. C. (2002). Thermal characteristics of InP, InAlAs, and A1GaAsSb metamorphic buffer layers used in $In_{0.52}Al_{0.48}As$/ $In_{0.53}Ga_{0.47}As$ heterojunction bipolar transistors grown on GaAs substrates, *J. Electron. Mater.*, **31**, pp. 196–199.

12. Bovolon, N., Baureis, P., Muller, J.-E., Zwicknagl, P., Schultheis, R., and Zanoni, E. (1998). A simple method for the thermal resistance measurement of AlGaAs/GaAs heterojunction bipolar transistors, *IEEE Trans. Electron Devices*, **45**, pp. 1846–1848.

13. Blakemore, J. S. (1982). Semiconducting and other major properties of gallium arsenide, *J. Appl. Phys.*, **53**, p. R123.

14. Joy, R. C., and Shlig, E. S. (1970). Thermal properties of very fast transistors, *IEEE Trans. Electron Devices*, **17**, pp. 596–594.

15. Walkey, D. J., Celo, D., and Smy, T. J. (2003). A simplified model for the effect of interfinger metal on maximum temperature rise in a multifinger bipolar transistor, *IEEE Trans. Comput. Aided Design Integr. Circuits Syst.*, **22**, pp. 15–25.

16. Licari, J. J., and Enlow, L. R. (1988). *Hybrid Microcircuit Technology Handbook* (Noyes Publications, New Jersey).

17. Rieh, J.-S., Johnson, J., Furkay, S., Greenberg, D., Freeman, G., and Subbanna, S. (2002). Structural dependence of the thermal resistance of trench-isolated bipolar transistors, *IEEE BCTM Tech. Dig.*, pp. 100–103.

18. Both, W., and Herrmann, F. P. (1982). Thermal resistivity of quaternary solid solution $Ga_xIn_{1-x}As_yP_{1-y}$ lattice-matched to InP and GaAs, *Cryst. Res. Technol.*, **17**, pp. K117–K122.

19. Adachi, S. (1982). Lattice thermal resistivity of III-V compound alloys, *J. Appl. Phys.*, **54**, pp. 1844–1848.

20. Nakwaski, W. (1988). Thermal conductivity of binary, ternary, and quaternary III-V compounds, *J. Appl. Phys.*, **64**, pp. 159–166.

21. Masana, F. N. (1996). A closed form solution of junction to substrate thermal resistance in semiconductor chips, *IEEE Trans. Compon. Packag. Manuf. Technol. A*, **19**, pp. 539–545.

22. Wakita, A. S., Rohdin, H., Robbins, V. M., Moll, N., Su, C-Y., Nagy, A., *et al.* (1998). Low-noise bias reliability of AlInAs/GaInAs MODFETs with linearly graded low-temperature buffer layers grown on GaAs substrates, *10th International Conference on Indium Phosphide and Related Materials (IPRM98)*, pp. 223–226.

23. Egawa, T., Murata, Y., Jimbo, T., and Umeno, M. (1997). Low-temperature continuous-wave operation of AlGaAs-GaAs vertical-cavity surface-emitting lasers on Si substrates, *IEEE Photon. Technol. Lett.*, **19**, pp. 872–874.

24. Wang, H., Ng, G. I., and Hopgood, A. A. (1998). Kinetics of non-radiative defect related degradation in GaAs/AlGaAs heterojunction bipolar transistors, *J. Phys. D: Appl. Phys.*, **31**, pp. 3168–3171.

25. Yang, H., Wang H., and Radhakrishnan, K. (2002). Metamorphic InP/InGaAs heterojunction bipolar transistors under high-current and high temperature stress, *14th International Conference on Indium Phosphide and Related Materials (IPRM02)*, pp. 369–372.

26. Gupta, A., Young A., and Bayraktaroglu, B. (2001). InGaP makes HBT reliability a non-issue, *GaAs Mantech Conf.*, pp. 203–206.

27. Wang, H. (2003). Development of InP-based metamorphic heterojunction bipolar transistors on GaAs substrates, Internal Report (School of EEE, Nanyang Technological University, Singapore).

28. Zhou, W., Tang, C. W., Zhu, J., Lau, K. M., Zeng, Y., Liu, H. G., *et al.* (2007). Metamorphic heterostructure InP/GaAsSb/InP HBTs on GaAs substrates by MOCVD, *IEEE Electron Device Lett.*, **28**, pp. 539–542.

Chapter 8

Metamorphic Quantum Well Lasers

Yuxin Song,[a] Shumin Wang,[a] Xiangjun Shang,[b] and Zhichuan Niu[b]

[a] *Department of Microtechnology and Nanoscience, Photonics Laboratory, Chalmers University of Technology, SE-41296 Gothenburg, Sweden*
[b] *State Key Laboratory for Superlattices and Microstructures, Institute of Semiconductors, Chinese Academy of Sciences, P.O. Box 912, Beijing 100083, China*
yuxin.song@chalmers.se

This chapter provides an overview of long wavelength metamorphic quantum well lasers. The idea of metamorphic growth is to compromise large lattice mismatch by utilizing a relaxed buffer layer. Structural design and growth optimization of metamorphic buffer layers are reviewed with emphasis on composition grading scheme and doping effects. Progress on long wavelength (>1.2 μm) metamorphic quantum well lasers is summarized.

8.1 Introduction

Metamorphic growth is a technique by which a virtual substrate template with a desired lattice constant and a low defect density can be obtained by growing a thin mismatched but relaxed buffer layer, typically a few micrometers or less, on a conventional substrate.

Lattice Engineering: Technology and Applications
Edited by Shumin Wang

ISBN 978-981-4316-29-3 (Hardcover), 978-981-4364-25-6 (eBook)
www.panstanford.com

It provides a means to monolithically grow lattice-mismatched heterostructures on large and low-cost substrates, or integrate two functional heteorsturctures with different lattice constants. The metamorphic growth technique has been successfully used in majority carrier electronic devices where the device size shrinks down to micrometer scale or less to enhance the device speed. This corresponds to an average defect density of $10^8 cm^{-2}$, which can be easily achieved in, e.g., arsenide alloys. For optoelectronic devices where the device size is typically in the scale of $10–10^3 \mu m$ or minority carrier devices where the device performance is very sensitive to carrier traps, the defect density should be further reduced, thus imposing significant technological challenges. The two main challenges are rough interface and a high density of threading dislocations (TDs) in the active region of the devices [1] that have direct impacts on device performances, lifetime, and yield. There has been steady progress on improving metamorphic materials leading to demonstrations of metamorphic long wavelength (>1.3 μm) lasers on GaAs at room temperature.

In this chapter, we review optimization of metamorphic heterostructures, using alloy graded InGaAs on GaAs for telecom wavelengths (1.3 and 1.55 μm) and the effect of doping on the material quality. We will then review long wavelength (>1.3 μm) metamorphic quantum well lasers on GaAs and Si.

8.2 Design of Metamorphic Buffer Layers

8.2.1 *Bulk-Like, Step-Graded, and Alloy-Graded Metamorphic Buffer Layers*

Metamorphic buffers can be in the form of a thick bulk-like [2, 3], a step-graded with distinct lattice mismatch in each step [4, 5] or alloy-graded where the lattice mismatch is continuously changed with the buffer thickness [6–8]. In addition, different dislocation filtering techniques, such as strain compensated superlattices [9], quantum dots (QDs) [10] and nano-voids [11], and use of strong bonding materials like dilute nitrides [12], can be applied in combination with the buffer design to minimize or even eliminate the TDs.

For most III-V semiconductors, a uniform bulk layer with a desired lattice constant is the simplest way to compose a metamorphic buffer layer and produce a virtual substrate template. If the lattice mismatch between the buffer and the substrate is less than 2%, strain relaxation occurs by formation of a network of 60° mixed type of misfit dislocations (MDs) at the interface when a critical thickness is reached [13]. Most initially generated MDs are accompanied by two TDs forming a half-loop. The TDs can glide in the (111) plane under the residual stress. As a result, TDs can glide to the wafer edges, be annihilated if the two TDs have the opposite Burgers vectors, form a new TD or be pinned by defects or a perpendicular MD. New MDs are generated with the buffer thickness [14] either via dislocation half-loops or via dislocation multiplications and the residual strain decreases. All the MDs are terminated at the buffer/substrate interface leaving a high density of TDs. The TD density decreases with the buffer thickness [15], and a very thick buffer is required to achieve a reasonably low defect density for device applications unless dislocation filtering techniques are used.

If the lattice mismatch between the buffer and the substrate is larger than 2%, surface undulation or QDs will be formed causing a rough surface. When large QDs are merged, edge-type dislocations are formed resulting in sessile 90° TDs [16]. Such 90° TDs are very difficult to be bent by strain or glide laterally as 60° TDs and will penetrate through the whole rest structures. For the $In_xGa_{1-x}As$ buffer layer on GaAs, design of uniform metamorphic buffer can only be practical when the In composition is less than about 0.28 in order to avoid any three dimensional features on the growth surface. The advantage of using this uniform buffer is that the surface is relatively smooth, but the TD density can be high.

In a step-graded buffer layer, lattice mismatch is introduced by using several steps. Strain relaxation occurs at several interfaces, and thus the MDs are distributed at different interfaces and the impedance on the glide of TDs by a MD along the perpendicular direction is reduced. The new deposited layer will exert excess stress to the underlying layer and increase the driving force for the TD gliding. The TD density can be significantly reduced in a step-graded buffer compared with that of a uniform buffer [5]. One drawback of this method is that the last interface is quite close to the surface than that of a uniform buffer, and thus the MDs at the last interface can have strong influence on the surface morphology leading to a pronounced cross-hatch pattern. The strain relaxation is not complete in a step-graded buffer, in particular in the last step. The necessary growth stops due to the time required to change the source temperatures during molecular-beam-epitaxy (MBE) growth may have some adverse effects such as a high-point defect density at the interfaces.

Alloy-graded buffer layer is up to now one of the most successful procedure for metamorphic optoelectronic devices and has been extensively investigated for several material combinations under different growth conditions [6, 16–19]. The lattice mismatch is introduced continuously by continuously changing the alloy compositions in the buffer layer. The depth profile of the alloy composition can be linear or arbitrary. The generated MDs and TDs are 60° mixed type if the change of the alloy composition is not too big. The MDs distribute spatially along the growth direction, thus reducing the chance for the TDs to be pinned by MDs running in the perpendicular direction during the gliding. There is continual stress on the top part of the growing buffer, yielding strong driving force for the TDs to glide. The TDs will be bent during the buffer growth and eventually be eliminated by forming MDs. No growth stop is required in this case. The drawback is the formation of a cross-hatch surface similar to the step-graded buffer. The alloy-graded metamorphic buffer is a focus of this chapter and will be further discussed in the following sections.

8.2.2 *"Dislocation-Free" Region and Residual Strain*

Figure 8.1 shows a cross-sectional TEM (XTEM) image of a linearly graded $In_xGa_{1-x}As$ buffer with $x = 0 \sim 0.42$ on a GaAs substrate. A clear boundary (marked by the broken line) can be found in the image leaving a nearly "dislocation-free" region above it. This is a general phenomenon. The "dislocation-free" region is not completely relaxed, and its thickness and the residual strain can

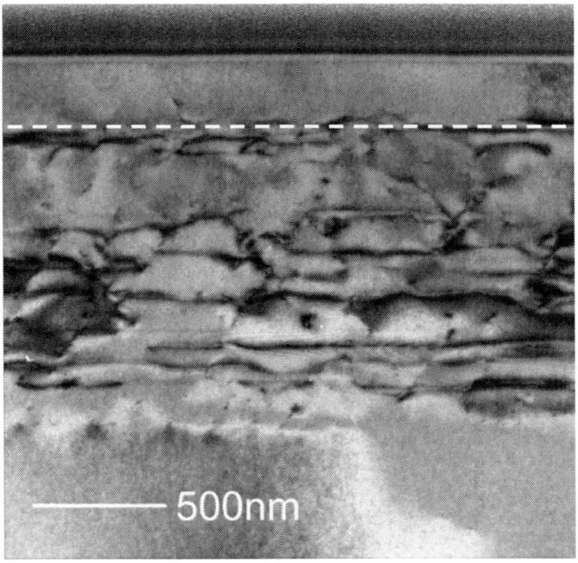

Figure 8.1. An XTEM image of a linearly graded $In_xGa_{1-x}As$ buffer on GaAs with $x = 0 \sim 0.42$.

be predicted by Tersoff model [20]. In this model, the equilibrium distribution of dislocations and residual strain along the growth direction are calculated by minimizing the sum of strain energy and dislocation energy. Theoretically, at the equilibrium, the material below this boundary is fully relaxed while the upper "dislocation-free" region is fully strained. For a linear grading profile, the thickness of the "dislocation-free" region, h, can be expressed by

$$h = \sqrt{2\lambda/bc\varepsilon'} \tag{8.1}$$

and the residual strain, ε, is

$$\bar{\varepsilon} = \sqrt{2\lambda\varepsilon'/bc} \tag{8.2}$$

where λ is the dislocation energy per unit length, b is the misfit component of the Burgers vector of the dislocation, c is an appropriate elastic constant for biaxial strain, and ε' is the grading slope of the lattice mismatch. For a fixed material system, λ, b, and c can be treated as constants. So the thickness of the "dislocation-free" region and the residual strain are only dependent on the grading slope. With increasing the alloy grading slope (thus also of the lattice mismatch), the thickness of the "dislocation-free" region decreases while the residual strain in the upper part of the buffer increases. The measured thickness of the "dislocation-free" region is always smaller than the calculated value because the strain below the boundary is only partially relaxed due to the dislocation interactions and the difference ranges from 10% to 40% from our experiments.

Freund found that there exists a minimum distance between a MD and the sample surface that allows the TD segments to overcome the impeding force when meeting a perpendicular MD [21]. Therefore a thick "dislocation-free" region is favorable for gliding of TDs. The thick "dislocation-free" region enables the newly generated dislocation half loops and the existing TDs to glide for a long distance before being pinned, leaving longer segments of MDs. This is consistent with TEM observations: a sample with longer average MD segments often has a lower TD density. A thick "dislocation-free" region requires a small alloy grading slope. On the other hand, a large alloy grading slope results in large residual strain and thus residual stress that is the driving force for 60° type TDs to glide [22]. Combining these two factors, there exists an optimal alloy grading slope for the metamorphic buffer design.

8.2.3 *Step-Like Strain Relaxation*

Another phenomenon found in Fig. 8.1 is that the MDs are formed at discrete levels, which indicates a step-like strain relaxation process even though the alloy grading is continuous. The Tersoff model is based on equilibrium assumptions and can't explain this phenomenon. We propose the following model. In the beginning of the growth, the epi-layer grows pseudomorphically until the thickness reaches a critical value and the dislocation half loops begin to form. The dislocation half loops glide toward the interface and thus relax part of the strain from the surface through the epi-layer. During the gliding process, the residual strain, which is equal to the difference between the lattice mismatch (position dependent) and the degree of strain relaxation (constant), decreases and so does the stress that drives the dislocation gliding. The stress is eventually balanced by the dislocation line tension and the Peierls-Nabarro friction force [23]. The gliding is then stopped and MDs are formed. No MDs will be generated in the region where the stress is smaller than the friction force. Now, the strain in the upper part of the layer becomes much smaller than before the relaxation. The growth continues, and new strain accumulates again. When it reaches the critical thickness, a similar relaxation process occurs again. New MDs will be generated at the places where the new force balance is established as discussed above. Thus, the strain is relaxed discretely. If the gliding of TDs is impeded by, e.g., the MDs in the perpendicular direction or crystal defects, the length of MD segments can be short and it is difficult to see the step-like distribution of MDs.

Figure 8.2 shows another example where the In source temperature is linearly varied, leading to an exponentially graded In composition shown on the right side of the figure. It can be clearly seen that in the beginning, the small alloy grading slope requires a large thickness to initiate the first strain relaxation. When the alloy grading slope becomes steeper, the space between the two discrete neighbor MD layers turns to be smaller. This experimental finding supports our model above.

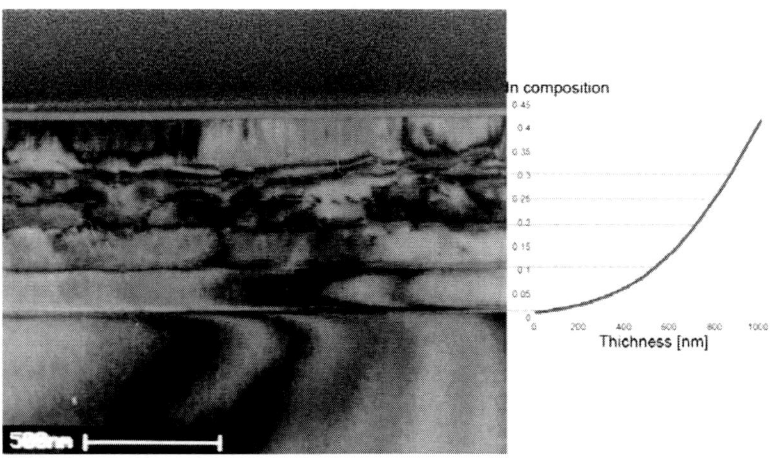

Figure 8.2. An XTEM image of an LTG $In_xGa_{1-x}As$ layer with $x = 0 \sim 0.42$. The figure on the right-hand side shows the In grading profile.

8.3 Doping in Alloy-Graded Buffer Layers

Alloy-graded buffer is one of the most successful metamorphic growth schemes for metamorphic devices and has been extensively investigated [6, 8, 24–31]. The effect of doping on the metamorphic buffer quality has to a large extent been ignored at the early time when developing metamorphic high electron mobility transistors, since in such devices the electron current runs horizontally in the "dislocation-free" region and the metamorphic buffer is usually undoped. For most optoelectronic devices where a back contact on a conducting substrate is employed to drive the current through the substrate and the metamorphic buffer, it is important to investigate the effect of doping on the metamorphic template quality. We found that different types of dopants (*n*- or *p*-type) have a significant effect on structural property and surface morphology in alloy-graded $In_xGa_{1-x}As$ buffer layers grown on GaAs. In this section, we will review this effect and the optimization of $In_xGa_{1-x}As$ metamorphic buffer layers in more detail.

8.3.1 *Doping Effects on Surface Morphology and Structural Property*

Since we intend to drive a current through a metamorphic buffer, we only consider the effect of a high $n-$ (Si at 4×10^{18} cm^{-3}) and *p*-type (Be at 3×10^{19} cm^{-3}) doping used as contact layers in a laser structure. We found that different types of dopants in alloy-graded InGaAs metamorphic buffers grown on GaAs can significantly influence surface morphology and structure quality [32]. This effect can be controlled and utilized to improve the quality of metamorphic structures. Three samples with an identical structure designed to be approximately the lower half of a laser structure with a total thickness of 1.85 μm on a 900 nm $In_xGa_{1-x}As$ buffer with x changing from 0.02 to 0.25 were grown at 450°C measured by a thermal couple using a linear temperature grading (LTG) in the EPI930 MBE system. The detailed structure and growth conditions can be found in Ref. [32].

It can be observed from AFM images in Fig. 8.3 that all samples have cross-hatch patterns. The surface of the Si-doped sample (b) is

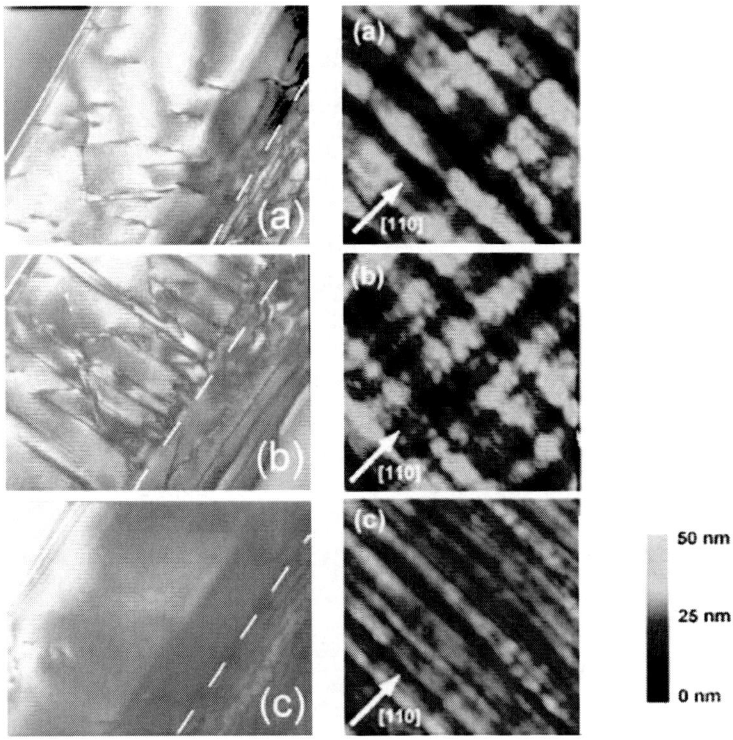

Figure 8.3. XTEM (left) and 20 × 20 μm^2 AFM images (right) of samples with undoped (a), Si-doped (b), and Be-doped (c) LTG $In_xGa_{10-x}As$ buffers with x changing from 0.02 to 0.25. Ref. [32], reprinted with permission from AIP.

more isotropic but rougher compared with the undoped sample (a), while the Be doping dramatically reduces the roughness for sample (c). From the TEM images, we can find that for the undoped sample (a), the majority of TDs are 60° type. The Si-doped sample (b) has a much higher density of TDs with both 60° type and 90° type. The Be-doped sample (c) has much fewer TDs than the other two samples. Photoluminescence (PL) intensity of sample (c) is approximately twice higher than that of sample (a) and six times than that of sample (b). In summary, Be doping reduces the TD density and the surface roughness, while Si doping has the opposite effects.

8.3.2 *Effects of Alloy Grading Profiles*

In the previous work on metamorphic lasers, we found that the adverse effects of Si doping are less serious when using a linear alloy grading (LAG) profile and therefore conducted a study on the doping effect in different alloy grading profiles [33].

Three different alloy grading profiles, shown in Fig. 8.4, are tested in samples consisting of a 0.75 μm thick graded $In_xGa_{1-x}As$ buffer layer with x from 0.02 to 0.42 grown at 380°C. All the samples are capped by 5 nm GaAs to protect the surface. Since the beam flux depends exponentially on the source temperature, a linear source temperature grading (LTG) results in a super-linear alloy grading curve with a slow initial increase of the In composition and a fast increase at the end. The second profile is a truly LAG realized by a carefully designed source temperature ramping program to compensate the exponential dependence. The last one is a mirror image of the LTG (M-LTG) profile with respect to the LAG showing a sub-linear curve. For each grading profile, three samples with non-doping, Be doping, and Si doping were grown. The doping levels are the same as those in the previous section.

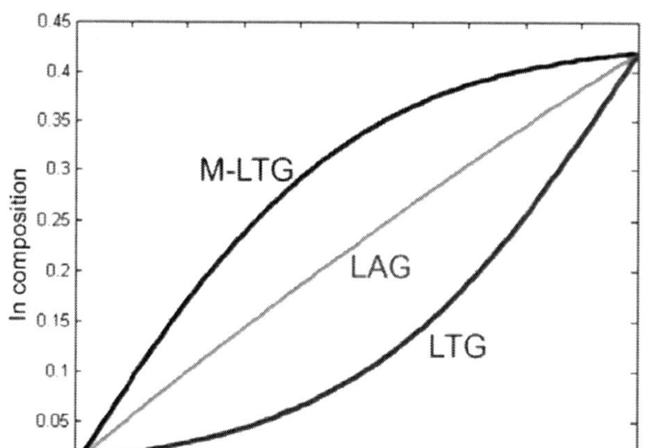

Figure 8.4. Three alloy grading profiles.

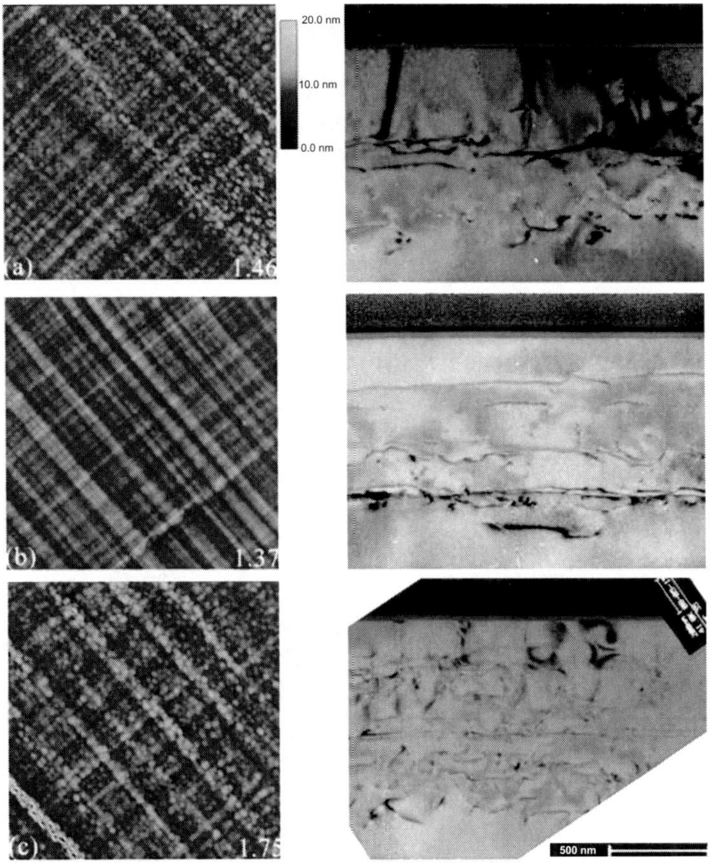

Figure 8.5. Surface morphology obtained by $10 \times 10\ \mu m^2$ AFM scan (left) and cross-sectional TEM images (right) of the undoped (a), the Be-doped (b), and the Si-doped (c) M-LTG $In_xGa_{1-x}As$ buffers with $x = 0 \sim 0.42$. The numbers at the lower-right corner of the AFM images are the RMS roughness values in nanometer.

AFM images with 1×1 and $10 \times 10\ \mu m^2$ were performed for all samples. The left column of Fig. 8.5 shows the AFM images of the M-LTG samples as an example. All the nine samples have a cross-hatch pattern on the surface and some have other features like 3D islands superimposed on the cross-hatch patterns as shown in Fig. 8.5a,c. For the Be-doped samples, the surface morphology is almost pure cross-hatch without 3D islands, while for all undoped and Si-

Table 8.1. RMS surface roughness in nm measured over $10 \times 10\ \mu m^2$ AFM scans

Profile	Undoped	Be-doped	Si-doped
M-LTG	1.46	1.37	1.75
LAG	1.25	1.34	1.49
LTG	1.95	0.80	2.20

Source: Ref. [33], reprinted with permission from AIP.

doped samples, 3D islands with different shapes are found. The shape of the 3D islands for the undoped and the Si-doped M-LTG samples in Fig. 8.5 are round. The 3D islands of the LAG samples have a similar shape as that of the M-LTG ones, while the islands on the LTG samples are shaped as elongate dashes separated by deep trenches giving a high roughness. Table 8.1 summarizes the root-mean-square (RMS) roughness over $10 \times 10\ \mu m^2$ scans.

A large portion of the roughness comes from the 3D islands. All Si-doped samples have strong 3D islands as well as large roughness. For both the undoped and the Si-doped samples, the roughness follows the trend that LAG $<$ M-LTG $<$ LTG. The large roughness of the LTG samples are caused by the elongated dash shaped islands. As mentioned above, the LTG and the M-LTG profile can be approximated by one small and one large grading slope at initial and final stage and vice versa. The surface morphology depends on the accumulated overall strain that is proportional to the grading slope and the surface smoothness. It was experimentally confirmed that formation of InAs dot-like or dash-like 3D islands on InGaAs depends on the surface roughness [34]. A smooth surface leads to dot-like formation, while a rough surface favors dash-like formation. For LTG samples, the initial strain accumulation is low and strain relaxation occurs via formation of MDs leading to a corrugated cross-hatch surface pattern. When the grading slope becomes high, dash-like 3D islands are formed on a rough cross-hatched surface. For LAG and M-LTG samples, the initial strain accumulation is high, and strain relaxation occurs via formation of dot-like 3D islands on a relatively smooth surface (wetting layer) rather than via formation of MDs. As the grading slope of M-LTG is larger than that of LAG, the 3D islands are formed at a much early stage and can subsequently merge to form large 3D islands, thus higher surface roughness.

The surface morphology is ultimately determined by the thermodynamics, but the surface kinetics plays an important role. The amount of In atoms at the growth front and the In surface diffusion length are two important parameters. The amount of In on the growth front consists of the incoming In atoms and the In atoms segregated from the underneath. The In segregation is a well-known effect and is important even at a growth temperature of 380°C [35]. A larger grading slope and a higher final In composition lead to more significant In segregation. The measured different surface morphologies with different doping indicate that doping has a strong effect on In segregation. The Be doping can suppress the segregation of In atoms to the surface and thus delay or avoid formation of 3D islands, while the Si doping is likely to enhance the In segregation leading to large 3D islands and roughness.

The right column of Fig. 8.5 shows the cross-sectional TEM images of the M-LTG samples. Some TDs are observed in the undoped sample. The Be-doped sample has few TDs at the top part and a clear boundary of "dislocation-free region", while more TDs are found in the Si-doped sample. The TD density can be correlated with the surface roughness such that the rougher sample also has more TDs. This observation is general for other grading profiles. The TD density follows the order of Be-doped < undoped < Si-doped samples. However, the dislocation types of the TDs are very different for different grading profiles. All the Be-doped samples despite of the grading profiles are almost free of TDs. For the undoped and the Si-doped samples, the TDs in the M-LTG samples are mainly 90° type with few 60° type ones, while for the other profiles most TDs are the common 60° type. This is expected as for small grading slopes, the strain relaxation occurs via formation of 60° mixed type dislocations, while for large grading slopes where 3D islands are formed, 90° dislocations are generated as a result of merger of 3D islands. The different types of TDs in the Si-doped samples can be clearly observed in Fig. 8.5. The formation of 90° type TDs in the M-LTG samples at the middle part of the growth supports the conclusion of the strong Si induced In segregation effect resulting in formation and coalescence of 3D islands.

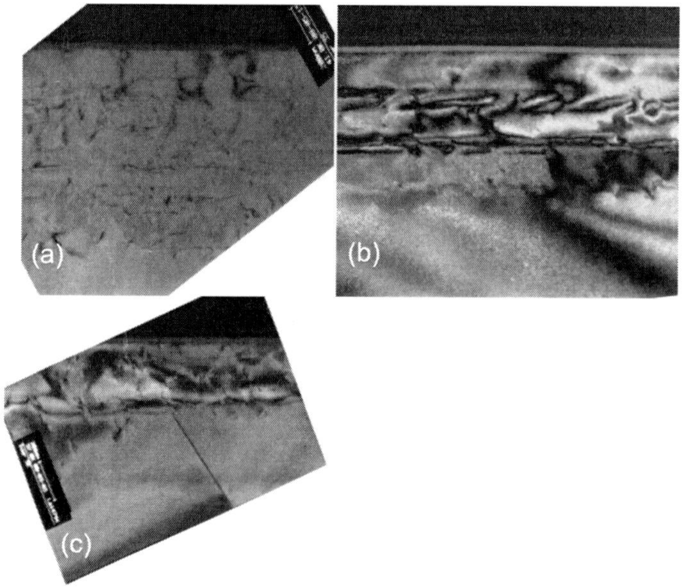

Figure 8.6. Cross-sectional TEM images Si-doped samples with M-LTG (a), the LAG (b), and the LTG (c) profiles.

One interesting feature in Si-doped samples known in Fig. 8.6 is that the dislocation type changes from 60° to 90° with increasing the grading slope and at a critical slope range shown in Fig. 8.6b, the top part has only a few TDs. At this particular grading slope, the difference in surface roughness and the TD density between the Si-doped LAG sample and the undoped and the Be-doped ones is relatively small. This offers an interesting design window for metamorphic devices where a Si-doped buffer must be used.

8.3.3 *Effects of Grading Slopes and Final In Compositions*

Based on the linear grading profile, the effect of doping is further investigated with different grading slopes, final In compositions as well as buffer thicknesses [36]. The sample structures and growth parameters are similar to the ones presented in the previous section. All the samples (A to E) are linearly graded with different In grading slopes, final In compositions, and thicknesses, as illustrated in

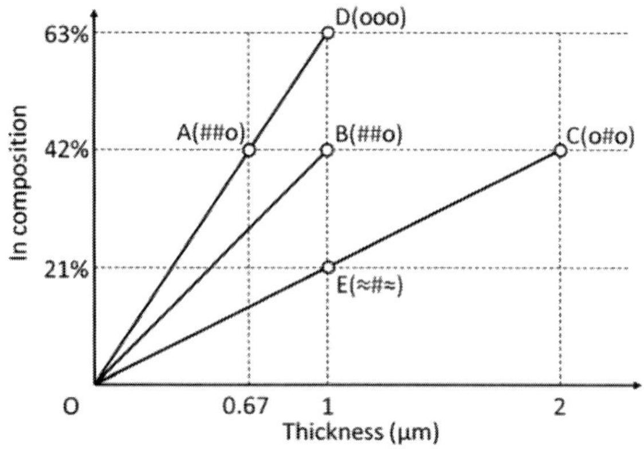

Figure 8.7. Sketch of In grading of the samples. The symbols in the bracket denote the surface morphology of relatively flat surface (#), round islands (o), and wire like islands (≈) in the sequence of undoped, Be-doped, and Si doped samples. Ref. [36], reprinted with permission from AIP.

Fig. 8.7. For each site, one undoped, one Be-doped, and one Si-doped samples were grown. The surface morphology was characterized by AFM over 1×1 and $10 \times 10\ \mu m^2$ areas and summarized also in Fig. 8.7. There are three typical surface features, quasi-2D flat surface (#), 3D islands (o), and wire-like (or dash-like) islands (≈). The height of the 3D islands is typically 3–5 nm. The wire-like islands are along the [1–10] direction and have a width of 60–70 nm, a length of 200–300 nm, and a height of 5 nm, which is slightly higher than that of the 3D islands. It was found from Fourier transform of surface heights that a lateral size of about 1 μm is a boundary below which the surface roughness is governed by the surface features rather than the cross-hatch. The RMS roughness values of the 15 samples measured from $1 \times 1\ \mu m^2$ images are summarized in Table 8.2.

Doping in the LAG $In_xGa_{1-x}As$ buffers has a distinguished effect on the surface morphology. Similar to the observed results in the previous section, the Be doping reduces surface roughness by suppressing the In segregation and thus formation of islands, while the Si doping has the opposite effect. For a thick buffer with a large grading slope (Sample D), there are still a few 3D islands found from the Be-doped D samples. However, the number of the islands is much less compared with the undoped counterpart. All the other

Table 8.2. RMS roughness in nanometer obtained from $1 \times 1\ \mu m^2$ AFM scans

	Undoped	Be-doped	Si-doped
A	0.77	0.67	1.16
B	1.06	1.12	1.07
C	1.27	0.98	1.44
D	1.85	1.10	1.86
E	1.47	0.57	2.55

Source: Ref. [36], reprinted with permission from AIP.

Be-doped samples have no islands at all, while all the Si-doped samples reveal strong 3D islands.

All the Be-doped samples show high-quality "dislocation free regions" with an average TD density of $10^8\ cm^{-2}$ or lower, i.e., below the detection limit. Many dislocations are observed in the Si-doped samples. However, the dislocation type and density are dependent on the grading slope. Figure 8.8 shows the TEM images of the Si-doped D, B, and E samples, and the results mimic those shown in Fig. 8.6 for different grading profiles. From Fig. 8.8a, some regular MDs can be found at the bottom part of the sample, but 90° TDs are seen in the upper part. Regular MDs can be found in almost all parts of the sample B, and a relatively clean region with a few TDs can also be seen in the upper part of the sample. The dislocations in the

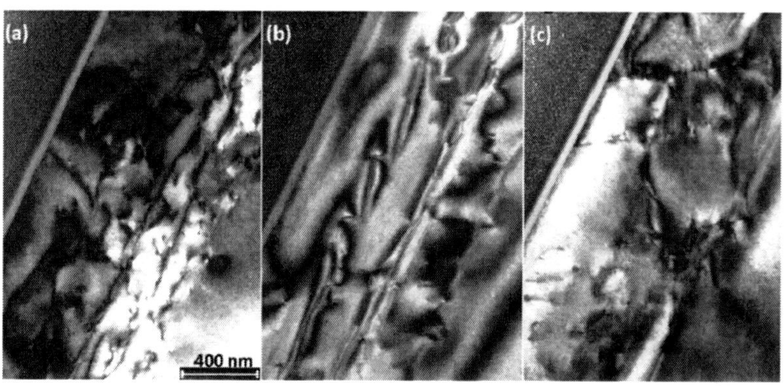

Figure 8.8. Cross-sectional TEM images of the Si-doped D (a), B (b), and E (c) samples. Ref. [36], reprinted with permission from AIP.

Table 8.3. Estimated density of TDs ($\times 10^8$ cm^{-2}) of the samples with different grading slopes and buffer thicknesses. The value "0" means that the density of TDs is below the detection limit of 1×10^8 cm^{-2}

	Undoped	Be-doped	Si-doped
A	11.5	0	20.5
B	0	0	4
C	3	0	29
D	2	3	35
E	4.5	3	18

Source: Ref. [36], reprinted with permission from AIP.

sample E are 60° type and distribute randomly through the buffer with a very high TD density in the top region of the sample. The estimated TD density is summarized in Table 8.3. The dislocation distribution of the Si-doped D sample is found to be very similar to that of the A sample, which has the same grading slope, while the Si-doped C and E samples are also similar. However, the TD density increases with the buffer thickness for the same grading slope when comparing the A and the D samples as well as the E and the C samples, in contrast to the undoped samples.

In summary, surface morphology and dislocations in LAG metamorphic buffers depend highly on the grading slop, buffer thickness, and the final In composition. A small grading slope facilitates formation of wire-like islands or dashes surrounded by deep trenches. Dislocations are typically 60° type. A large grading slope leads to formation of 3D islands that roughens the surface, in particular when the final In composition is high. 90° TDs are generated as a result of merger of 3D islands. Doping has a strong effect on the surface and structure properties of the LAG $In_xGa_{1-x}As$ buffers on GaAs. Beryllium doping can reduce both the surface roughness by suppressing the In segregation and the TD density by reducing the pinning effect and enhancing the gliding of the 60° TDs. Si doping has the opposite effect resulting in sessile 90° TDs. A moderate In grading slope (about 42%/μm) is preferable to minimize the negative effects on both the surface roughness and the TD density caused by the Si doping.

8.3.4 *Design of Alloy-Graded Buffers for Optoelectronic Devices*

The observed experimental findings have strong impact on design of metamorphic optical electronic devices such as lasers using InGaAs as buffer layers. Conventional lasers are grown on *n*-type substrates and employ backside contacts for current injection. In this case, a Si-doped metamorphic buffer is needed to drive the current through the device. The grading slope should be optimized to minimize the TD density as well as interface roughness in the laser active region. On the other hand, a Be-doped graded buffer reveals less TDs in the active region and smooth interface regardless of the other parameters like grading profile and grading slopes. The optical quality of such metamorphic materials is expected to be high. The devices employing a Be-doped buffer need to be grown on a *p*-type substrate or require a co-planar contact scheme for current injection when grown on a semi-insulation substrate. This may lead to thermal problems related to high resistance or a non-uniform current injection. A thick bottom contact layer above the Be-doped buffer and better controlled device processes are necessary in this case.

8.4 Metamorphic Quantum Well Lasers

Quest for metamorphic quantum well lasers has two motivations. The first is to achieve long wavelength infrared (IR) lasers (>1.2 μm) on cheap substrates like GaAs or Si to reduce cost or realize monolithic integration of lasers with electronics. Near IR lasers at telecom wavelengths (1.3 and 1.55 μm) are nowadays grown on InP substrates, while mid-IR (MIR 2–5 μm) lasers are often grown on GaSb substrates. The second motivation is that metamorphic templates with new lattice constants other than the existing commercial ones offer large design freedoms for laser design that can possess better device performance than those designed on conventional substrates. For example, one challenge for telecom lasers is to achieve uncooled high-speed laser operation at a high ambient temperature up to 85°C. Commercial InP-based lasers have a low characteristic temperature, T_0, of about 60 K due to the bad carrier confinement in shallow QWs. It is theoretically predicated [37] that the laser design on a new substrate with a lattice constant corresponding to $In_{0.3}Ga_{0.7}As$ would give maximum conduction band offset and thus a high T_0-value. For this reason, ternary $In_{0.3}Ga_{0.7}As$ substrates have been prepared and excellent 1.3 μm lasers have been demonstrated on these ternary substrates [38]. One drawback for the ternary substrate is that heat dissipation can significantly influence the laser performance as the thermal conductivity becomes worse with increase of the In content. The metamorphic growth provides a more flexible technique to prepare high quality templates with an arbitrary lattice constant. In this section, we will review telecom and MIR metamorphic QW lasers. It should be noted that in early 1980s, a very thick (>10 μm) buffer was usually employed to suppress TDs, and telecom lasers on Si have been demonstrated at 300 K [39]. We will focus on the metamorphic QW lasers employing a thin buffer (<5 μm) below.

8.4.1 *Metamorphic Telecom Lasers*

The first metamorphic InGaAs/GaAs telecom laser on GaAs using a thin buffer was reported by Uchida *et al.* from Fujitsu in Japan in 1994 [40]. They employed *n*-type 2.4 μm thick LAG $In_xGa_{1-x}As$ layer and 1 μm $In_{0.3}Ga_{0.7}As$ layer as the buffer. The waveguide is formed by 1 μm $In_{0.8}Ga_{0.2}P$ on both sides of the separate confinement layer (SCH) consisted of an $In_{0.4}Ga_{0.6}As$ QW sandwiched between 0.1 μm $In_{0.4}Ga_{0.6}As_{0.8}P_{0.2}$. By using 90% high reflectivity coating on both facets, they demonstrated 1.27 μm lasers at 300 K with a threshold current density, J_{th}, of 500 A/cm^2. The measured T_0-value is 100 K in the temperature range of 10–60°C. Later the same group compared use of AlGaInAs and InGaAsP in the SCH region [41]. Theoretical calculations reveal that the optical confinement of AlGaInAs is higher than that of InGaAsP. The characteristic temperature from ridge waveguide lasers (RWG) with a 5 μm ridge width is 85–110 K when using AlGaInAs and 60–80 K when using InGaAsP. Figure 8.9a,b shows the T_0-values for different SCH materials and the P–I curves, respectively.

In 2003, the Ioffe group in Russia reported 1.3 μm $In_{0.4}Ga_{0.6}As$ metamorphic QW lasers on GaAs using a 1 μm n+ $In_{0.2}Ga_{0.8}As$

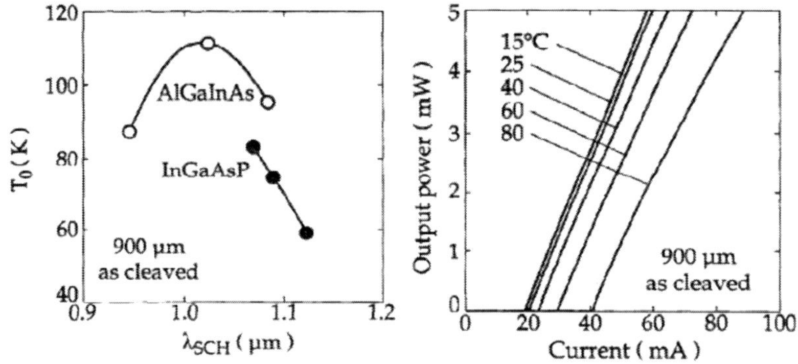

Figure 8.9. (a) Dependence of the characteristic temperature on different SCH materials and (b) laser output power as a function of injection current (P–I curve) at different temperatures. Ref. [41], reprinted with permission from IEEE.

uniform buffer [42]. The metamorphic buffer was grown by MBE at 400°C, while the rest of the structures were grown at 500°C. The threshold current density for broad area lasers (100 × 1500 μm^2) is 3.3 kA/cm^2 at 290 K. Later Tångring *et al.* reported 1.27 μm metamorphic $In_{0.4}Ga_{0.6}As$ QW lasers on GaAs using a 0.8 μm Be-doped InGaAs buffer grown by MBE [43]. A Si-doped n+ $In_{0.18}Ga_{0.82}As$ bottom contact layer is then grown and a coplanar contact scheme is used for current injection as discussed in the Section 8.3.4. The sample was annealed by rapid thermal annealing (RTA) at 650°C for 10 s before being processed. A 4 μm wide RWG laser with a cavity length of 1.5 mm shows a J_{th} of 1 kA/cm^2 under pulsed excitation. The characteristic temperature is 70 K at 10–50°C for 1 mm cavity lasers with high reflectivity coating [44]. Later, Wu *et al.* reported 1.34 μm metamorphic InGaAs QW lasers at 300 K under continuous wave (cw) operation [45]. A 1 μm LAG InGaAs buffer is grown by MBE to provide an $In_{0.18}Ga_{0.82}As$ template. The active region consists of a 7 nm thick $In_{0.47}Ga_{0.53}As$ single QW surrounded by 100 nm $In_{0.18}Ga_{0.82}As$ barriers on both sides, while the waveguide consists of 1.5 μm $In_{0.18}Al_{0.36}Ga_{0.5}As$ doped by Si and Be, respectively. The sample was thermally annealed at 850°C for 10 s, which resulted in seven times reduction of the threshold current

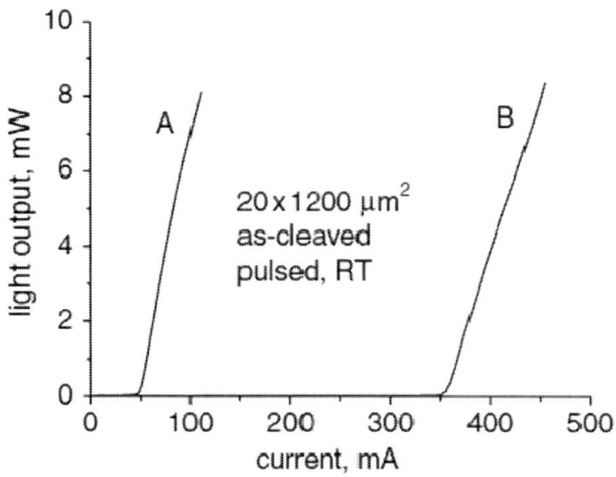

Figure 8.10. *P–I* curves from metamorphic lasers with (A) and without (B) rapid thermal annealing. Ref. [45], reprinted with permission from IET.

as shown in Fig. 8.10. For a cavity of 20×1200 μm^2, the threshold current density is only 205 A/cm^2 under cw operation.

Impressive 1.26 μm metamorphic InGaAs QW lasers on GaAs have been recently demonstrated by the NTT group in Japan [3, 46–48]. A 1.6 μm thick $n+$ $In_{0.18}Ga_{0.82}As$ buffer was grown by MOVPE followed by a 100 nm thick $In_{0.1}Ga_{0.9}As$. The AFM shows a flat surface with an RMS roughness value of 0.97 nm over a 25×25 μm^2 region. An asymmetric waveguide is employed comprising of 1.5 μm $n+$ $In_{0.09}Al_{0.61}Ga_{0.3}As$ as the lower cladding layer and 1.5 μm $p+$ $In_{0.58}Ga_{0.42}P$ as the upper cladding layer. Use of an InGaP cladding layer ensures formation of an inverted mesa by wet-etching with a large top area for making an Ohmic contact, see the scanning electron microscopy (SEM) image shown in Fig. 8.11. The ridge width at the neck is about 2 μm, which is crucial for achieving single transverse mode lasing. The SCH region consists of 2–3 $In_{0.4}Ga_{0.6}As$ QWs embedded in 120 nm $In_{0.1}Ga_{0.9}As$ barriers on each side. The TD density estimated from TEM is below 4×10^6 cm^{-2}. Figure 8.12 shows the *P–I* curves of the 1.26 μm laser under cw excitation. The maximum operation temperature is 173 and 188°C for a 1200 μm cavity laser under pulsed and cw excitation, respectively. The measured T_0-value is 81 K in 25–145°C. For a short cavity of 300 μm,

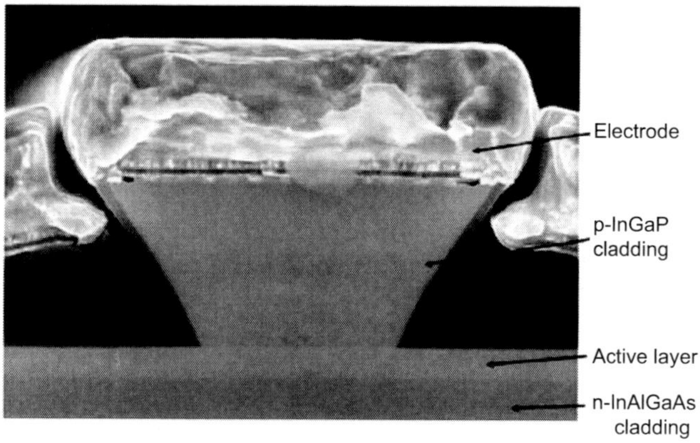

Figure 8.11. SEM image of a metamorphic RWG laser with an inverted mesa. Ref. [48], reprinted with permission from IEEE.

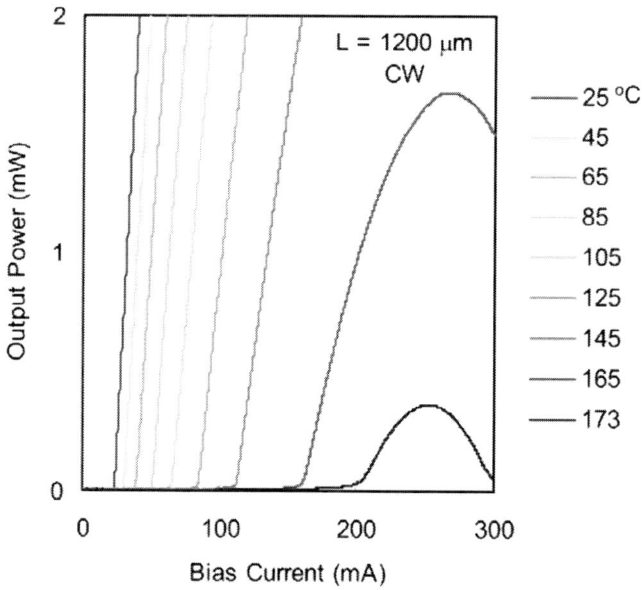

Figure 8.12. P–I curves of metamorphic RWG lasers at different temperatures. Ref. [48], reprinted with permission from IEEE.

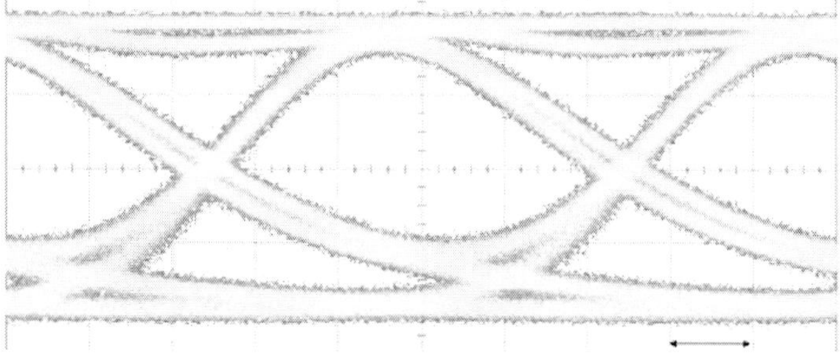

Figure 8.13. 10 Gbit/s eye diagram of a 200 μm RWG laser at 85°C. Ref. [48], reprinted with permission from IEEE.

a threshold current of 8 mA and a differential quantum efficiency of 38% are demonstrated. To achieve a high-speed modulation bandwidth, high reflectivity coating (94%) was deposited on one facet of a 200 μm long laser. The measured maximum relaxation oscillation frequency was 15 and 6 GHz at 25 and 85°C, respectively. Uncooled 10 Gbit/s operation was demonstrated up to 85°C from a 200 μm metamorphic InGaAs laser on GaAs, see Fig. 8.13.

For 1.55 μm metamorphic QW lasers, high In composition is needed in both the metamorphic template and the QW. It is thus very challenging for design and optimization of metamorphic growth. There are two recent reports on 1.55 μm metamorphic QW lasers on GaAs from our groups [7, 49]. The metamorphic buffer consists of 1.6 μm linear alloy graded InGaAs with a final In of 0.39 grown by MBE. The grading slope is between that of B and C samples shown in Fig. 8.7. The growth temperature was linearly ramped down from 500 to 400°C during growth of the metamorphic buffer and was then kept at 440°C for the rest of the structures. To suppress In accumulation on the growth surface, a thin 3 nm GaAs layer was grown at several interfaces up to the QW. In addition, a low flux of Sb with a Sb/As ratio of 1/500 was exposed to serve as surfactant to reduce surface roughness. The RMS surface roughness value measured from AFM is 0.5 and 5.6 nm over 1 × 1 and 10 × 10 μm^2, respectively. The laser structure consists of a waveguide composed of two 1.5 μm thick $In_{0.32}Al_{0.3}Ga_{0.38}As$ cladding layers, an

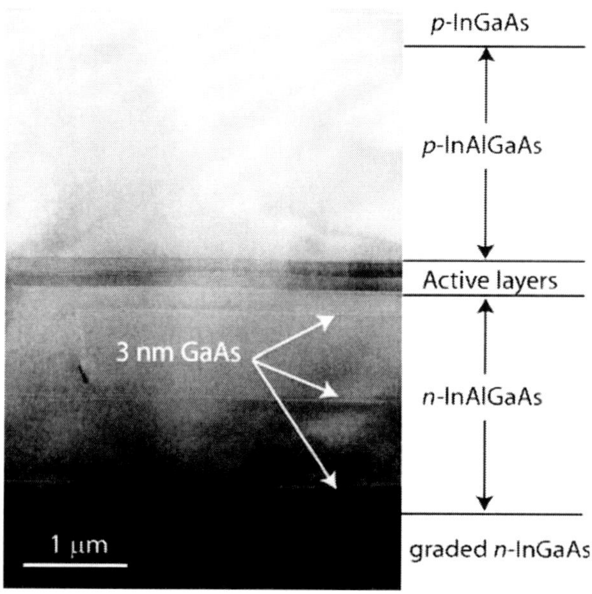

Figure 8.14. Cross-sectional TEM image of the 1.58 μm metamorphic laser structure. Ref. [7], reprinted with permission from AIP.

SCH region of 130 nm thick $In_{0.32}Ga_{0.68}As$ barriers on both sides of a 7 nm thick $In_{0.4}Ga_{0.6}As$ QW. Figure 8.14 shows a cross-sectional TEM image of the laser structure, and the TD density is estimated to be less than 10^7 cm^{-2}. Post growth thermal annealing at 625°C for 30 s is necessary to enhance the PL intensity by one order of magnitude. Figure 8.15 shows $P-I$ and $V-I$ curves for a 50×1250 μm^2 laser at 293 K, and the lasing wavelength is 1.58 μm. The lowest threshold current density is 490 A/cm^2 with a slope efficiency of 0.18 W/A. These values are comparable with the best 1.55 μm GaInNAs lasers on GaAs [50].

8.4.2 *Mid-Infrared Metamorphic Quantum Well Lasers*

Mid-IR (2–5 μm) lasers have potential applications for next generation gas and biochemical sensors and therefore have attracted great attention currently. These lasers are usually grown on GaSb or InP substrates. There are, however, a number of reports on MIR

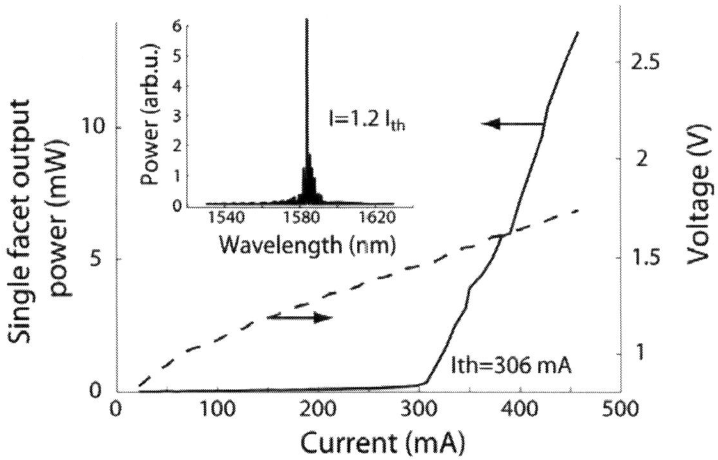

Figure 8.15. *P–I* and *V–I* curves of the 1.58 μm metamorphic laser at 293 K. The inset shows the lasing wavelength. (Ref. [7], reprinted with permission from AIP).

metamorphic QW lasers on GaAs, InP, and Si substrates. The buffer design and strain relaxation mechanism of AlGaSb on GaAs and Si with large lattice mismatch are very different from an InGaAs buffer discussed in the previous section. Under certain growth conditions of AlGaSb on GaAs and Si, 3D islands are suppressed and initial strain relaxation occurs via formation of 90° misfit dislocation arrays at the interface, the so-called interfacial misfit array [51]. In contrast to formation of 3D islands or 60° mixed type MDs that usually generate a large amount of TDs, such a strain relaxation mode results in few TDs and a flat surface that are essential for making metamorphic lasers.

Pease *et al.* reported optically pumped metamorphic GaInSb/AlGaInSb QW lasers on GaAs in the wavelength range of 2.5–3.5 μm in 2003 [52]. They used 0.5 μm AlSb followed by a multi-step AlInSb metamorphic buffer with an increment of In composition of 9% in each step grown at 500°C by MBE. Each step is composed of an $Al_{1-x}In_xSb/Al_{1-y}In_ySb$ ($x, y = 0$, 0.3, and 0.5) superlattice with a period of 10 nm to filter out TDs. The laser structure consists of 2 μm $Al_{1-x}In_xSb$ cladding layers on each side of the AlGaInSb/GaInSb QWs. The In composition in the cladding layer, barrier and QWs vary

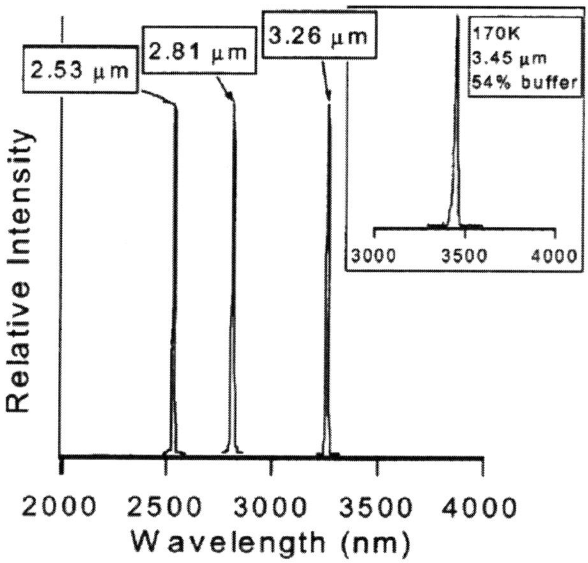

Figure 8.16. Lasing spectra of optically pumped GaInSb/AlGaInSb metamorphic QW lasers on GaAs. Ref. [52], reprinted with permission from AIP.

to achieving different emission wavelength between 2.58 and 3.45 μm. Surface roughness was found to increase from 2.4 to 15.2 nm with the final In composition of the metamorphic buffer from 0.27 to 0.54, which signifies a strong In segregation effect. Optically pumped lasers are demonstrated using a 980 nm stacked diode array under pulsed excitation (50 μs). Figure 8.16 shows lasing spectra up to 3.26 μm at 300 K and 3.45 μm at 170 K. The threshold pump density is 459, 169, and 651 W/cm^2 at 2.5, 2.8, and 3.3 μm, respectively.

Electrically pumped MIR metamorphic lasers on GaAs were demonstrated using GaInSb/AlGaInSb QWs by Nash *et al.* [53]. An AlInSb buffer was directly grown on GaAs substrates followed by an 8 μm thick AlInSb bottom cladding layer. The total laser structure excluding the metamorphic buffer exceeds 15 μm. Lasing was achieved at 3.5 μm up to 95 K. Later, the same group reported improved laser performance with a maximum operating temperature of 220 K at 3.3 μm [54].

Hill and Yang reported interband cascade lasers at 4 μm on GaAs [55]. Interband cascade lasers combine advantages of quantum

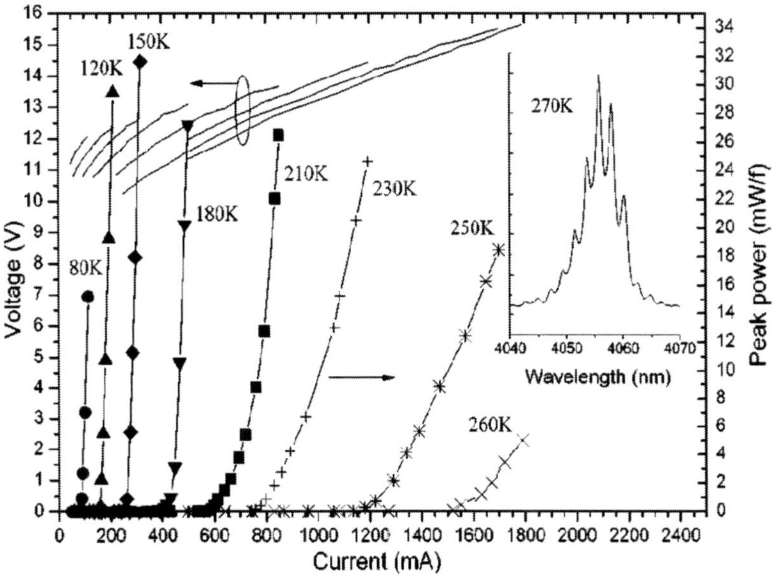

Figure 8.17. *P–I* and *V–I* curves of metamorphic interband cascade lasers on GaAs at different temperatures. Ref. [55], reprinted with permission from AIP.

cascade lasers and type-II QW lasers and have potential to achieve MIR lasers operating above room temperature [56]. The growth was initiated by first depositing 10 nm AlSb at 550°C followed by a 0.5 μm GaSb buffer. Then the metamorphic buffer template and a GaSb substrate are loaded to a multi-wafer MBE system. The laser structure consists of InAs/AlSb bottom and top superlattice cladding layers served as a waveguide and 23 cascade stages designed for 4 μm laser emission. Figure 8.17 shows *P–I* and *V–I* curves of the laser at different temperatures. Lasing under pulsed condition at 4.05 μm is maintained up to 270 K. The corresponding laser grown on the GaSb substrate lases up to 300 K under pulsed excitation and up to 145 K under cw excitation.

Impressive 2.2 μm laser performance has been demonstrated on both GaAs [57] and Si substrates [58] by Rodriguez *et al.*, using the interfacial misfit growth method recently. This growth method has previously led to room temperature lasing at 1.65 μm [59] and 1.82 μm [60], both under pulsed excitation. The metamorphic buffer

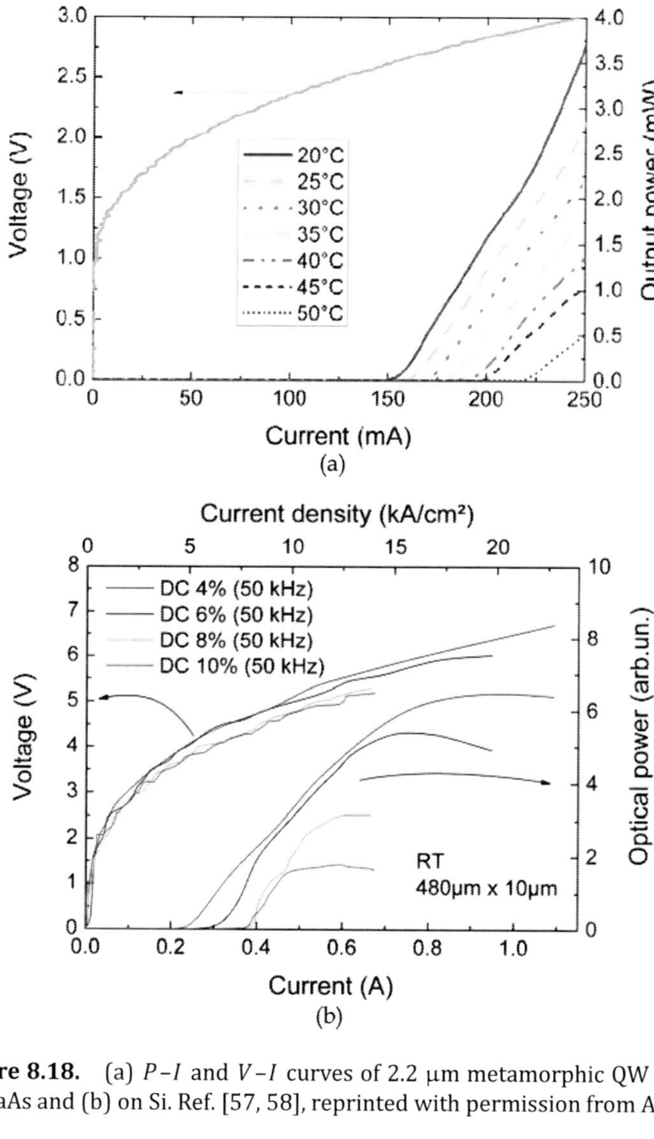

Figure 8.18. (a) P–I and V–I curves of 2.2 μm metamorphic QW lasers on GaAs and (b) on Si. Ref. [57, 58], reprinted with permission from AIP.

consists of 1 μm GaSb on GaAs and 5 nm AlSb followed by 1 μm GaSb on 4° miscut Si grown at 510°C by MBE. The laser design is similar in both cases and consists of 1.5 μm AlGaAsSb cladding layers, 30 nm AlGaAsSb barriers, and two 9 nm thick $Ga_{0.68}In_{0.32}As_{0.06}Sb_{0.94}$ QWs. Figure 8.18 shows P–I and V–I curves for the lasers on GaAs

(a) and on Si (b). Lasing under cw excitation up to 50°C has been demonstrated on GaAs substrates with a threshold current density in the range of 1.5–2.5 kA/cm^2 for a 10 μm ridge width. This is about 10 times of the J_{th}-value from a similar laser grown on GaSb. The characteristic temperature in the temperature range of 20–50°C is 86 K, which is comparable to the state-of-the-art values from lasers grown on GaSb [61]. The lasers fabricated on Si are only operable under pulsed excitation with a threshold current density in the range of 4.5–8.2 kA/cm^2.

One drawback for MIR lasers employing Sb-containing compounds on either GaSb substrates or metamorphic templates is that the device process technology is much less mature as compared with GaAs and InP based devices, especially how to avoid fast performance degradation due to oxidation of AlSb-containing layers. For this reason, growth of Al-free MIR lasers on InP substrates is an attractive alternative. Kirch *et al.* recently studied growth of metamorphic InAsP templates using MOVPE [62]. The buffer design consists of 15 layers of 100 nm thick step-graded InAsP capped by a 500 nm thick $InAs_{0.68}P_{0.32}$ layer. Two material combinations for the cladding/SCH are designed to achieve MIR emission: $InAs_{0.66}P_{0.34}/In_{0.85}Ga_{0.15}As$ and $InP_{0.8}Sb_{0.2}/InAs_{0.66}P_{0.34}$. The latter design has a higher optical confinement factor and shows higher room temperature PL intensity due to the better carrier confinement than the former one. However, the large miscibility gap of InPSb and strict partial pressure constraints to prevent hillock formation during MOVPE growth make it difficult to grow high quality InPSb. Metamorphic 2.45 μm $InAs_{0.66}P_{0.34}/In_{0.85}Ga_{0.15}As$ QW lasers were demonstrated at 77 K with a J_{th}-value in the range of 290–500 A/cm^2 for a 30 × 1000 μm^2 cavity.

In summary, there have been some demonstrations of metamorphic telecom and MIR lasers on GaAs, InP, or even Si substrates, with some devices operating at above room temperature. For large lattice mismatch (7% or more), high quality AlGa(In)Sb buffers can be grown on GaAs and Si substrates using interfacial misfit method. This method is able to reduce the TD density down to 10^5 cm^{-2} or lower [63] and is thus very promising for making metamorphic lasers. For medium lattice mismatch, conventional alloy grading is often used to provide a metamorphic template. Laser performance is highly related to the template quality, namely the TD density and the surface roughness. It should be noted that development of metamorphic lasers is at its early stage. There are still many challenges that need to be solved, for example, yield, reliability, lifetime, and thermal management, etc., before the technology becomes attractive for industry. Therefore further improving metamorphic template quality is crucial to push this technology forward.

References

1. Wang, S. M., Tangring, I., Gu, Q. F., Sadeghi, M., Larsson, A., Wang, X. D., *et al.* (2006). Metamorphic InGaAs quantum wells for light emission at 1.3–1.6 μm, *Thin Solid Films*, **515**, pp. 4348–4351.
2. Sugo, M., Mori, H., Sakai, Y., and Itoh, Y. (1992). Stable cw operation at room temperature of a 1.5-μm wavelength multiple quantum well laser on a Si substrate, *Appl. Phys. Lett.*, **60**, pp. 472–473.
3. Arai, M., Tadokoro, T., Fujisawa, T., Kobayashi, W., Nakashima, K., Yuda, M., *et al.* (2009). Uncooled (25–85 degrees C) 10 Gbit/s operation of 1.3 μm-range metamorphic Fabry-Perot laser on GaAs substrate, *Electron. Lett.*, **45**, p. 359.
4. Chen, J., Fernández, J. M., Chang, J. C. P., Kavanagh, K. L., and Wieder, H. H. (1992). Modulation-doped $In_{0.3}Ga_{0.7}As/In_{0.29}Al_{0.71}As$ heterostructures grown on GaAs by step grading, *Semicond. Sci. Technol.*, **7**, p. 601.
5. Lubyshev, D., Fastenau, J. M., Fang, X.-M., Wu, Y., Doss, C., Snyder, A., *et al.* (2004). Comparison of As- and P-based metamorphic buffers for high performance InP heterojunction bipolar transistor and high electron mobility transistor applications, *J. Vac. Sci. Technol. B*, **22**, p. 1565.
6. Molina, S. I., Pacheco, F. J., Araujo, D., Garcia, R., Sacedon, A., Calleja, E., *et al.* (1994). Strain relief in linearly graded composition buffer layers: A design scheme to grow dislocation-free ($< 10^5$ cm^{-2}) and unstrained epilayers, *Appl. Phys. Lett.*, **65**, pp. 2460–2462.
7. Tångring, I., Ni, H. Q., Wu, B. P., Wu, D. H., Xiong, Y. H., Huang, S. S., *et al.* (2007). 1.58 μm InGaAs quantum well laser on GaAs, *Appl. Phys. Lett.*, **91**, p. 221101.
8. Lee, K. E., and Fitzgerald, E. A. (2010). High-quality metamorphic compositionally graded InGaAs buffers, *J. Cryst. Growth*, **312**, pp. 250–257.

9. Fritz, I. J., Gourley, P. L., Dawson, L. R., and Schirber, J. E. (1988). Electrical and optical studies of dislocation filtering in InGaAs/GaAs strained layer superlattices, *Appl. Phys. Lett.*, **53**, p. 1098.
10. Mi, Z., Yang, J., Bhattacharya, P., and Huffaker, D. L. (2006). Self-organised quantum dots as dislocation filters: the case of GaAs-based lasers on silicon, *Electron. Lett.*, **42**, p. 121.
11. Ledentsov, N. N., Kovsh, A. R., Shchukin, V. A., Mikhrin, S. S., Krestnikov, I. L., Kozhukhov, A. V., *et al.* (2005). QD lasers: physics and applications, *Proc. SPIE,* **5624**, pp. 335–344.
12. Song, Y., Wang, S., Lai, Z., and Sadeghi, M (2010). Enhancement of optical quality in metamorphic quantum wells using dilute nitride buffers, *Appl. Phys. Lett.*, **97**, p. 91903.
13. Matthews, J. W., and Blakeslee, A. E. (1974). Defects in epitaxial multilayers. I. Misfit dislocations, *J. Cryst. Growth*, **27**, pp. 118–125.
14. Dodson, B. W., Tsao, J. Y., and Taylor, P. A. (1988). Stability and metastability of semiconductor strained-layer structures, *Superlattices Microstruct.*, **4**, pp. 417–422.
15. Ayers, J. E. (2007). Heteroepitaxy of semiconductors: theory, growth and characterization, CRC Press, Boca Raton, USA.
16. Guha, S., Madhukar, A., and Rajkumar, K. C. (1990). Onset of incoherency and defect introduction in the initial stages of molecular beam epitaxical growth of highly strained $In_xGa_{1-x}As$ on GaAs(100), *Appl. Phys. Lett.*, **57**, pp. 2110–2112.
17. Jeong, Y. (2007). Invalidity of graded buffers for InAs grown on GaAs (0 0 1) — A comparison between direct and graded-buffer growth, *J. Cryst. Growth*, **301**, pp. 235–239.
18. Bertoli, B., Suarez, E. N., Ayers, J. E., and Jain, F. C. (2009). Misfit dislocation density and strain relaxation in graded semiconductor heterostructures with arbitrary composition profiles, *J. Appl. Phys.*, **106**, pp. 73517–73519.
19. Capotondi, F., Biasiol, G., Ercolani, D., Grillo, V., Carlino, E., Romanato, F., *et al.* (2005). Strain induced effects on the transport properties of metamorphic InAlAs/InGaAs quantum wells, *Thin Solid Films*, **484**, pp. 400–407.
20. Tersoff, J. (1993). Dislocations and strain relief in compositionally graded layers, *Appl. Phys. Lett.*, **62**, p. 693.
21. Freund, L. B. (1990). A criterion for arrest of a threading dislocation in a strained epitaxial layer due to an interface misfit dislocation in its path, *J. Appl. Phys.*, **68**, pp. 2073–2080.

22. Freund, L. B. (1990). The driving force for glide of a threading dislocation in a strained epitaxial layer on a substrate, *J. Mech. Phys. Solids*, **38**, pp. 657–679.
23. Peierls, E. (1940). The size of a dislocation, *Proc. Phys. Soc.*, **52**, pp. 34–37.
24. Hayakawa, T., Suyama, T., Kondo, M., Takahashi, K., Yamamoto, S., and Hijikata, T. (1986). Improvements in AlGaAs laser diodes grown by molecular beam epitaxy using a compositionally graded buffer layer *Appl. Phys. Lett.*, **49**, pp. 191–193.
25. Fitzgerald, E. A., Kim, A. Y., Currie, M. T., Langdo, T. A., Taraschi, G., and Bulsara, M. T. (1999). Dislocation dynamics in relaxed graded composition semiconductors, *Mater. Sci. Eng., B*, **67**, pp. 53–61.
26. Ferrari, C., Rossetto, G., and Fitzgerald, E. A. (2002). Misfit dislocation and threading dislocation distributions in InGaAs and GeSi/Si partially relaxed heterostructures, *Mater. Sci. Eng., B*, **91–92**, pp. 437–440.
27. Choy, H. K. H., and Fonstad, J. C. G. (2005). Effects of substrate temperature on the growth of InGaAs compositionally graded buffers and on quantum well structures grown above them, *J. Vac. Sci. Technol. B*, **23**, pp. 2109–2113.
28. Bertoli, B., Suarez, E. N., Ayers, J. E., and Jain, F. C. (2009). Misfit dislocation density and strain relaxation in graded semiconductor heterostructures with arbitrary composition profiles, *J. Appl. Phys.*, **106**, pp. 073519-7.
29. Choi, H., Jeong, Y., Cho, J., and Jeon, M. H. (2009). Effectiveness of non-linear graded buffers for In(Ga, Al)As metamorphic layers grown on GaAs (0 0 1), *J. Cryst. Growth*, **311**, pp. 1091–1095.
30. Lee, K. E., and Fitzgerald, E. A. (2009). Digital metamorphic alloys, *J. Appl. Phys.*, **106**, pp. 074911–074912.
31. Sidoti, D., Xhurxhi, S., Kujofsa, T., Cheruku, S., Reed, J., Bertoli, B., *et al.* (2010). Critical layer thickness in exponentially graded heteroepitaxial layers, *J. Electron. Mater.*, **39**, pp. 1140–1145.
32. Tangring, I., Wang, S. M., Zhu, X. R., Larsson, A., Lai, Z. H., and Sadeghi, M. (2007). Manipulation of strain relaxation in metamorphic heterostructures, *Appl. Phys. Lett.*, **90**, p. 071904.
33. Tångring, I., Song, Y. X., Lai, Z. H., Wang, S. M., Sadeghi, M., and Larsson, A. (2009). A study of the doping influence on strain relaxation of graded composition InGaAs layers grown by molecular beam epitaxy, *J. Cryst. Growth*, **311**, pp. 1684–1687.

34. Sritirawisarn, N., van Otten, F. W. M., Eijkemans, T. J., and Nötzel, R. (2007). Surface morphology induced InAs quantum dot or dash formation on InGaAsP/InP (1 0 0), *J. Cryst. Growth*, **305**, pp. 63–69.
35. Moison, J. M., Guille, C., Houzay, F., Barthe, F., and Rompay, M. V. (1989). Surface segregation of third-column atoms in group III-V arsenide compounds: Ternary alloys and heterostructures, *Phys. Rev. B*, **40**, p. 6149.
36. Song, Y. X., Wang, S. M., Tangring, I., Lai, Z., and Sadeghi, M. (2009). Effects of doping and grading slope on surface and structure of metamorphic InGaAs buffers on GaAs substrates, *J. Appl. Phys.*, **106**, pp. 123531–123537.
37. Ishikawa, H. (1993). Theoretical gain of strained quantum well grown on an InGaAs ternary substrate, *Appl. Phys. Lett.*, **63**, p. 712.
38. Otsubo, K. Shoji, H. Kusunoki, T. Suzuki, T. Uchida, T. Nishjima, Y., *et al.* (1997). High T_0 (140 K) and low-threshold long-wavelength strained quantum well lasers on InGaAs ternary substrates, *Electron. Lett.* **33**, pp. 1795–1797.
39. Razeghi, M., Defour, M., Blondeau, R., Omnes, F., Maurel, P., Acher, O., *et al.* (1988). First CW operation of a $Ga_{0.25}In_{0.75}As_{0.5}P_{0.5}$-InP laser on a silicon substrate, *Appl. Phys. Lett.*, **53**, pp. 2389–2390.
40. Uchida, T., Kurakake, H., Soda, H., and Yamazaki, S. (1994). 1.3 μm InGaAs/GaAs strained quantum well lasers with InGaP cladding layer, *Electron. Lett.*, **30**, pp. 563–565.
41. Kurakake, H., Uchida, T., Higashi, T., Ogita, S., and Kobayashi, M. (1996). 1.3 μm high T0 strained MQW laser with AlGaInAs SCH layers on a hetero-epitaxial InGaAs buffer layer, *Semicond. Laser Conf., 1996., 15th IEEE International*, pp. 71–72.
42. Zhukov, A. E., Kovsh, A. R., Mikhrin, S. S., Semenova, E. S., Maleev, N. A., Vasil'ev, A. P., *et al.* (2003). Metamorphic lasers for 1.3-μm spectral range grown on GaAs substrates by MBE, *Semiconductors*, **37**, pp. 1119–1122.
43. Tånging, I., Wang, S., Sadeghi, M., and Larsson, A. (2006). A.1.27 mum metamorphic InGaAs quantum well lasers on GaAs substrates, *Electron. Lett.* **42**, pp. 691–693.
44. Tångring, I., Wang, S. M., Sadeghi, M., Larsson, A., and Wang, X. D. (2007). Metamorphic growth of 1.25~1.29 mm InGaAs quantum well lasers on GaAs by molecular beam epitaxy, *J. Cryst. Growth*, **301–302**, pp. 971–974.
45. Wu, D., Wang, H., Wu B., Ni, H., Huang, S., Xiong, Y., *et al.* (2008). Low threshold current density 1.3 mum metamorphic InGaAs/GaAs quantum well laser diodes, *Electron. Lett.*, **44**, pp. 474–475.

46. Arai, M., Nakashima, K., Fujisawa, T., Tadokoro, T., Kobayashi, W., Yuda, M., *et al.* (2008). High temperature operation of 1.26 μm ridge waveguide laser with InGaAs metamorphic buffer on GaAs substrate, *IEEE 21st Int. Semicond. Laser Conf.*, pp. 57–58.

47. Arai, M., Fujisawa, T., Kobayashi, W., Nakashima, K., Yuda, M., and Kondo, Y. (2008). High-temperature operation of 1.26 μm Fabry-Perot laser with InGaAs metamorphic buffer on GaAs substrate, *Electron. Lett.* **44**, p. 1359.

48. Arai, M., Nakashima, K., Fujisawa, T., Tadokoro, T., Kobayashi, W., Yuda, M., *et al.* (2009). High-temperature operation of 1.26-μm ridge waveguide laser with InGaAs metamorphic buffer on GaAs substrate, *IEEE J. Sel. Topics Quantum Electron.* **15**, pp. 724–730.

49. Wang, H. L., Wu, D. H., Wu, B. P., Ni, H. Q., Huang, S. S., Xiong, Y. H., *et al.* (2009). Metamorphic InGaAs quantum well laser diodes at 1.5 μm on GaAs grown by molecular beam epitaxy, *Chin. Phys. Lett.* **26**, p. 014214.

50. Ha, W., Gambin, V., Bank, S., Wistey, M., Yuen, H., Kim, S., *et al.* (2002). Long-wavelength GaInNAs(Sb) lasers on GaAs, *IEEE J. Quantum Electron.*, **38**, pp. 1260–1267.

51. Qian, W., Skowronski, M., Kaspi, R., De Graef, M., and Dravid, V. P. (1997). Nucleation of misfit and threading dislocations during epitaxial growth of GaSb on GaAs(001) substrates, *J. Appl. Phys.* **81**, pp. 7268–7272.

52. Pase, E. A., Dawson, L. R., Vaughn, L. G., Rotella, P., and Lester, L. F. (2003). 2.5–3.5 μm optically pumped GaInSb/AlGaInSb multiple quantum well lasers grown on AlInSb metamorphic buffer layers, *J. Appl. Phys.*, **93**, pp. 3177–3181.

53. Nash G. R., Smith, S. J., Coomber, S. D., Przeslak, S., Andreev, A., Carrington, P., *et al.* (2007). Midinfrared GaInSb/AlGaInSb quantum well laser diodes grown on GaAs, *Appl. Phys. Lett.*, **91**, p. 131118.

54. Nash, G. R., Przeslak, S., Smith, S. J., de Valicourt, G., Andreev, A. D., Carrington, P., *et al.* (2009). Midinfrared GaInSb/AlGaInSb quantum well laser diodes operating above 200 K, *Appl. Phys. Lett.*, **94**, p. 091111.

55. Hill, C. J., and Yang, R. Q. (2004). Interband cascade lasers grown on GaAs substrates lasing at 4 microns, *Appl. Phys. Lett.*, **85**, p. 3014.

56. Vurgftman, I., Canedy, C. L., Kim, C. S., Kim, M., Bewley, W. W., Lindle, J. R., *et al.* (2009). Mid-infrared interband cascade lasers operating at ambient temperatures, *New J. Phys.*, **11**, p. 125015.

57. Rdriguez, J. B., Cerutti, L., and Tournie, E. (2009). GaSb-based, 2.2 μm type-I laser fabricated on GaAs substrate operating continuous wave at room temperature, *Appl. Phys. Lett.*, **94**, p. 023506.

58. Rriguez, J. B., Cerutti, L., Grech, P., and Tournie, E. (2009). Room-temperature operation of a 2.25 μm electrically pumped laser fabricated on a silicon substrate, *Appl. Phys. Lett.*, **94**, p. 061124.

59. Mehta, M., Jallipalli, A., Tatebayashi, J., Kutty, M. N., Albrecht, A., Balakrishnan, G., *et al.* (2007). Room-temperature operation of buffer-free GaSb–AlGaSb quantum-well diode lasers grown on a GaAs platform emitting at 1.65 μm, *IEEE Photon. Technol. Lett.*, **19**, pp. 1628–1630.

60. Tateayashi, J., Jallipalli, A., Kutty, M. N., Huang, S. H., Balakrishnan, G. Dawson, L. R., *et al.* (2007). Room-temperature lasing at 1.82 μm of GaInSb/AlGaSb quantum wells grown on GaAs substrates using an interfacial misfit array, *Appl. Phys. Lett.*, **91**, p. 141102.

61. Salhi, A., Rouillard, Y., Angellier, J., and Garcia, M. (2004). Very-low-threshold 2.4-μm GaInAsSb-AlGaAsSb laser diodes operating at room temperature in the continuous-wave regime, *IEEE Photon. Technol. Lett.*, **16**, pp. 2424–2426.

62. Kirch, J. Garrod, T., Kim, S., Park, J. H., Shin, J. C., Mawst, L. J., *et al.* (2010). $InAs_yP_{1-y}$ metamorphic buffer layers on InP substrates for mid-IR diode lasers, *J. Cryst. Growth*, **312**, pp. 1165–1169.

63. Jallipalli, A., Balakrishnan, G., Huang, S. H., Rotter, T. J., Nunna, K., Liang, B. L., *et al.* (2009). Structural analysis of highly relaxed GaSb grown on GaAs substrates with periodic interfacial array of 90° misfit dislocations, *Nanoscale Res. Lett.*, **4**, pp. 1458–1462.

Chapter 9

High-Performance Metamorphic In(Ga)As/GaAs Quantum Dot Lasers on GaAs and Si

Zetian Mi,[a] Jun Yang,[b] and Pallab Bhattacharya[b]

[a] *Department of Electrical and Computer Engineering, McGill University, 3480 University Street, Montreal, Quebec H3A 2A7, Canada*
[b] *Solid-State Electronics Laboratory, Department of Electrical Engineering and Computer Science, University of Michigan, Ann Arbor, MI 48109-2122, USA*
zetian.mi@mcgill.ca

This chapter describes the recent developments of high-performance metamorphic In(Ga)As quantum dot lasers on GaAs and Si substrates. The epitaxial growth and materials characteristics of metamorphic buffer layers and quantum dot heterostructures on GaAs and Si are first discussed. The use of multiple layers of self-organized In(Ga, Al)As/GaAs quantum dots as three-dimensional (3-D) dislocation filter to reduce dislocation density in metamorphic heterostructures on Si is also presented. The design, epitaxial growth, fabrication, and performance characteristics of self-organized In(Ga)As quantum dot lasers on both GaAs and Si, with emission wavelengths in the range of ~1.0–2.0 μm, are subsequently described, followed by discussions on the future prospects and challenges of metamorphic quantum dot lasers.

Lattice Engineering: Technology and Applications
Edited by Shumin Wang

ISBN 978-981-4316-29-3 (Hardcover), 978-981-4364-25-6 (eBook)
www.panstanford.com

9.1 Introduction

Semiconductor lasers, with the incorporation of self-organized In(Ga)As quantum dots as active regions, have been extensively investigated, due to the 3-D carrier confinement of the dots and the resulting near-discrete density of states [1–21]. Such quantum dot heterostructures are formed on a two-dimensional (2-D) wetting layer in the Stranski–Krastanow growth mode [4, 22–23], which offers an unprecedented control over the structural, electronic, and optical properties of the dot layers, compared to other fabrication techniques. Self-organized In(Ga)As/GaAs quantum dot lasers can exhibit nearly ideal characteristics, including ultralow threshold current ($J_{th} < 20$ A/cm^2) [11, 24], nearly temperature invariant operation [25–27], high output power (14 W) [28], large modulation bandwidth ($f_{-3dB} = 24.5$ GHz) [17], and near-zero chirp, and α-parameter [17, 29–30]. The operation wavelengths of such pseudomorphic devices on GaAs, however, are generally in the range of 0.9–1.3 μm. To achieve 1.55 μm quantum dot lasers on GaAs, metamorphic heterostructures have been investigated [1, 18–20, 31–33]. This approach allows for an unprecedented control of the emission wavelengths of the dots by simply varying the In composition and, therefore the bandgap of the surrounding InGaAs barriers, in addition to the dot growth conditions. Metamorphic quantum dot lasers may also promise significant cost and performance advantages for optical communications, compared to conventional 1.55 μm semiconductor lasers with the incorporation of InGaAsP/InP double heterostructures or GaInNAs(Sb) quantum wells, which characteristically have high threshold current, temperature-sensitive operation ($T_0 \sim 40$–50 K), and large values of chirp and α-parameter [34–36].

Additionally, for application in chip-level optical interconnects, it is imperative to develop high-performance electrically injected semiconductor lasers that can be monolithically integrated with Si-electronics in a CMOS compatible process. Although significant advances have been made in Si-optoelectronics, the achievement of an electrically injected all-Si laser has remained elusive [37–38]. Alternatively, the integration of III-V semiconductor materials and devices on Si that involves the use of direct epitaxial growth, wafer bonding, or compliant substrates have been explored [39–43]. To date, highly promising results have been realized using relaxed and graded SiGe layers, AlSb buffer, evanescent coupling, and self-organized quantum dots as the gain region [21, 39, 41, 44–48]. Owning to the near-discrete density of states and the resulting large gain and differential gain, metamorphic quantum dot lasers on Si promise significantly improved performance, compared to conventional III-V quantum well lasers on Si. It has also been recognized that the coherently strained quantum dots are capable of generating large strain fields in the surrounding materials, which can inhibit the propagating dislocations, leading to almost defect-free quantum dots and low defect densities in device active regions [21, 41, 49].

In this chapter, we provide an overview of the recent developments of metamorphic quantum dot lasers, with a special focus on the growth kinetics and materials characteristics of metamorphic quantum dot laser heterostructures on both GaAs and Si substrates. In Section 9.2, we discuss the epitaxial growth and characterization of high-quality InGaAs metamorphic buffer layers on GaAs as well as the use of self-organized InAs quantum dots as highly effective dislocation filters in achieving low defect density GaAs heterostructures on Si substrates. The epitaxial growth and characterization of In(Ga)As/GaAs metamorphic quantum dot heterostructures on GaAs and Si are presented in Section 9.3. The design, growth, fabrication, and characterization of metamorphic quantum dot lasers on GaAs and on Si are discussed in Sections 9.4 and 9.5, respectively. Finally, conclusions and future prospects are presented in Section 9.6.

9.2 Epitaxial Growth of High-Quality Metamorphic In(Ga)As Buffer Layers on GaAs and Si Substrates

The performance of metamorphic lasers on GaAs and Si has been largely limited by the presence of high densities of dislocations, due to the large differences in lattice constants and thermal expansion coefficients between the device active regions and the underlying substrates [39–40, 45, 50–51]. Additionally, III-V lasers on Si suffer severely from the formation of antiphase domains related to the polar and nonpolar surface incompatibility between the two materials. In this regard, low defect density metamorphic buffer layers have been extensively investigated with the use of strained layer superlattices [50, 52], relaxed and graded SiGe layers [39], thermal cycle annealing [19, 53], selective area growth on patterned substrates [54], special buffer layers including AlSb [45] and $SrTiO_3$ [55], different compositional grading schemes in InGa(Al)As buffer layers [56–62], and more recently quantum dot dislocation filters [21, 48]. In this section, we describe the epitaxial growth and characterization of low defect density InGaAs metamorphic buffer layers on GaAs, followed by discussions on the achievement of high-quality metamorphic GaAs heterostructures on Si with the use of self-organized InAs quantum dots as dislocation filters.

9.2.1 *InGaAs Metamorphic Buffer Layers on GaAs*

The epitaxial growth and properties of InGaAs metamorphic buffer layers on GaAs, with the use of different grading schemes, including linear source temperature grading, linear alloy grading, and step grading profiles, have been intensively investigated [56–62]. Such metamorphic buffer layers have been used to realize high-performance metamorphic electronic devices, including high electron mobility transistors [63] and heterojunction bipolar transistors [64], and, more recently, metamorphic quantum dot and well lasers on GaAs [1, 19–20, 32, 65–68]. In what follows, we discuss the special growth techniques, including a number of critical growth/annealing steps involved in achieving nearly defect-free single- and multiple-step-graded InGaAs metamorphic buffer layers on GaAs.

The growth of nearly defect-free $In_{0.15}Ga_{0.85}As$ metamorphic buffer layers on GaAs is first described. After oxide desorption and the growth of a GaAs buffer layer at $\sim$600°C on GaAs substrate, a 0.3 μm $In_{0.15}Ga_{0.85}As$ buffer layer was grown at relatively low temperatures ($\sim$400°C) [19], which can effectively suppress the propagation of dislocations [69–70]. An *in situ* thermal annealing was performed to further improve the crystalline quality of the

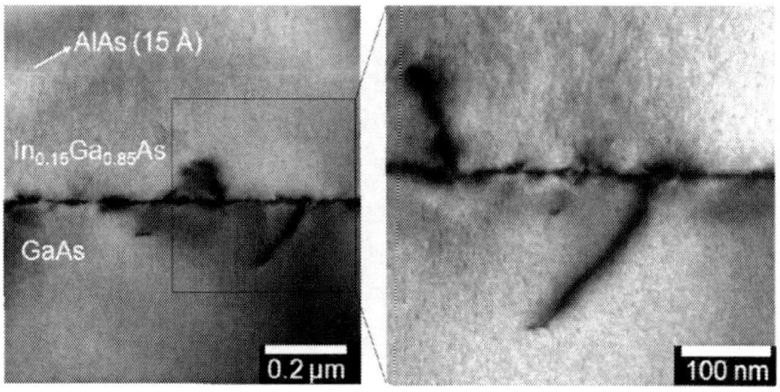

Figure 9.1. Cross-sectional transmission electron microscopy image of the $In_{0.15}Ga_{0.85}As$ metamorphic buffer layers grown on GaAs substrate. Ref. [1]. Reprinted with permission from IEEE.

InGaAs metamorphic buffer layers [71]. Prior to the annealing, a thin (~15 Å) AlAs layer was grown to effectively protect the surface. The cross-sectional transmission electron microscopy image of the metamorphic InGaAs buffer layer grown on GaAs is shown in Fig. 9.1 [1]. The thin (~15 Å) AlAs layer, where an *in situ* thermal annealing process was performed, can be identified. It is evident that dislocations are largely confined at the InGaAs/GaAs misfit interface and the propagation of dislocations is effectively suppressed. To enhance the strain relaxation and to further eliminate residual dislocations in the metamorphic buffer layers, multiple steps of low-temperature growth and *in situ* annealing were conducted. Under optimized growth conditions, the resulting metamorphic InGaAs layers can exhibit smooth surface morphology with faint crosshatch and are relatively free of microstructural roughness or stacking faults.

To achieve high-quality InGaAs metamorphic buffer layers with higher In compositions, various alloy grading schemes have been utilized [32, 56–58, 60, 62]. We investigated step-graded metamorphic buffer layers consisting of 0.3 μm $In_{0.16}Ga_{0.84}As$, $In_{0.20}Ga_{0.80}As$, and $In_{0.23}Al_{0.35}Ga_{0.42}As$ layers [20]. All these metamorphic layers were grown at relatively low temperatures (~390°C) to suppress the propagation of dislocations. The growth techniques were similar to those described above for single step-graded $In_{0.15}Ga_{0.85}As$ metamorphic buffer layers. Briefly, a thin (~15 Å) AlAs layer was grown after each 0.3 μm metamorphic buffer layer, which was followed by an *in situ* thermal annealing at ~700°C to enhance the strain relaxation and to reduce the defect densities. Recently, high-quality $In_{0.25}Ga_{0.75}As$ metamorphic layers have also been demonstrated on GaAs using an appropriate defect reduction technique [32]. In this approach, the InGaAs metamorphic layer is capped by a thin GaAs-AlAs layer and the dislocated region can be selectively removed by annealing.

It is also important to note that doping plays an important role in the strain relaxation of metamorphic buffer layers [61, 66, 72]. It was observed that both dislocation densities and surface roughness can be reduced by Be doping. On the other hand, higher threading dislocation densities have been measured in Si-doped InGaAs metamorphic buffer layers grown on GaAs substrates. The underlying mechanism has been investigated. It is generally believed that Be atoms can act as a surfactant, which helps suppress 3-D growth by reducing In segregation and the strain at the growth front. To improve the quality of Si-doped metamorphic InGaAs buffer layers, a relatively low Si doping and/or a moderate alloy grading profile are generally employed to minimize the generation of dislocations.

9.2.2 *Self-Organized Quantum Dot Dislocation Filters*

It is of interest to investigate the use of self-organized quantum dots as dislocation filters, due to the fact that the material within/surrounding quantum dots has large strain and a strained layer is capable of bending propagating dislocations [21, 48]. The bending of threading dislocations enhances the chance of their annihilation or propagation to the sample edge, which can lead to a reduction of dislocation densities in the device active regions [44, 52, 73–74]. Dislocation filters consisting of 2-D strained-layer superlattice have been utilized to realize III-V quantum well lasers on Si since the 1980s [50, 52]. Compared to conventional 2-D strained-layer superlattice, self-organized quantum dots provide larger and anisotropic strain for dislocation bending. In addition, the localized strain field surrounding the quantum dots can trap point defects of opposite strain due to the energy minimization. Therefore, quantum dots incorporated in the buffer layer are expected to be

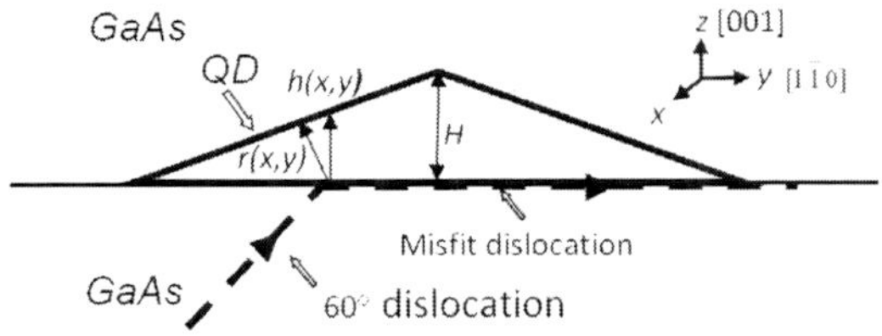

Figure 9.2. Cross-sectional schematic description of the bending of a 60° dislocation by a quantum dot. Ref. [21]. Reprinted with permission from IEEE.

highly effective in blocking dislocations and defects. Defect-density reduction in GaN grown on sapphire substrates has been observed in the presence of quantum dots [75].

9.2.2.1 Theoretical model

A quasi-3-D model based on strain relaxation has been used to analyze dislocation bending by quantum dots [21]. The self-organized quantum dots are coherently strained and nearly free of dislocations [4, 76]. They are also assumed to have a pyramidal shape. Figure 9.2 schematically shows that a dislocation, generated at the lattice mismatched interface, propagates upwards and is bent beneath the island. The bending of the dislocation will generate a segment of misfit dislocation. Bending occurs only when the strain energy released due to the generation of the misfit dislocation (ΔE_{rel}) is greater than the dislocation self-energy (ΔE_{dis}) [77–78], where ΔE_{rel} and ΔE_{dis} for the dislocation of unit length are described by:

$$\Delta E_{rel} = \frac{2G_{dot}(1+v)}{(1-v)} f_{eff} h b \cos\alpha \tag{9.1}$$

$$\Delta E_{dis} = \frac{1}{2\pi} \frac{G_{buff} G_{dot}}{G_{buff} + G_{dot}} b^2 \left(\frac{1 - v\cos^2\beta}{1-v} \right) \left[\ln\left(\frac{2r}{b} \right) + 1 \right] \tag{9.2}$$

Here, ΔE_{rel} and ΔE_{dis} are determined by the material properties, including the lattice mismatch f, Poisson ratio v, and Young's modules G_{buff} and G_{dot} (the subscripts of *buff* and *dot* denote GaAs buffer and quantum dots, respectively). They also depend on the type, magnitude, and orientation of misfit dislocations, characterized by Burgers vector b and its orientation angles α and β with respect to the layer interface and dislocation line, respectively. Due to the island geometry and anisotropic strain distribution of quantum dots, ΔE_{rel} and ΔE_{dis} vary with position (x, y) where the dislocation encounters the island. For simplicity, the following assumptions are used in the model: (i) the island has uniform strain distribution with an effective lattice mismatch of f_{eff}, (ii) the thickness of the strain-released layer is the local height $h(x, y)$ of the island, and (iii) the outer cut-off radius of the dislocation strain field is the distance $r(x, y)$ to the nearest facet of the island. Based on these assumptions, the bending area underlying the quantum dot, within which a propagating dislocation is bent when $\Delta E_{rel} > \Delta E_{dis}$, can be estimated. The bending area ratio, defined as the bending area of a single quantum dot divided by its base area, has been calculated for InAs, $In_{0.5}Ga_{0.5}As$, and $In_{0.6}Al_{0.4}As$ quantum dots. As shown in Fig. 9.3a, InAs dots, due to the relatively large dot size and strain field, are much more effective in bending dislocations compared to the other two types of dots. More specifically, InAs dots, which have a typical base width of 20–30 nm, exhibit an effective bending-area ratio of 70–80%. With a dot density of 2–4 $\times$ 10^{10}cm^{-2}, a single layer of InAs quantum dots has a bending-area ratio of up to 10–20%.

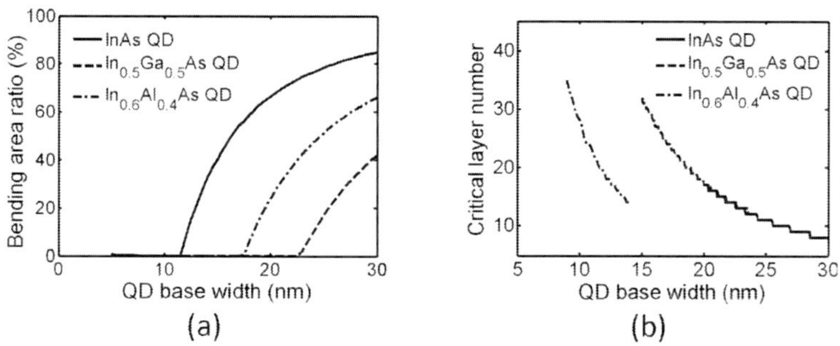

Figure 9.3. (a) Bending area ratio versus quantum dot base width for InAs, $In_{0.5}Ga_{0.5}As$, and $In_{0.6}Al_{0.4}As$ dots; (b) Calculated critical layer number for the generation of single-kink dislocation loop for InAs, $In_{0.5}Ga_{0.5}As$, and $In_{0.6}Al_{0.4}As$ quantum dot multilayers. Ref. [21]. Reprinted with permission from IEEE.

It is also expected that multiple quantum dot layers can further enhance dislocation filtering. However, the quantum dot barrier thickness and layer number have to be carefully optimized in order to avoid the generation of single-kink or double-kink misfit dislocation loops due to the accumulated strain. Single-kink dislocations depend on the depth of the buried dot layers and will preferentially occur first due to a lower energy barrier. Using the excess stress model [79], one can estimate the critical number of quantum dot layers before single-kink dislocations are generated. Calculations are performed by assuming the quantum dot as an equivalent 2-D uniformly strained layer with a thickness of $h = H/3$ and equal strain energy per area. As shown in Fig. 9.3b, the critical layer numbers, depending on the dot sizes, are 10–15 for InAs dots, 20–30 for $In_{0.5}Ga_{0.5}As$ dots and 15–35 for $In_{0.6}Al_{0.4}As$ dots, with 50nm GaAs barrier layers in between.

9.2.2.2 Experimental characterization

The effect of dislocation filtering can be further revealed by studying the photoluminescence properties of active $In_{0.5}Ga_{0.5}As$ quantum dots in laser heterostructures grown on Si with the incorporation of 10 layers of InAs, $In_{0.5}Ga_{0.5}As$, and $In_{0.6}Al_{0.4}As$ quantum dots as dislocation filters [21]. As shown in Fig. 9.4a, the luminescence intensities measured from the $In_{0.5}Ga_{0.5}As$ active quantum dots vary greatly, depending on the type of quantum dot buffer layer in the sample. The difference in luminescence intensities is a direct consequence of the difference in the density of defects and other nonradiative recombination centers in the quantum dot active region. It is observed that the sample with InAs quantum dot buffer layer exhibits the highest photoluminescence intensity. The full-width at half-maximum (FWHM) is $\sim$60 meV, which is comparable to that measured for similar laser heterostructures grown on GaAs substrates. InAlAs quantum dots can further induce surface roughness, and therefore the photoluminescence emission from the top $In_{0.5}Ga_{0.5}As$ quantum dots is very weak and broad. It is evident that InAs quantum dots are most effective in improving the luminescent properties of the active quantum dot region, confirming that they can block propagating dislocations more efficiently, due to

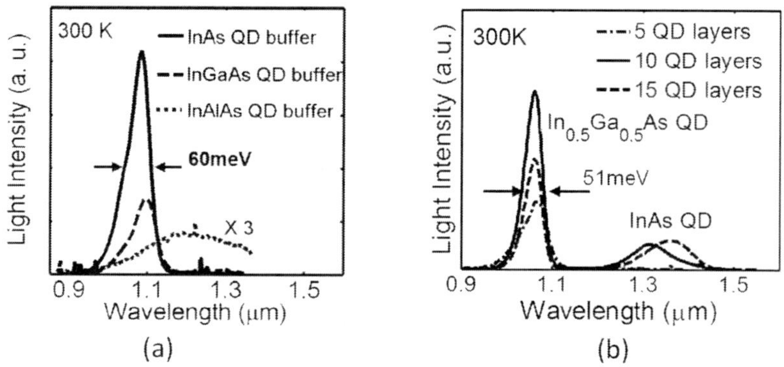

Figure 9.4. Photoluminescence spectra measured at 300 K from $In_{0.5}Ga_{0.5}As$ quantum dots grown on Si: (a) with different quantum dot buffer layers grown at 460°C; and (b) with 5–, 10–, and 15-layer InAs quantum dots in the buffer layers grown at 510°C. Ref. [21]. Reprinted with permission from IEEE.

the large dot sizes and strain field. The effects of growth temperature and the number of quantum dot layers in the buffer region on dislocation filtering were further investigated. Shown in Fig. 9.4b is the photoluminescence spectrum measured from an $In_{0.5}Ga_{0.5}As$ quantum dot heterostructure grown on Si with 10 layers of InAs quantum dots as the buffer and the growth temperature for these dots at a higher temperature of ~510°C. In this case, luminescence emission from the InAs quantum dot buffer region can also be clearly observed, in spite of the defects around this layer. Also shown in Fig. 9.4b are the photoluminescence spectra from active $In_{0.5}Ga_{0.5}As$ quantum dots grown on Si with the use of 5 and 15 layers of InAs quantum dots as the buffer. It is evident that 10 quantum dot layers are optimum, which is in agreement with the theoretical studies described above.

The propagation of dislocations in the InAs quantum dot buffer layers on Si was further studied using cross-sectional transmission electron microscopy [21]. Images of dislocations were recorded under various diffraction conditions along the [110] axis, specifically for g = $[2\bar{2}0]$, $[1\bar{1}1]$, $[\bar{1}11]$, [004], and $[1\bar{1}3]$, as shown in Fig. 9.5. Due to the invisibility criterion of $g \times b = 0$, there are generally two types of threading dislocations for GaAs on Si, including pure-edge

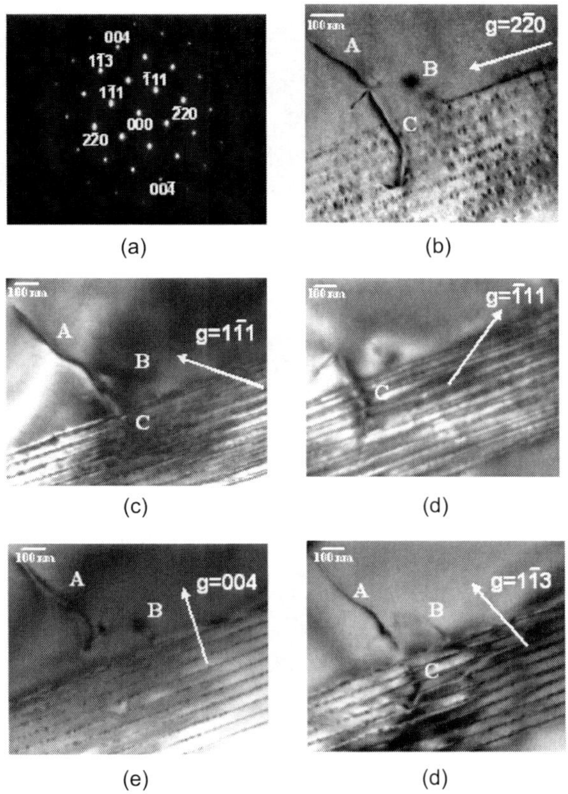

Figure 9.5. Electron diffraction pattern (a), and cross-sectional transmission electron microscopy image of dislocation propagation in the 10-layer InAs quantum dots buffer layer under various diffraction conditions: (b) $g = \langle 2\bar{2}0 \rangle$, (c) $g = \langle 1\bar{1}1 \rangle$, (d) $g = \langle \bar{1}11 \rangle$, (e) $g = \langle 004 \rangle$, and (f) $g = \langle 1\bar{1}3 \rangle$. The zone axis is $\langle 110 \rangle$. Ref. [21]. Reprinted with permission from IEEE.

dislocations with Burger's vector $b = \pm(\bar{1}10)$ (labeled as C in Fig. 9.5) and 60°-mixed dislocations with $b = \pm(01\bar{1})$ or $\pm(101)$ (labeled as A and B in Fig. 9.5). It is observed that InAs quantum dot buffer layers can efficiently bend 60°-mixed dislocations (labeled as B). More importantly, pure-edge dislocations (labeled as C) can be terminated at the surface of quantum dots, which was not possible by 2-D strained-layer superlattice [44]. This may be related to the formation of either a dislocation loop at the quantum dot surface due to stronger and anisotropic stress surrounding the nanoisland or a dislocation of reverse Burger's vector, which annihilates the pre-existing dislocation.

9.3 Growth and Characterization of Metamorphic Quantum Dot Heterostructures

9.3.1 *Metamorphic Quantum Dot Heterostructures on GaAs*

The growth of high-quality InAs quantum dots embedded in a metamorphic heterostructure has proved to be difficult, compared to their pseudomorphic counterparts, due to the complicated surface diffusion kinetics at a metamorphic InGa(Al)As growth front. In what follows, the epitaxial growth and characterization of long wavelength (1.45–2.0 μm) InAs metamorphic quantum dot heterostructures are described.

The achievement of high-quality 1.45 μm InAs metamorphic quantum dots on $In_{0.15}Ga_{0.85}As$ metamorphic buffer layers on GaAs is first discussed. As Illustrated in Fig. 9.6a, the metamorphic quantum dot heterostructure consists of ~2.9 ML InAs, capped

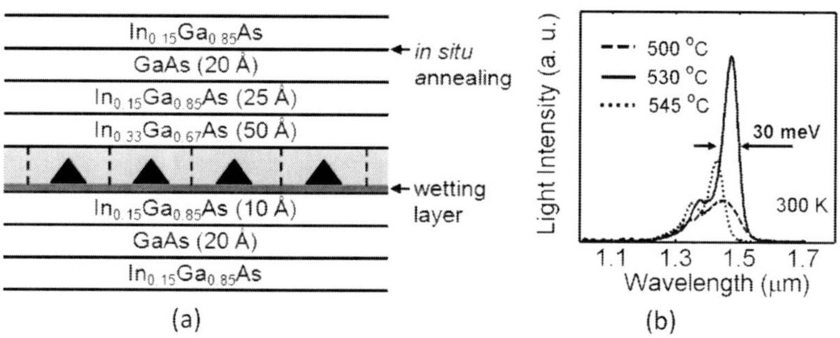

Figure 9.6. (a) Schematic of the metamorphic quantum dot heterostructure showing the formation of highly uniform InAs quantum dot layers due to the enhanced indium adatom surface migration. The dashed lines illustrate the adatom surface migration around the dots. (b) Room temperature photoluminescence emission spectra for metamorphic InAs quantum dot samples grown at various temperatures. Ref. [1]. Reprinted with permission from IEEE.

by a 50 Å $In_{0.33}Ga_{0.67}As$ strain-reducing layer [1]. Such a dot-in-a-well heterostructure has been commonly used to extend the emission wavelengths of InAs quantum dots [80]. Thin (20 Å) GaAs layers were also grown prior to and after the InAs quantum dot formation, which can smoothen the growth front and minimize phase separation. The incorporation of such GaAs barrier layers provides additional benefits, including effective strain compensation as well as a stronger carrier confinement for the quantum dots.

The growth temperature plays a crucial role in determining the optical quality of the metamorphic InAs quantum dot heterostructures. Shown in Fig. 9.6b are the room-temperature photoluminescence spectra of InAs quantum dots grown at various temperatures. The best quality InAs metamorphic quantum dots, in terms of both the size uniformity and photoluminescence emission efficiency, is achieved at a growth temperature of $\sim$530°C. The measured photoluminescence intensity and spectral FWHM of InAs metamorphic quantum dots grown in this temperature range are comparable to those of 1.3 μm pseudomorphic InAs quantum dots. Illustrated in Fig. 9.7a is the atomic force microscopy image of an uncapped InAs metamorphic quantum dot layer grown under optimized conditions. The quantum dots exhibit an average height

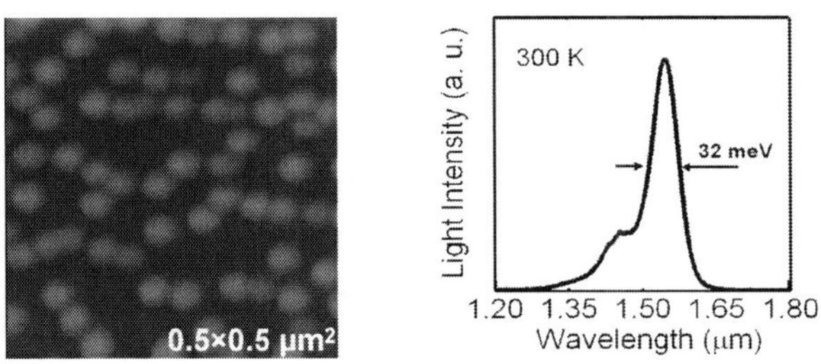

(a) (b)

Figure 9.7. (a) Atomic force microscopy image of an uncapped InAs metamorphic quantum dot layer grown under optimized conditions. (Ref. [18]. Reprinted with permission from Elsevier.) (b) Photoluminescence spectra measured at 300 K for an InAs metamorphic quantum dot heterostructure grown on multiple-step-graded InGaAs buffer layers on GaAs. Ref. [20]. Reprinted with permission from American Institute of Physics.

of $\sim$14 nm, based width of $\sim$45 nm, and areal density of $\sim$2.6 $\times$ 10^{10} cm^{-2}. It is also evident that the dots are highly uniform, which is consistent with the measured very narrow spectral FWHM ($\sim$30 meV) at room temperature. Additionally, metamorphic InAs quantum dots with a very high surface areal density of $\sim 10^{11}$ cm^{-2} can be achieved by depositing $\sim$2.5 ML InAs on $In_{0.2}Ga_{0.8}As$ metamorphic buffer [32, 81]. When capped by a 4 nm $In_{0.4}Ga_{0.6}As$ strain reducing layer, the dots exhibit emission wavelengths in the range of $\sim$1.4–1.5 μm.

To further extend the emission wavelength to $\sim$1.55 μm and beyond, InAs quantum dot heterostructures grown on $In_xGa_{1-x}As$ buffer layers with x larger than 0.2 have been investigated. Shown in Fig. 9.7b is the room temperature photoluminescence spectrum of InAs quantum dots grown on an $In_{0.23}Ga_{0.77}As$ metamorphic buffer layer on GaAs [20]. Each InAs quantum dot layer consists of $\sim$3.0 ML InAs, capped by a 50 Å $In_{0.40}Ga_{0.60}As$ strain-reducing layer. Thin ($\sim$20 Å) GaAs layers were also incorporated before and after each quantum dot layer. Under optimized growth conditions, such InAs quantum dot heterostructures exhibit strong photoluminescence emission, with a peak wavelength at $\sim$1.55 μm and a relatively narrow ($\sim$32 meV) spectral FWHM at 300 K.

Compared to the 1.0–1.1 μm pseudomorphic InAs/GaAs quantum dot heterostructures, long wavelength InAs/InGaAs metamorphic quantum dots generally exhibit smaller maximum gain, due to the smaller dot density and larger dot size. It is therefore essential to incorporate multiple InAs quantum dot layers in the laser active region in order to achieve high-performance metamorphic quantum dot lasers. However, the large strain field associated with InAs dots and the resulting strain-driven In adatom migration toward the top of the dots can lead to severe surface undulations in the subsequently grown InGaAs capping layers [1, 82]. The surface undulations may also adversely affect the formation of the following dot layers, even when relatively thick InGaAs ($\sim$50 nm) barrier layers are incorporated, which lead to the formation of dislocations in the barrier and dot layers in regions where excess strain builds up. As a consequence, directly grown multiple 1.5 μm metamorphic InAs quantum dot layers generally exhibit very poor optical quality. It has been reported that the formation of dislocations during the growth of multiple layers of 1.3 μm quantum dots can be drastically reduced by using the technique of high growth temperature (~600°C) spacer layers (HGTSLs) [83]. This growth technique, however, cannot be implemented during the growth of 1.55 μm metamorphic quantum dot heterostructures, due to its incompatibility with the optimum growth temperatures (~500–530°C) used for the metamorphic InGaAs barrier layers. An alternate approach is to employ a simple *in situ* annealing process after the growth of each InAs quantum dot and the InGaAs and GaAs capping layers [1, 19]. The thermal annealing can greatly minimize surface undulations and therefore smoothen the growth front, due to the significantly enhanced adatom surface migration at elevated temperatures. The GaAs layer can also avoid any significant indium desorption from the surface during the annealing process. Subsequently, InGaAs barrier layers, with a thickness of ~50 nm, were grown at ~530°C. The annealing conditions, including the temperature and duration, as well as the thickness of the GaAs capping layer have to be carefully optimized. From extensive studies, it was determined that an *in situ* annealing at ~580–600°C for ~8 minutes after the 20 Å GaAs capping layer enabled the achievement of 1.5 μm metamorphic quantum dots with the best optical and structural quality. Shown in Fig. 9.8 is the transmission electron microscopy image of multiple InAs metamorphic quantum dot layers grown under optimized conditions, which are nearly free of

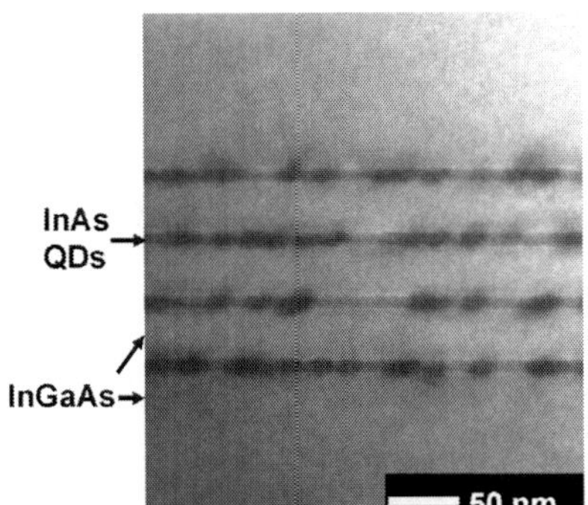

Figure 9.8. Cross-sectional transmission electron microscopy image of multiple InAs metamorphic quantum dot layers.

dislocations. The present experiments confirm the role of an *in situ* thermal annealing in achieving superior quality multiple layers of InAs quantum dot heterostructures. It may also be noted that the highly effective, often used HGTSLs technique actually performed an unintentional *in situ* annealing process when the substrate temperature was ramped up and the barrier layers were grown at elevated temperatures. It has also been reported that proper heat treatment can evaporate defect regions in InGaAs/GaAs quantum dot heterostructures, leading to improved crystal quality [32, 84].

Recently, InAs quantum dots (dashes), with emission wavelengths at ~1.6–2.0 μm, have been grown on GaAs substrates using AlGaAsSb metamorphic buffer layers [31, 85]. In this approach, multiple-step-graded AlGaAsSb metamorphic buffer layers were first grown on GaAs substrates, followed by the growth of a 2 μm $Al_{0.5}Ga_{0.5}As_{0.54}Sb_{0.46}$ layer. Using this buffer scheme, the lattice constant of the growth matrix is increased from 5.653 to 5.869 Å. The resulting lattice mismatch between InAs and the surrounding metamorphic barriers is only ~3.2%, which leads to the formation of quantum dash structures. The metamorphic quantum dashes exhibit a relatively high density ($\sim 8 \times 10^{10}$ cm^{-2}). Room temperature emission wavelength at ~2.02 μm was measured.

While emission wavelengths of InAs quantum dots on GaAs substrates can be tuned to ~1.5–2.0 μm using the afore-described metamorphic approaches, the luminescence efficiency decreases considerably at longer wavelengths. This is due, in part, to the reduced carrier confinement associated with the smaller energy band discontinuity between the dots and the surrounding InGaAs metamorphic matrix with high In compositions. The resulting thermal escape of carriers and their subsequent recombination in the wetting layer and barrier states can lead to a significantly quenched emission at or near room temperature [86]. Better carrier confinement can be achieved by embedding InAs quantum dots in GaAs barriers, as described above, or by incorporating InAlAs barrier layers [87–89], which led to significantly improved luminescence efficiency for metamorphic InAs quantum dots with emission wavelengths in the range of 1.55 μm and beyond [1, 87]. In this regard, the use of metamorphic heterostructures provides a greater degree of flexibility in engineering both the emission wavelength and the carrier confinement of the dots, compared to their pseudomorphic counterpart.

9.3.2 *Metamorphic Quantum Dot Heterostructures on Si*

The afore-described growth techniques are also applicable to the development of high-quality metamorphic quantum dot heterostructures on Si substrates. In these studies, (001)-oriented Si substrates misoriented 2–6° toward [111] or [110] were often utilized to eliminate the formation of antiphase domains and stacking faults at the III-V and Si misfit interfaces [90]. Self-organized quantum dot layers can be directly nucleated on Si substrates or grown on suitable buffer layers on Si [41, 91]. In the latter case, GaAs or SiGe metamorphic buffer layers were first grown, followed by the deposition of quantum dot heterostructures. (In,Ga)As quantum dots grown on Si, with emission wavelengths from ~1.0 to 1.3 μm, have been realized [21, 92–94]. Their characteristics, including the size, density, uniformity, and emission wavelengths, are comparable to those grown on GaAs substrates and can be precisely controlled by varying the growth conditions. The photoluminescence emission spectrum of $In_{0.5}Ga_{0.5}As$ quantum dots grown on Si substrates with the incorporation of InAs quantum dot dislocation filters is shown in Fig. 9.9. Also shown in the figure are the photoluminescence spectra of similar quantum dots grown on GaAs and on Si without the use of InAs quantum dot buffer layer. It is seen that the optical quality of InGaAs metamorphic quantum dots grown on Si is comparable, in terms of both the photoluminescence intensity and spectral FWHM, to that of InGaAs pseudomorphic quantum dots on GaAs substrates. InAs metamorphic quantum dot heterostructures, with emission wavelengths in the 1.3 μm range, have also been grown on Si substrates using thin GaAs or relaxed SiGe buffer layers [93–95].

9.4 Metamorphic Quantum Dot Lasers on GaAs

Low threshold current, high speed, and zero chirp semiconductor lasers with operation wavelength λ at ~1.55 μm are required for long-haul optical communication systems. In what follows, we

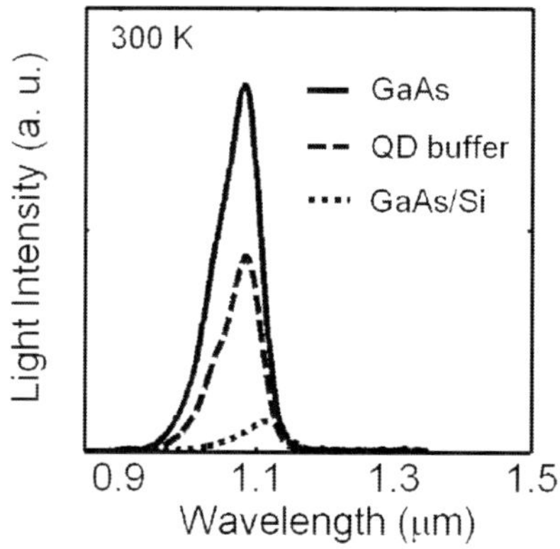

Figure 9.9. Photoluminescence spectra measured at 300 K for InGaAs quantum dots grown on GaAs substrates and on Si with and without InAs quantum dot buffer layers. Ref. [21]. Reprinted with permission from IEEE.

describe the epitaxial growth, fabrication, and characterization of long wavelength (1.45–2.0 μm) metamorphic quantum dot lasers on GaAs substrates that promise such nearly ideal characteristics. It has been recognized that conventional quantum dot lasers suffer from hot-carrier related problems, due to the presence of wetting layer states and closely spaced hole energy levels [13–15]. Recently, the special techniques of p-doping and tunnel injection have been demonstrated to be highly effective in alleviating these problems [13, 15–17]. In the scheme of tunnel injection, "cold" carriers are injected directly into the dots from an injector well through a tunnel barrier. This reduces carrier and phonon heating and leads to increased gain and differential gain. Very large modulation bandwidth has been measured in 1.0 μm tunnel injection quantum dot and quantum well lasers [17, 96]. In the technique of p-doping, each quantum dot layer is modulation doped p-type, which provides holes in the ground state of the dots. Deleterious effects due to a thermally broadened hole distribution can be largely eliminated. With the incorporation of these special techniques, metamorphic quantum dot lasers can exhibit a very low threshold ($J_{th} \sim 60$ A/cm^2), extremely high temperature stability ($T_0 = 620$ K), large frequency response ($f_{-3dB} = 8$ GHz), and very small linewidth enhancement factor and chirp [1, 19]. Additionally, metamorphic

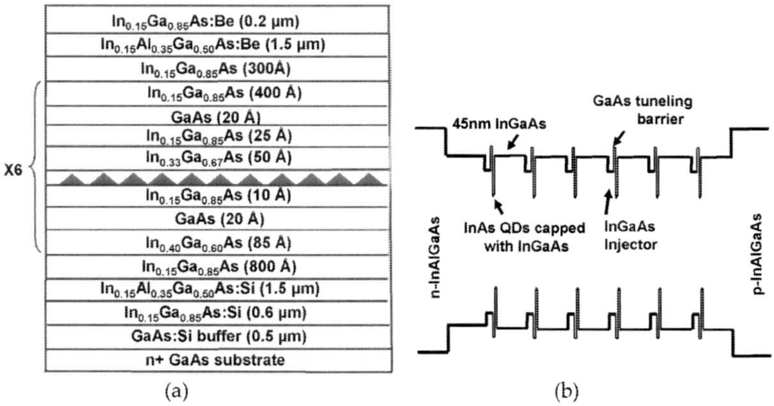

Figure 9.10. (a) Metamorphic InAs tunnel injection quantum dot laser heterostructures grown on GaAs by molecular beam epitaxy and (b) schematic energy band diagram of the laser active region. Ref. [19]. Reprinted with permission from American Institute of Physics.

quantum dot lasers with high output power ($\sim$7 W) and long-term reliability have been demonstrated [20, 32].

The metamorphic InAs quantum dot tunnel injection laser is first described. The $In_{0.15}Ga_{0.85}As/In_{0.15}Al_{0.35}Ga_{0.50}As$ separate confinement heterostructure quantum dot laser, illustrated in Fig. 9.10a, was grown on an $In_{0.15}Ga_{0.85}As$ metamorphic buffer layer on n^+-GaAs (001) substrates [19]. The 1.5 μm thick $In_{0.15}Al_{0.35}Ga_{0.50}As$ n- and p-cladding layers were doped with Si and Be, respectively. The device active region consists of six stacks of coupled well and dot tunnel heterostructures, separated by $\sim$45 nm $In_{0.15}Ga_{0.85}As$ barrier layers. The energy band diagram of the laser active region is schematically shown in Fig. 9.10b. Each coupled well-dot layer consists of 8.5 nm $In_{0.4}Ga_{0.6}As$ injector well, a 2 nm GaAs tunnel barrier, and an InAs quantum dot layer. The photoluminescence peak wavelengths of the injector well and quantum dot layer are carefully tuned to $\sim$1.17 and 1.45 μm, respectively, such that the electron ground state in the well is approximately one phonon energy above the first excited states of the dots. The measured photoluminescence emission spectrum at room temperature is illustrated in Fig. 9.11, wherein the emission peaks from the injector well and the ground and first excited states of the dots are identified.

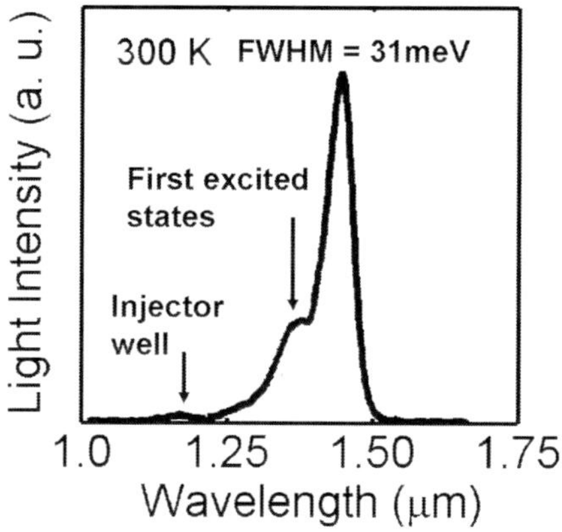

Figure 9.11. Photoluminescence spectrum measured from the metamorphic tunnel injection laser active region at 300 K. Ref. [19]. Reprinted with permission from American Institute of Physics.

The output characteristics of 1.45 μm metamorphic InAs quantum dot tunnel injection lasers are described. Shown in Fig. 9.12a is the measured light-current characteristics of a 1000 $\times$ 80 μm^2 device with ~95% and 80% high reflectivity coatings on both facets under pulsed mode (1% duty cycle) operation [19]. The laser threshold is ~63 A/cm^2, which is comparable to 1.3 μm pseudomorphic InAs quantum dot lasers on GaAs and considerably smaller than the commonly reported values for InAs quantum dot/dash lasers on InP substrates [30, 83, 97–99]. The output spectrum at an injection current of 80 mA is shown in the inset. The lasers exhibit an internal quantum efficiency of ~72% and a cavity loss of ~2.8 cm^{-1}.

The dynamic characteristics of a 600 $\times$ 3 μm^2 near-single mode device were investigated under pulsed mode operation (1% duty cycle) to minimize heating effect [19] and 95% and 80% facet coatings were employed on the two facet ends, respectively. Illustrated in Fig. 9.12b are the small-signal modulation characteristics measured at various injection currents at room temperature. A maximum 3 dB bandwidth of ~8 GHz was measured at an injection current of 55 mA, which suggests that metamorphic quantum dot lasers are suitable for 10 Gbps operation for high-speed optical communications. The modulation efficiency is ~1.06 $GHz/mA^{1/2}$.

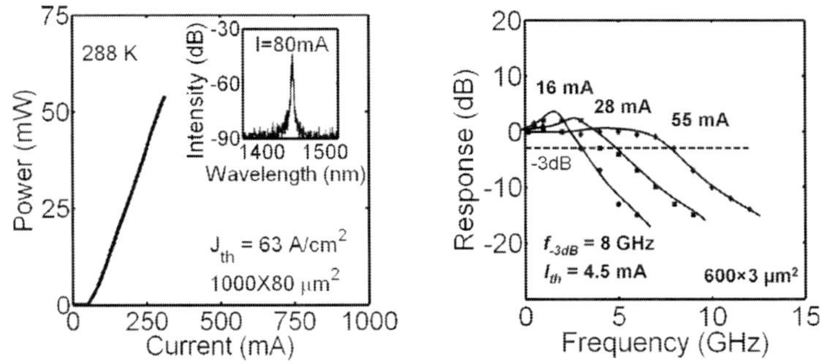

(a) (b)

Figure 9.12. (a) Light-current characteristics and output spectrum (inset) of a broad area metamorphic InAs tunnel injection quantum dot laser under pulsed mode (1% duty cycle) operation. (b) Small-signal modulation response of the tunnel injection metamorphic lasers. Ref. [19]. Reprinted with permission from American Institute of Physics.

The linewidth enhancement factors (α-parameters) as a function of frequency were obtained by measuring the subthreshold emission spectra of the nearly single-mode ridge waveguide lasers and by calculating the Fabry-Pérot mode peak-to-valley ratios and shifts between two different bias currents. The values of the α-parameters are essentially zero. It is important to note that for a conventional quantum dot laser, the measured α-parameter under subthreshold bias conditions can be very different from that at high bias conditions, due to carrier occupation of wetting layer and barrier states [100]. In tunnel injection quantum dot lasers, however, the α-parameter is expected to be weakly dependent on injection currents, due to the injection of "cold" electrons directly into the lasing states [101]. Therefore, the α-parameter above threshold in tunnel injection quantum dot lasers is expected to be close to that measured under subthreshold bias conditions. The chirp of the metamorphic quantum dot lasers was measured by measuring the broadening of a single longitudinal mode using an optical spectrum analyzer. The measurements were done as a function of frequency of the modulating current. The chirp is very low, ~0.1 Å. The low values of α parameter and chirp are attributed to the reduction in hot-carrier effects due to tunnel injection.

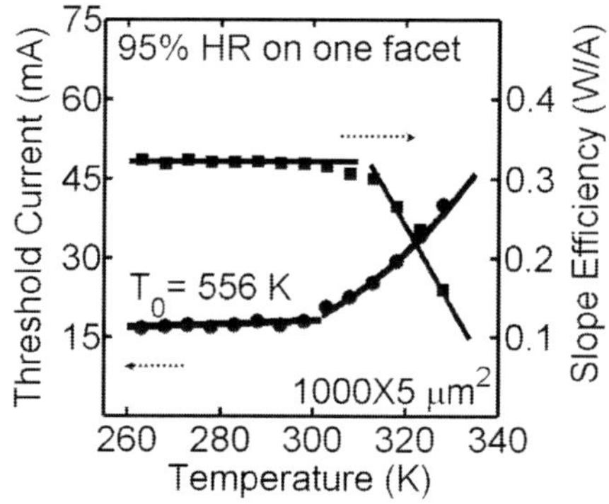

Figure 9.13. Variation of threshold current with temperature of the p-doped metamorphic quantum dot laser. Ref. [18]. Reprinted with permission from Elsevier.

The growth and characteristics of p-doped InAs metamorphic quantum dot lasers have also been investigated. The laser active region consists of 8 quantum dot layers, which are modulation doped p-type using Be (~20 holes per dot) [18, 20]. The devices exhibit a large T_0 of 556 K and constant output slope efficiency (~0.31 W/A) in the temperature range of 263–305 K, shown in Fig. 9.13 [18]. Very large T_0 (up to ∞) have also been measured in p-doped pseudomorphic quantum dot lasers on GaAs and InP substrates [25, 30, 102–103]. One possible explanation for the observed very large T_0 is the increase in Auger recombination in quantum dots upon p-doping and its decrease with increasing temperature, which offsets the increasing trends of other recombination currents [25, 102]. The large T_0 in p-doped quantum dot lasers may also be explained by considering the temperature dependence of the gain of p-doped quantum dot lasers [103].

In order to achieve emission wavelengths at ~1.55 μm, $In_{0.23}Ga_{0.77}As/In_{0.23}Al_{0.35}Ga_{0.42}As$ separate confinement heterostructure metamorphic quantum dot lasers grown on multiple-step-graded InGaAs buffer layers have been developed. Illustrated in Fig. 9.14a, the laser active region consists of six InAs metamorphic quantum dot layers, separated by ~40 nm $In_{0.23}Ga_{0.77}As$ barrier layers [20]. The lasers exhibit an emission wavelength of ~1.52 μm and a relatively low threshold current of ~460 A/cm^2, shown in

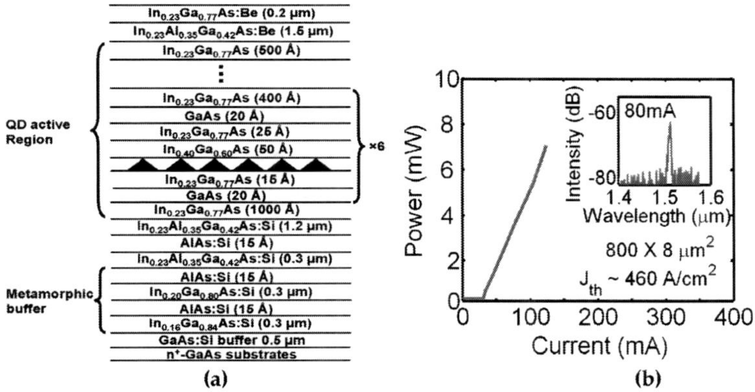

Figure 9.14. (a) Metamorphic InAs quantum dot heterostructures grown on multiple-step-graded InGaAs buffer layers. (b) Light-current characteristics and output spectrum (inset) of 1.52 μm InAs metamorphic quantum dot lasers grown on GaAs. Ref. [20]. Reprinted with permission from American Institute of Physics.

Fig. 9.14b. The output efficiency, however, has remained relatively low, ~0.1 W/A, potentially due to the presence of defects in the device heterostructures.

InAs metamorphic quantum dot lasers on GaAs have also been demonstrated with the use of AlGaAsSb buffer layers [31]. A minimum threshold current of ~300 A/cm^2 was achieved under pulsed operation for a 7.7 mm cavity length. The emission wavelengths vary in the range of 1.27 to 1.34 μm, which corresponds to the first excited state transition of the active dots.

Metamorphic quantum dot lasers can exhibit relatively large modal gain. From detailed analysis of the spontaneous emission efficiency of a 1.5 μm metamorphic laser consisting of 10 InAs dot layers, it was derived that the modal gain can be possibly as high as 190 cm^{-1} [104]. It is therefore expected that metamorphic quantum-dot lasers are suitable for high power operation. Over 7W and 800mW output power values were measured from broad-area (1500 × 100 μm^2) and narrow-stripe (2000 × 6 μm^2) metamorphic quantum dot lasers under pulsed mode operation, respectively [32].

The reliability of metamorphic quantum dot lasers has also been investigated. A lifetime of nearly 2000 h was measured for a 1.45 μm metamorphic quantum dot ridge waveguide laser at an output power of 10 mW under quasi-continuous bias conditions (50% duty cycle) at 300 K [20]. A lifetime of ~800 h was also reported for 1.46 μm metamorphic quantum dot lasers at 10°C under continuous wave operation [105]. It may be noted that these measurements were performed on unpackaged laser diodes and the intrinsic laser lifetime may be significantly longer.

9.5 Metamorphic Quantum Dot Lasers on Si

The growth and characterization of metamorphic quantum dot lasers on Si substrates have also been investigated [21, 41, 48–49, 92, 95, 106]. Linder *et al.* reported the first InGaAs quantum dot lasers on Si [49]. Although the device was operated at ~80 K and exhibited very large threshold, it was found that the laser active region had substantially reduced defect densities and the dots themselves may be defect-free, which was attributed to the bending of dislocations, due to the strong strain field in and around the dots, as discussed in Section 9.2.2. This observation is also consistent with the experimental results reported by Kazi *et al.* that showed an improved lifetime for semiconductor lasers on Si incorporating quantum dot like active regions, compared to similar quantum well lasers grown on Si [92]. In this section, we describe the recent achievement of room-temperature operational InGaAs metamorphic quantum dot lasers on Si with emission wavelengths in the range of 1.0–1.1 μm [1, 21, 41, 95, 107].

GaAs/$Al_{0.7}Ga_{0.3}As$ separate confinement heterostructure quantum dot lasers, illustrated in Fig. 9.15, were grown on (001)-oriented Si substrates misoriented 4° toward [111] [1, 21, 48]. A thin (<2 μm) GaAs buffer layer was first grown by metal organic vapor phase epitaxy. Subsequently, a 10-layer InAs quantum dot buffer, separated by 500 Å GaAs barrier layers, was grown at ~515°C to serve as the dislocation filters. The bending of a 60° dislocation induced by the strain field of a quantum dot layer is schematically shown in Fig. 9.15. The laser active region consists of three coupled $In_{0.5}Ga_{0.5}As$ quantum dot layers. The p- and n-cladding layers consist of 1.0 μm $Al_{0.7}Ga_{0.3}As$ layers, doped with Be

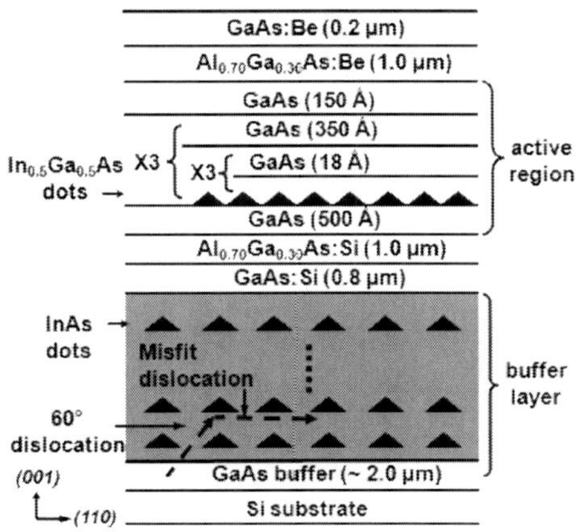

Figure 9.15. Schematic of the self-organized $In_{0.5}Ga_{0.5}As$ quantum dot laser heterostructure grown on Si with the incorporation of InAs quantum dots as dislocation filters. The bending of a 60° dislocation induced by the strain field of a dot layer is illustrated. Ref. [1]. Reprinted with permission from IEEE.

and Si, respectively. The quantum dot active region was grown at ~510°C, and the rest of the laser structure was grown at ~620°C. p-Doped quantum dot lasers were also grown on Si, with each quantum dot layer modulation doped using Be, averaging ~20 holes per dot. The room temperature photoluminescence emission of the metamorphic quantum dot laser heterostructures on Si is shown in Fig. 9.4b, which exhibits a strong peak intensity and relatively narrow (~50 meV) spectral FWHM, comparable to that from quantum dot heterostructures grown on GaAs substrates. The emission peak at ~1.3 μm corresponds to the luminescence emission from the quantum dot dislocation filter region, indicating that the InAs quantum dot buffer layer, or part of it, exhibits reasonably high optical quality.

Light-current measurements were performed on broad area (600 × 80 μm^2) lasers under pulsed bias conditions (10 KHz, 1% duty cycle) at room temperature. As illustrated in Fig. 9.16a, the device exhibits a relatively low (J_{th} ~ 900 A/cm^2) threshold current. The output spectrum measured at a drive current of 1.3 × J_{th} is shown in the inset. It is also important to note that the measured threshold current is much lower than that of similar

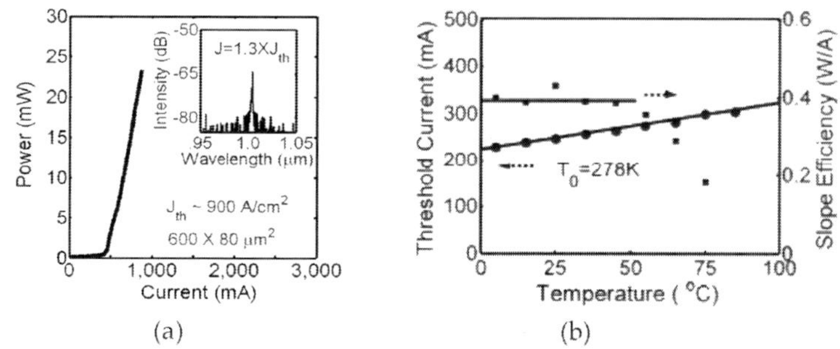

Figure 9.16. (a) Threshold and (b) slope efficiency versus temperature under pulsed mode (1% duty cycle of 100 μs) for a p-doped $In_{0.5}Ga_{0.5}As$ quantum dot laser on Si. Ref. [21]. Reprinted with permission from IEEE.

quantum dot lasers grown directly on Si without the incorporation of InAs quantum dot buffer layers, which generally exhibit threshold currents of $\geq$1500 A/cm^2. Relatively high output power (~150 mW) has also been measured from metamorphic InAs quantum dot lasers on Si. Additionally, a small signal modulation bandwidth of ~5.5 GHz was measured in near single-mode ridge waveguide lasers, which is comparable to that of conventional InGaAs quantum dot lasers grown GaAs substrates.

Performance characteristics of p-doped quantum dot lasers have also been studied. Shown in Fig. 9.16b is the variation of threshold current and output slope efficiency with temperature for p-doped quantum dot lasers on Si. We derived a very large value of T_0 (278 K) in the temperature range of 5–85°C, compared to a value of ~100 K for the undoped devices [106]. Additionally, it is observed that the output slope efficiency is nearly independent of temperature in the range of 5–50°C. The very high temperature stability is attributed to the improved gain and reduced carrier leakage associated with p-doping. Such performance characteristics are highly desired for the on-chip integration of quantum dot lasers with other optoelectronic and electronic components on a Si platform.

Self-organized quantum dot lasers on Si, with emission wavelengths in the range of 1.3–1.55 μm have also been investigated [95]. However, preliminary studies showed that these heterostructures exhibited a lasing wavelength at ~1.1 μm, which corresponds to the excited states of the dots. To achieve lasing at the ground state, it is essential to minimize the cavity loss by reducing dislocations and surface roughness and by optimizing the facet reflectivity.

9.6 Summary and Future Prospects

In summary, self-organized In(Ga)As metamorphic quantum dot lasers on GaAs and Si, with emission wavelengths in the range of ~1.0–2.0 μm have been developed. Excellent performance characteristics, including ultralow threshold current ($J_{th} \approx 60$ A/cm^2), extremely high temperature stability ($T_0 \sim 500$–600 K), and high output power (~7 W) have been measured for 1.5 μm metamorphic quantum dot lasers on GaAs. Self-organized In(Ga)As quantum dot lasers grown on Si also exhibit reasonable performance, including relatively low threshold (~900 A/cm^2) and high temperature stability ($T_0 = 278$ K). The achievement of such high-performance metamorphic optoelectronic devices has been attributed to the advancement of epitaxial growth, including the use of low temperature growth, *in situ* thermal annealing, and multiple InAs quantum dot layers as effective dislocation filters, and to the incorporation of the special techniques of tunnel injection and p-doping in the laser active region.

Recently, the monolithic integration of quantum dot lasers, waveguides, and modulators on a Si platform has been demonstrated [95, 107–108]. Future developments include the realization of 1.3–1.55 μm metamorphic quantum dot lasers on Si and further minimization of the laser threshold and improvement of the device reliability. The achievement of high-quality metamorphic quantum dot heterostructures will also find important applications in a range of optoelectronic and electronic devices, including low cost, high-efficiency multi-junction solar cells, mid-infrared photonics, and nanoscale III-V transistors on Si. More recently, metamorphic heterostructures have been employed for achieving single quantum dot emission at ~1.55 μm on GaAs substrates [109], providing a highly promising approach for realizing single photon emitters at the telecommunication wavelengths.

References

1. Mi, Z., and Bhattacharya, P. (2008). Pseudomorphic and metamorphic quantum dot heterostructures for long-wavelength lasers on GaAs and Si (Invited paper), *IEEE J. Sel. Top. Quantum Electron.*, **14**(4), pp. 1171–1179.
2. Asada, M., Miyamoto, Y., and Suematsu, Y. (1986). Gain and threshold of three-dimensional quantum box-lasers, *IEEE J. Quantum Electron.*, **22**(9), pp. 1915–1921.
3. Arakawa, Y., and Sakaki, H. (1982). Multidimensional quantum well laser and temperature dependence of its threshold current, *Appl. Phys. Lett.*, **40**(11), pp. 939–941.
4. Leonard, D., Krishnamurthy, M., Reaves, C. M., *et al.* (1993). Direct formation of quantum-sized dots from uniform coherent islands of InGaAs on GaAs surfaces, *Appl. Phys. Lett.*, **63**(23), pp. 3203–3205.
5. Kirstaedter, N., Ledentsov, N. N., Grundmann, M., *et al.* (1994). Low threshold, large T_0 injection laser emission from (InGa)As quantum dots, *Electron. Lett.*, **30**(17), pp. 1416–1417.
6. Kamath, K., Bhattacharya, P., Sonowski, T., *et al.* (1996). Room-temperature operation of $In_{0.4}Ga_{0.6}As$/GaAs self-organized quantum dot lasers, *Electron. Lett.*, **32**(15), pp. 1374–1375.
7. Mirin, R., Gossard, A. and Bowers, J. (1996). Room temperature lasing from InGaAs quantum dots, *Electron. Lett.*, **32**(18), pp. 1732–1734.
8. Shoji, H., Nakata, Y., Mukai, K., *et al.* (1996). Room temperature CW operation at the ground state of self-formed quantum dot lasers with multi-stacked dot layer, *Electron. Lett.*, **32**(21), pp. 2023–2024.
9. Huffaker, D. L., Park, G., Zou, Z., *et al.* (1998). 1.3 μm room-temperature GaAs-based quantum-dot laser, *Appl. Phys. Lett.*, **73**(18), pp. 2564–2566.
10. Fafard, S., Hinzer, K., Raymond, S., *et al.* (1996). Red-emitting semiconductor quantum dot lasers. *Science*, **274**(5291), pp. 1350–1353.
11. Eliseev, P. G., Li, H., Liu, G. T., *et al.* (2001). Ground-state emission and gain in ultralow-threshold InAs-InGaAs quantum-dot lasers, *IEEE J. Sel. Top. Quantum Electron.*, **7**(2), pp. 135–142.
12. Bimberg, D., Kirstaedter, N., Ledentsov, N. N., *et al.* (1997). InGaAs-GaAs quantum-dot lasers. *IEEE J. Sel. Top. Quantum Electron.*, **3**(2), pp. 196–205.

13. Bhattacharya, P., Ghosh S., Pradhan, S., *et al.* (2003). Carrier dynamics and high-speed modulation properties of tunnel injection InGaAs–GaAs quantum-dot lasers, *IEEE J. Quantum Electron.*, **39**(8), pp. 952–962.
14. Shchekin, O. B., and Deppe, D. G. (2002). 1.3 μm InAs quantum dot laser with $T_o = 161$ K from 0 to 80°C. *Appl. Phys. Lett.*, **80**(18), pp. 3277–3279.
15. Miyamoto, Y., Miyake, Y., Asada, M., *et al.* (1989). Threshold current density of GaInAsP/InP quantum-box lasers, *IEEE J. Quantum Electron.*, **25**(9), pp. 2001–2006.
16. Shchekin, O. B., Ahn, J., and Deppe, D. G. (2002). High temperature performance of self-organised quantum dot laser with stacked p-doped active region, *Electron. Lett.* **38**(14), pp. 712–713.
17. Fathpour, S., Mi, Z., and Bhattacharya, P. (2005). High-speed quantum dot lasers, *J. Phys. D: Appl. Phys.*, **38**(13), pp. 2103–2111.
18. Mi, Z., Yang, J., and Bhattacharya, P. (2007). Molecular beam epitaxial growth and characteristics of ultra-low threshold 1.45 μm metamorphic InAs quantum dot lasers on GaAs, *J. Cryst. Growth*, **301–302**, pp. 923–926.
19. Mi, Z., Bhattacharya, P., and Yang, J. (2006). Growth and characteristics of ultralow threshold 1.45 μm metamorphic InAs tunnel injection quantum dot lasers on GaAs, *Appl. Phys, Lett.*, **89**(15), art. no. 153109.
20. Mi, Z., Wu, C., Yang, J., *et al.* (2008), Molecular beam epitaxial growth and characteristics of 1.52 μm metamorphic InAs quantum dot lasers on GaAs., *J. Vacuum Sci. Technol.*, **26**(3), pp. 1153–1156.
21. Yang, J., Bhattacharya, P., and Mi, Z. (2007). High-performance $In_{0.5}Ga_{0.5}As$/GaAs quantum-dot lasers on silicon with multiple-layer quantum-dot dislocation filters. *IEEE Trans. Electron. Dev.*, **54**(11), pp. 2849–2855.
22. Madhukar, A., Xie, Q., Chen, P., *et al.* (1994). Nature of Strained InAs three-dimensional island formation and distribution on GaAs(100), *Appl. Phys. Lett.*, **64**(20), pp. 2727–2729.
23. Siverns, P. D., Malik, S., McPherson, G., *et al.* (1998). Scanning transmission-electron microscopy study of InAs/GaAs quantum dots, *Phys. Rev. B*, **58**(16), pp. R10127–R10130.
24. Freisem, S., Ozgur, G., Shavritranuruk, K., *et al.* (2008). Very-low-threshold current density continuous-wave quantum-dot laser diode. *Electron. Lett.*, **44**(11), pp. 679–681.

25. Fathpour, S., Mi, Z., Bhattacharya, P., *et al.* (2004). The role of Auger recombination in the temperature-dependent output characteristics ($T_0 = \infty$) of p-doped 1.3 μm quantum dot lasers, *Appl. Phys. Lett.*, **85**(22), pp. 5164–5166.
26. Marko, I. P., Mass, N. F., Sweeney, S. J., *et al.* (2005). Carrier transport and recombination in p-doped and intrinsic 1.3 μm InAs/GaAs quantum-dot lasers. *Appl. Phys. Lett.*, **87**(21), art. no. 211114.
27. Novikov, I. I., Gordeev, N. Y., Karachinskii, L. Y., *et al.* (2005). Effect of p-doping of the active region on the temperature stability of InAs/GaAs QD lasers, *Semiconductors*, **39**(4), pp. 477–480.
28. Zhukov, A. E., Kovsh, A. R., Mikhrin, S. S., *et al.* (1999). 3.9W CW power from sub-monolayer quantum dot diode laser, *Electron. Lett.*, **35**(21), pp. 1845–1847.
29. Kondratko, P. K., Chuang S. L., Walter, G., *et al.* (2003). Observations of near-zero linewidth enhancement factor in a quantum-well coupled quantum-dot laser, *Appl. Phys. Lett.*, **83**(23), pp. 4818–4820.
30. Mi, Z., Bhattacharya, P., and Fathpour S. (2005). High-speed 1.3 μm tunnel injection quantum-dot lasers. *Appl. Phys. Lett.*, **86**(15), art. no. 153109.
31. Xin, Y. C., Vaughn, L. G., Dawson, L. R., *et al.* (2003). InAs quantum-dot GaAs-based lasers grown on AlGaAsSb metamorphic buffers, *J. Appl. Phys.*, **94**(3), pp. 2133–2135.
32. Ledentsov, N. N., Shchukin, V. A., Kettler, T., *et al.* (2005). MBE-grown metamorphic lasers for applications at telecom wavelengths, *J. Cryst. Growth*, **301–302**, pp. 914–922.
33. Karachinsky, L. Y., Kettler, T., Gordeev, N. Y., *et al.* (2005). Continuous-wave lasing of single-mode metamorphic quantum dot lasers for the 1.5-μm spectral region. *Semiconductor*, **39**(12), pp. 1415–1419.
34. Temkin, H., Coblentz, D., Logan, R. A., *et al.* (1993). High temperature characteristics of InGaAsP/InP laser structures, *Appl. Phys. Lett.*, **62**(19), pp. 2402–2404.
35. Thijs, P. J. A., Van Dongen, T., Tiemeijer, L. F., *et al.* (1994). High-performance $\lambda = 1.3$ μm InGaAsP-InP strained-layer quantum well lasers, *J. Lightw. Technol.*, **12**(1), pp. 28–37.
36. Bank, S. R., Goddard, L. L., Wistey, M. A., *et al.* (2005). On the temperature sensitivity of 1.5-μm GaInNAsSb lasers, *IEEE J. Sel. Top. Quantum Electron.* **11**(5), pp. 1089–1098.
37. Walters, R. J., Bourianoff, G. I., and Atwater, H. A. (2005). Field-effect electroluminescence in silicon nanocrystals, *Nature Mater.*, **4**(2), pp. 143–146.

38. Rong, H. S., Jones, R., Liu, A. S., *et al.* (2005). A continuous-wave Raman silicon laser, *Nature*, **433**(7027), pp. 725–728.

39. Groenert, M. E., Leitz, C. W., Pitera, A. J., *et al.* (2003). Monolithic integration of room-temperature cw GaAs/AlGaAs lasers on Si substrates via relaxed graded GeSi buffer layers, *J. Appl. Phys.,* **93**(1), pp. 362–367.

40. Fang, A. W., Park, H., Cohen, O., *et al.* (2006). Electrically pumped hybrid AlGaInAs-silicon evanescent laser, *Opt. Express,* **14**(20), pp. 9203–9210.

41. Mi, Z., Bhattacharya, P., Yang, J., *et al.* (2005). Room-temperature self-organised $In_{0.5}Ga_{0.5}As$ quantum dot laser on silicon, *Electron. Lett.,* **41**(13), pp. 742–744.

42. Pei, C. W., Turk, B., Wang, W. I., *et al.* (2001). Mechanism of the reduction of dislocation density in epilayers grown on compliant substrates, *J. Appl. Phys,* **90**(12), pp. 5959–5962.

43. Tatebayashi, J., Jallipalli, A., Kutty, M. N., *et al.* (2009). Monolithically integrated III-Sb-based laser diodes grown on miscut Si substrates, *IEEE J. Sel. Top. Quantum Electron.,* **15**(3), pp. 716–723.

44. El-Masry, N., Tarn J. C. L., Humphreys, T. P., *et al.* (1987). Effectiveness of strained-layer superlattices in reducing defects in GaAs epilayers grown on silicon substrates, *Appl. Phys. Lett.,* **51**(20), pp. 1608–1610.

45. Balakrishnan, G., Huang, S., Dawson, L. R., *et al.* (2005). Growth mechanisms of highly mismatched AlSb on a Si substrate, *Appl. Phys. Lett.,* **86**(3), art. no. 034105.

46. Currie, M. T., Samavedam, S. B., Langdo, T. A., *et al.* (1998). Controlling threading dislocation densities in Ge on Si using graded SiGe layers and chemical-mechanical polishing, *Appl. Phys. Lett.,* **72**(14), pp. 1718–1720.

47. Fischer, R., Morkoc, H., Neuman, D. A., *et al.* (1986). Material properties of high-quality GaAs epitaxial layers grown on Si Substrates, *J. Appl. Phys.,* **60**(5), pp. 1640–1647.

48. Mi, Z., Yang, J., Bhattacharya, P., *et al.* (2006). Self-organised quantum dots as dislocation filters: the case of GaAs-based lasers on silicon, *Electron. Lett.,* **42**(2), pp. 121–123.

49. Linder, K. K., Phillips, J., Qasaimeh, O., *et al.* (1999). Self-organized $In_{0.4}Ga_{0.6}As$ quantum-dot lasers grown on Si substrates, *Appl. Phys. Lett.,* **74**(10), pp. 1355–1357.

50. Fischer, R., Kopp, W., Morkoc, H., *et al.* (1986). Low threshold laser operation at room-temperature in GaAs/(Al,Ga)As structures grown directly on (100) Si, *Appl. Phys. Lett.,* **48**(20), pp. 1360–1361.

51. Egawa, T., Soga, T., Jimbo T., *et al.* (1991). Room-temperature continuous-wave operation of AlGaAs-GaAs single-quantum-well lasers on Si by metalorganic chemical-vapor deposition using AlGaAs-AlGaP intermediate layers, *IEEE J. Quantum Electron.*, **27**(6), pp. 1798–1804.
52. Sakai, S., Soga, T., Takeyasu, M., *et al.* (1986). Room-temperature laser operation of AlGaAs-GaAs double heterostructures fabricated on Si substrates by metalorganic chemical vapor-deposition, *Appl. Phys. Lett.*, **48**(6), pp. 413–414.
53. Egawa, T., Hasegawa, Y., Jimbo T., *et al.* (1992). Effects of dislocation and stress on characteristics of GaAs-based laser grown on Si by metalorganic chemical vapor deposition, *Jpn. J. Appl. Phys. Part 1*, **31**(3), pp. 791–797.
54. Yamaguchi, M., Tachikawa, M., Sugo, M., *et al.* (1990). Analysis for dislocation density reduction in selective area grown GaAs films on Si substrates, *Appl. Phys. Lett.*, **56**(1), pp. 27–29.
55. Eisenbeiser, K., Finder, J. M., Yu, Z., *et al.* (2000). Field effect transistors with $SrTiO_3$ gate dielectric on Si, *Appl. Phys. Lett.*, **76**(10), pp. 1324–1326.
56. Choy, H. K. H., and Fonstad C. G. (2005). Effects of substrate temperature on the growth of InGaAs compositionally graded buffers and on quantum well structures grown above them, *J. Vac. Sci. Technol. B*, **23**(5), pp. 2109–2113.
57. Molina, S. I., Pacheco, F. J., Araujo, D., *et al.* (1994). Strain relief in linearly graded composition buffer layers — a design scheme to grow dislocation-free ($<10^5$ cm^{-2}) and unstrained epilayers, *Appl. Phys. Lett.*, **65**(19), pp. 2460–2462.
58. Bertoli, B., Suarez, E. N., Jain, F. C., *et al.* (2009). Dislocations and strain relief in reverse-graded semiconductor layers, *Semiconductor Sci. Technol.*, **24**(12), art. no. 125006.
59. Kidd, P., Dunstan, D. J., Colson, H. G., *et al.* (1996). Comparison of the crystalline quality of step-graded and continuously graded InGaAs buffer layers, *J. Cryst. Growth*, **169**(4), pp. 649–659.
60. Jeong, Y., Choi, H., and Suzuki, T. (2007). Invalidity of graded buffers for InAs grown on GaAs (001) — a comparison between direct and graded-buffer growth, *J. Cryst. Growth*, **301**, pp. 235–239.
61. Song, Y., Wang S., Tangring, I., *et al.* (2009). Effects of doping and grading slope on surface and structure of metamorphic InGaAs buffers on GaAs substrates, *J. Appl. Phys.*, **106**(12), art. no. 123531.

62. Bertoli, B., Suarez, E. N., Ayers, J. E., *et al.* (2009). Misfit dislocation density and strain relaxation in graded semiconductor heterostructures with arbitrary composition profiles, *J. Appl. Phys.*, **106**(7), art. no. 073519.

63. Behet, M., Van Der Zanden, K., Borghs, G., *et al.* (1998). Metamorphic InGaAs/InAlAs quantum well structures grown on GaAs substrates for high electron mobility transistor applications, *Appl. Phys. Lett.*, **73**(19), pp. 2760–2762.

64. Zheng, H., Radhakrishnan, K., Wang, H., *et al.* (2000). Metamorphic InP/InGaAs double-heterojunction bipolar transistors on GaAs grown by molecular-beam epitaxy, *Appl. Phys. Lett.*, **77**(6), pp. 869–871.

65. Arai, M., Nakashima, K., Fujisawa, T., *et al.* (2009). High-temperature operation of 1.26-μm ridge waveguide laser with InGaAs metamorphic buffer on GaAs substrate, *IEEE J. Sel. Top. Quantum Electron.*, **15**(3), pp. 724–730.

66. Tangring, I., Wang, S., Sadeghi, M., *et al.* (2007). Metamorphic growth of 1.25–1.29 μm InGaAs quantum well lasers on GaAs by molecular beam epitaxy, *J. Cryst. Growth*, **301–302**, pp. 971–974.

67. Tangring, I., Ni, H., Wu, B., *et al.* (2007). 1.58 μm InGaAs quantum well laser on GaAs, *Appl. Phys. Lett.*, **91**(22), art. no. 221101.

68. Wu, D., Wang, H., Wu, B., *et al.* (2008). Low threshold current density 1.3 μm metamorphic InGaAs/GaAs quantum well laser diodes, *Electron. Lett.*, **44**(7), pp. 474–475.

69. Hoke, W. E. (1999). Molecular beam epitaxial growth and device performance of metamorphic high electron mobility transistor structures fabricated on GaAs substrates, *J. Vac. Sci. Technol., B*, **17**(3), pp. 1131–1135.

70. Zhukov, A. E., Kovsh, A. R., Mikhrin, S. S., *et al.* (2003). Metamorphic lasers for 1.3-μm spectral range grown on GaAs substrates by MBE, *Semiconductors*, **37**(9), pp. 1119–1122.

71. Baskar, K., Soga, T., Jimbo, T., *et al.* (1996). Growth of high quality $Al_{0.22}Ga_{0.78}As$ layers on Si substrates by metalorganic chemical vapor deposition, *J. Appl. Phys.*, **80**(7), pp. 4112–4115.

72. Tangring, I., Song, Y., Lai, Z., *et al.* (2009). A study of the doping influence on strain relaxation of graded composition InGaAs layers grown by molecular beam epitaxy, *J. Cryst. Growth*, **311**(7), pp. 1684–1687.

73. Fischer, R., Neuman, D., Zabel, H., *et al.* (1986). Dislocation reduction in epitaxial GaAs on Si (100), *Appl. Phys. Lett.*, **48**(18), pp. 1223–1225.

74. Matthews, J. W., and Blakeslee, A. E. (1974). Defects in epitaxial multilayers: I. Misfit dislocations, *J. Cryst. Growth*, **27**(1), pp. 118–125.

75. Huang, D., Reshchikov, M. A., Yun, F., *et al.* (2002). Defect reduction with quantum dots in GaN grown on sapphire substrates by molecular beam epitaxy, *Appl. Phys. Lett.*, **80**(2), pp. 216–218.

76. Berger, P. R., Chang, K., Bhattacharya, P., *et al.* (1988). Role of strain and growth-conditions on the growth front profile of $In_xGa_{1x}As$ on GaAs during the pseudomorphic growth regime, *Appl. Phys. Lett.*, **53**(8), pp. 684–686.

77. Tillmann, K., and Forster, A. (2000). Critical dimensions for the formation of interfacial misfit dislocations of $In_{0.6}Ga_{0.4}As$ islands on GaAs(001). *Thin Solid Films*, **368**(1), pp. 93–104.

78. Ovid'ko, I. A. (2002). Relaxation mechanisms in strained nanoislands, *Phys. Rev. Lett.*, **88**(4), 046103.

79. Tsao, J. Y., and Dodson, B. W. (1988). Excess stress and the stability of strained heterostructures, *Appl. Phys. Lett.*, **53**(10), pp. 848–850.

80. Ustinov, V. M., Maleev, N. A., Zhukov, A. E., *et al.* (1999). InAs/InGaAs quantum dot structures on GaAs substrates emitting at 1.3 μm, *Appl. Phys. Lett.*, **74**(19), pp. 2815–2817.

81. Ledentsov, N. N., Kovsh, A. R., Zhukov, A. E., *et al.* (2003). High performance quantum dot lasers on GaAs substrates operating in 1.5 μm range, *Electron. Lett.*, **39**(15), pp. 1126–1128.

82. Le Ru, E., Howe, P., Jones T. S., *et al.* (2003). Strain-engineered InAs/GaAs quantum dots for long-wavelength emission, *Phys. Rev. B*, **67**(16), art. no. 165303.

83. Liu, H. Y., Sellers, I. R., Badcock, T. J., *et al.* (2004). Improved performance of 1.3 μm multilayer InAs quantum-dot lasers using a high-growth-temperature GaAs spacer layer, *Appl. Phys. Lett.*, **85**(5), pp. 704–706.

84. Sizov, D., Maximov, M. V., Tsatsulnikov, A. F., *et al.* (2002). The influence of heat treatment conditions on the evaporation of defect regions in structures with InGaAs quantum dots in the GaAs matrix, *Semiconductors*, **36**(9), pp. 1020–1026.

85. Balakrishnan, G., Huang, S., Rotter, T. J., *et al.* (2004). 2.0 μm wavelength InAs quantum dashes grown on a GaAs substrate using a metamorphic buffer layer, *Appl. Phys. Lett.*, **84**(12), pp. 2058–2060.

86. Seravalli, L., Trevisi, G., Frigeri, P., *et al.* (2009). The role of wetting layer states on the emission efficiency of InAs/InGaAs metamorphic quantum dot nanostructures, *Nanotechnology*, **20**(27), art. no. 275703.

87. Seravalli, L., Frigeri, P., Trevisi, G., *et al.* (2008). 1.59 μm room temperature emission from metamorphic InAs/InGaAs quantum dots grown on GaAs substrates. *Appl. Phys. Lett.*, **92**(21), art. no. 213104.

88. Zhang, Z. Y., Xu, B., Jin, P., *et al.* (2002). Photoluminescence study of self-assembled InAs/GaAs quantum dots covered by an InAlAs and InGaAs combination layer, *J. Appl. Phys.*, **92**(1), pp. 511–514.

89. Liu, H. Y., and Hopkinson, M. (2003). Tuning the structural and optical properties of 1.3-μm InAs/GaAs quantum dots by a combined InAlAs and GaAs strained buffer layer, *Appl. Phys. Lett.*, **82**(21), pp. 3644–3646.

90. Ueda, O., Soga, T., Jimbo T., *et al.* (1989). Direct evidence for self-annihilation of antiphase domains in GaAs/Si heterostructures, *Appl. Phys. Lett.*, **55**(5), pp. 445–447.

91. Choi, B. H., Park, C. M., Song, S. H., *et al.* (2001). Selective growth of InAs self-assembled quantum dots on nanopatterned SiO_2/Si substrate, *Appl. Phys. Lett.*, **78**(10), pp. 1403–1405.

92. Kazi, Z. I., Egawa, T., Umeno, M., *et al.* (2001). Growth of $In_xGa_{1-x}As$ quantum dots by metal-organic chemical vapor deposition on Si substrates and in GaAs-based lasers, *J. Appl. Phys.*, **90**(11), pp. 5463–5468.

93. Li, L., Guimard, D., Rajesh, M., *et al.* (2008). Growth of InAs/Sb: GaAs quantum dots on silicon substrate with high density and efficient light emission in the 1.3 μm band, *Appl. Phys. Lett.*, **92**(26), art no. 263105.

94. Tanoto, H., Yoon, S. F., Ngo, C. Y., *et al.* (2008). Structural and optical properties of stacked self-assembled InAs/InGaAs quantum dots on graded $Si_{1-x}Ge_x$/Si substrate. *Appl. Phys. Lett.*, **92**(21), art. no. 213115.

95. Yang, J., Bhattacharya, P., Mi, Z., *et al.* (2008). Quantum dot lasers and integrated optoelectronics on silicon platform, *Chin. Opt. Lett.*, **6**(10), pp. 727–731.

96. Bhattacharya, P., Singh, J., Yoon, H., *et al.* (1996). Tunneling injection lasers: a new class of lasers with reduced hot carrier effects, *IEEE J. Quantum Electron.*, **32**(9), pp. 1620–1629.

97. Marko, I. P., Sweeney, S. J., Adams, A. R., *et al.* (2004). Recombination mechanisms in InAs/InP quantum dash lasers studied using high hydrostatic pressure, *Phys. Status Solidi B*, **241**(14), pp. 3427–3431.

98. Mi, Z., and Bhattacharya, P. (2006). DC and dynamic characteristics of p-doped and tunnel injection 1.65-mu m InAs quantum-dash lasers grown on InP (001), *IEEE J. Quantum Electron.*, **42**(11–12), pp. 1224–1232.

99. Caroff, P., Paranthoen, C., Platz, C., *et al.* (2005). High-gain and low-threshold InAs quantum-dot lasers on InP, *Appl. Phys. Lett.*, **87**(24), art. no. 243107.
100. Melnik, S., and Huyet, G. (2006). The linewidth enhancement factor alpha of quantum dot semiconductor lasers, *Opt. Exress*, **14**(7), pp. 2950–2955.
101. Mi, Z., and Bhattacharya, P. (2007). Analysis of the linewidth-enhancement factor of long-wavelength tunnel-injection quantum-dot lasers, *IEEE J. Quantum Electron.*, **43**(5–6), pp. 363–369.
102. Ghosh, S., Bhattacharya, P., Stoner, E., *et al.* (2001). Temperature-dependent measurement of Auger recombination in self-organized $In_{0.4}Ga_{0.6}As$/GaAs quantum dots, *Appl. Phys. Lett.*, **79**(6), pp. 722–724.
103. Sandall, I. C., Smowton, P. M., Thomson, J. D., *et al.* (2006). Temperature dependence of threshold current in p-doped quantum dot lasers, *Appl. Phys. Lett.*, **89**(15), art. no.151118.
104. Novikov, I. I., Gordeev, N. Y., Maximov, M. V., *et al.* (2005). Ultrahigh gain and non-radiative recombination channels in 1.5 μm range metamorphic InAs-InGaAs quantum dot lasers on GaAs substrates, *Semiconduct. Sci. Technol.*, **20**(1), pp. 33–37.
105. Kettler, T., Karachinsky, L. Y., Ledentsov, N. N., *et al.* (2006). Degradation-robust single mode continuous wave operation of 1.46 μm metamorphic quantum dot lasers on GaAs substrate, *Appl. Phys. Lett.*, **89**(4), art. no. 041113.
106. Mi, Z., Yang, J., Bhattacharya, P., *et al.* (2006). High performance self-organized InGaAs quantum dot lasers on silicon, *J. Vac. Sci. Technol. B*, **24**(3), pp. 1519–1522.
107. Mi, Z., Yang, J., Bhattacharya, P., *et al.* (2009). High-performance quantum dot lasers and integrated optoelectronics on Si, *Proc. IEEE*, **97**(7), pp. 1239–1249.
108. Yang, J., Mi, Z., and Bhattacharya, P. (2007). Groove-coupled In-GaAs/GaAs quantum dot laser/waveguide on silicon, *J. Lightw. Technol.*, **25**(7), pp. 1826–1831.
109. Semenova, E. S., Hostein, R., Patriarche, G., *et al.* (2008). Metamorphic approach to single quantum dot emission at 1.55 μm on GaAs substrate, *J. Appl. Phys.*, **103**(10), art. no. 103533.

Chapter 10

Nitride-Based Light-Emitting Diodes and Nitride-Based Laser Diodes by Plasma-Assisted Molecular Beam Epitaxy

Czeslaw Skierbiszewski
Institute of High Pressure Physics, Polish Academy of Sciences, Warsaw, 01-142, Poland
czeslaw@unipress.waw.pl

We present recent progress in the growth of nitride-based laser diodes (LDs) and light-emitting diodes (LEDs) made by plasma-assisted MBE (PAMBE). This technology is ammonia free, and nitrogen for the growth is activated by RF plasma source from nitrogen molecules. The recent demonstration of continuous wave (CW) blue InGaN LDs has opened a new perspective for PAMBE in optoelectronics. The LDs were fabricated on low threading dislocation density bulk GaN substrates at low growth temperatures 550–750°C. In this work, we describe the nitride growth fundamentals, the peculiarities of InGaN growth as well as properties of LEDs and LDs made by PAMBE.

Lattice Engineering: Technology and Applications
Edited by Shumin Wang

ISBN 978-981-4316-29-3 (Hardcover), 978-981-4364-25-6 (eBook)
www.panstanford.com

10.1 Introduction

The potential associated with nitride-based LEDs and LDs for solid-state lighting has generated a continuing interest in group III-N materials and their alloys. Until very recently, the key achievements and developments in the field of InGaN LDs have been made by the metal-organic vapour-phase epitaxy (MOVPE) technique [1, 2]. In spite of many potential advantages for MBE growth, such as *in situ* monitoring techniques [3–5], the low quantum efficiency of MBE-grown optoelectronic structures compared to MOVPE led many researchers to conclude that MBE cannot compete with MOVPE. It was also commonly believed that the only solution for MBE would be to bring the growth conditions as close as possible to the MOVPE; i.e. to apply high growth temperatures ($\sim$1050°C for GaN) and high nitrogen precursor overpressure. Indeed, ammonia MBE, which uses atomic beams for group III elements coupled with a large excess supply of NH_3 as the nitrogen precursor [3, 4], has been successful in improving the optical quality of nitride films and recently produced the first room temperature CW operation of a 405 nm LDs [6]. However, the corrosive nature of ammonia compounded by its large flows creates additional hazards and technological challenges, as well as leads to undesirable high hydrogen background during the epitaxial process. In the more widely employed PAMBE [5], purified nitrogen gas is activated using an RF-plasma and supplied to the growth surface at typical flow rates of 1–2 sccm. Early experimental results [7] showed that unlike ammonia MBE, PAMBE requires group III–rich conditions to achieve good material quality. Much progress has been made in both theoretical [8, 9] and experimental [10–13] understanding of the growth kinetics for such metal rich conditions. In spite of the relatively low growth temperatures, 550–750°C, state of the art GaN/AlGaN heterostructures with record high mobilities of two-dimensional electron gas [14–16] have been grown with PAMBE, making it the technique of choice for the production of electronic devices. Although sustained refinement of the PAMBE growth conditions on GaN/sapphire MOVPE templates allowed the demonstration of promising LED devices [17], it was the introduction of high quality GaN substrates which led to truly dramatic improvements in the optical quality of PAMBE grown structures, resulting in the room temperature high-power pulsed and CW blue-violet LDs [18, 19]. This chapter reports on the progress made by PAMBE in the area of optoelectronic devices, which demonstrates the potential of this technology.

10.2 Experimental

The growth of all nitride structures presented in this paper was performed in a customized VG90 Oxford MBE reactor equipped with two Veeco RF plasma sources, operating at 240–450 W for 0.8–2 sccm of N_2 flow. The pressure during growth was 1.5×10^{-5} to 7 Torr. The substrates used were either high pressure grown bulk GaN [20] or bulk GaN made by hydrate vapour phase epitaxy (HVPE). The epi-ready bulk substrates were prepared initially in a three-step process of mechanical polishing, dry etching, and deposition of a 2 μm GaN:Si buffer layer in the MOVPE reactor. Recently, two-step procedure is employed, which consists of mechanical polishing and mechano-chemical polishing. For growth, samples with different shapes were attached by indium to two inch GaN/Al_2O_3 templates. The back surfaces of the two inch GaN/Al_2O_3 templates were coated with a 0.7 μm molybdenum layer to improve the thermal coupling for radiative heating. The typical size of GaN bulk high pressure substrates was 4 mm × 5 mm, while bulk HVPE substrates were 10 mm × 10 mm. Here, we mainly discuss the growth of GaN, InGaN, and InAlGaN on the (0001) Ga polarity side except the section related to the growth of GaN, where growth on m-plane was demonstrated.

10.3 Growth of GaN

The analysis of experimental data for GaAs or Si crystals indicates that the optimum growth temperature, T_G, for epitaxy (step flow growth mode) of these materials equals about half of the melting temperature, T_M [21, 22]. This rule is also valid for GaN growth, where the GaN melting point lies somewhere between 2540 K (experimentally determined for 6 GPa [23]) and 2800 K

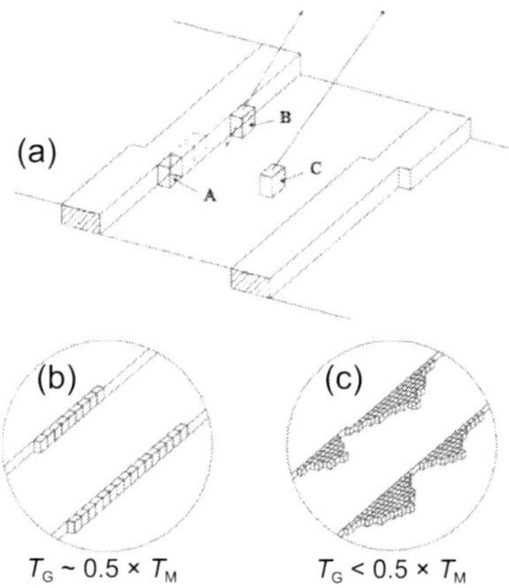

Figure 10.1. The two-dimensional step flow epitaxy model (a). Adatom attachments to the atomic kinks at growth temperature around half of the melting temperature of the crystal T_M (b) and much below (c).

(theoretically calculated [24]) and the optimum growth temperature used in MOVPE is in the range 1050–1100°C (1320–1370K). On a microscopic scale, at $T = 0.5 \times T_M$ atomic kinks start to become active; i.e. atoms can detach from an atomic kink [21, 22] at a rate comparable to typical attachment rates during deposition. This rule is strict for Kossel crystal and is also fulfilled for real crystal structures. During growth, atoms which reach the growth surface can be bound on the terraces (atom C on Fig. 10.1a), atomic ledges (atom B), or atomic kinks (atom A). The binding energies for the atoms on the terrace, on the ledge, and on the kink position increase as decrease the number of dangling bonds of adatoms C, B, and A. Therefore, adatom A has the highest binding energy among adatoms A, B, and C. The step-flow growth mode takes place when all atoms are finally bound on the atomic kinks (like adatom A), while those which did not reach the atomic kinks (adatoms B and C) are desorbed from the growth surface (Fig. 10.1). The mechanism of formation of flat (for $T_G \sim 0.5 \times T_M$) and rough (for $T_G < 0.5 \times T_M$) growth front is shown in Figs. 10.1b,c, respectively.

The adatoms at the kink positions have a half-filled number of dangling bonds, while atoms inside the crystal have no dangling bonds. Therefore the binding energy of the adatom at the atomic kink is equal to half the binding energy of an atom inside the crystal. The difficulty for GaN growth in MBE is that for temperature 0.5 $\times T_M = 1320$ K, *a high nitrogen overpressure (about 60 bars for* N_2*) is required to prevent GaN decomposition* [25–27]. The nitrogen overpressure required to stabilize InN is even higher. For more chemically active N precursors, like ammonia, the situation is better, but still an overpressure in the range of 0.1–2 bar must be used. Such overpressures are not compatible with MBE technology, which relies on high vacuum conditions for the delivery of atoms from the effusion cells to the growing layer. In fact, for typical MBE growth conditions, GaN begins to decompose rapidly at temperatures in excess of 800°C, restricting the epitaxy to temperatures much below the optimum point. Indeed, due to the arguments given above and the low diffusivity of N adatoms, the early efforts to grow GaN in MBE at T_G below 800°C, using group V-rich conditions typical for III-V epitaxy, gave unsatisfactory results. These early difficulties, seemingly well grounded in the simple thermodynamics of the processes involved, led many to believe that the only path for successful growth of nitrides in MBE reactors is to push the growth conditions as close as possible to those present in MOVPE reactors.

The breakthrough in the study of growth kinetics in PAMBE came with the finding that in Ga rich conditions, it is possible to grow relatively smooth layers at low growth temperatures [7, 10–16]. Further study of the Ga auto-surfactant effect revealed that high-quality growth is possible in a very narrow range of Ga/N values: low enough to be just below the formation of the droplets, but high enough to ensure the formation of a metallic Ga bilayer on the Ga polarity surface [10–14, 16]. As an example, we demonstrate in Figure 10.2 the PAMBE grown GaN surface morphology for N-rich (rough surface — three-dimensional growth) and Ga-rich (smooth surface — step-flow growth) conditions and growth temperature 710°C.

This apparent conundrum of finding step-flow growth at temperatures significantly lower than 0.5 $\times T_M$ can be resolved by considering that the decomposition of GaN (and therefore the activity of atomic kinks) is controlled by a kinetic barrier on (0001) c-plane (see Fig. 10.3). Schoonemaker *et al.* [28] demonstrated over 40 years ago that the height of this barrier is strongly reduced when

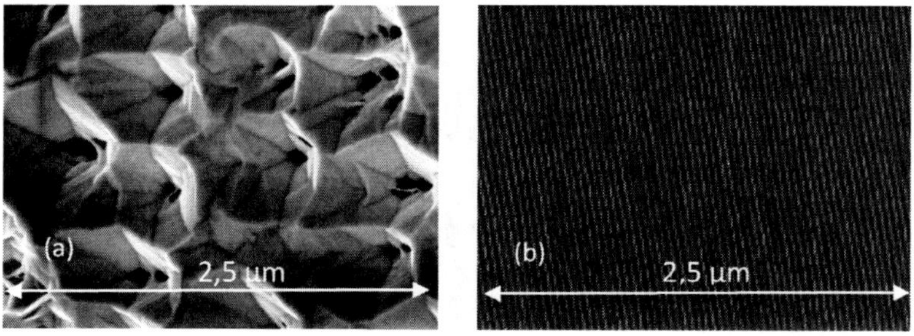

Figure 10.2. The Scanning Electron Microscope image of GaN layer grown by PAMBE in Nitrogen rich conditions — 3D growth mode (a) and Atomic Force Microscope image of GaN grown in Ga rich conditions — step-flow growth mode (b).

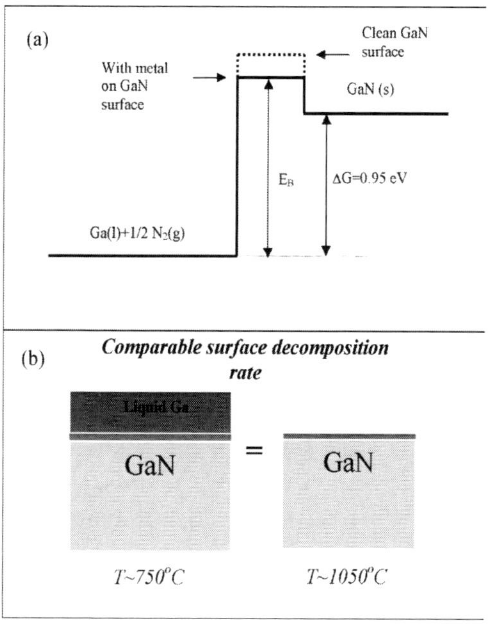

Figure 10.3. A schematic diagram showing reaction of decomposing GaN with and without catalyst (after Newman, [30]).

GaN is covered by liquid Ga (Fig. 10.3). It was also experimentally evidenced and discussed by others [29, 30] that for a given temperature, the presence of a metal layer on a GaN surface accelerates etching of GaN in comparison to the exposed surface. Therefore, in the metal-rich regime for PAMBE, the formation energies of the kink sites are reduced, substantially lowering the minimum growth temperature necessary to sustain the desirable step-flow growth mode. We can call it liquid phase epitaxy (LPE)-like conditions.

The kinetic barrier for GaN decomposition mentioned above depends on the surface properties — then it should be different for other GaN surfaces. Indeed, confirmation of this idea can be found in our recent experiments of growth of GaN on non-polar m-plane surface [31]. It is known that m-plane GaN surface is *less stable* than inert (0001) c-plane (e.g. chemical etching of m-plane is possible [32]). This was a starting point for experiments of GaN growth at N-rich conditions on m-plane surface by PAMBE. We achieved using *classical N-rich conditions* step-flow growth of GaN at temperature as low as 730°C. In Fig. 10.4, we show AFM images of m-plane GaN surface with parallel atomic steps indicating step-flow growth. For comparison'0001) c-plane substrate which has been attached to 2 inch GaN/Al_2O_3 template in the same growth run shows rough three-dimensional morphology.

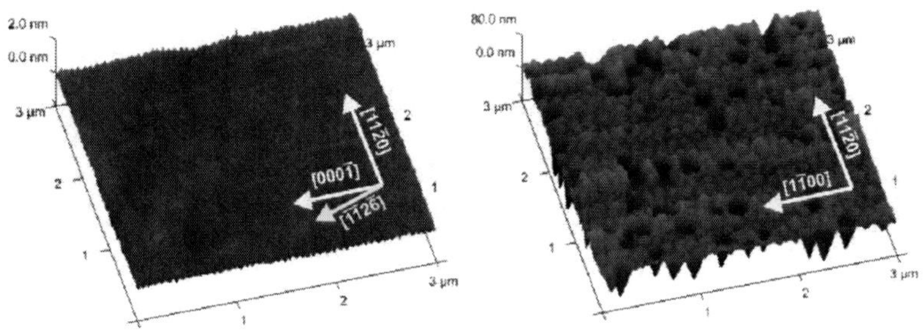

Figure 10.4. Parallel atomic steps on m-plane GaN layer grown in N-rich conditions at 730° C. Local substrate miscut is about 0.25° towards [11–26], which is the step-flow preferable direction (a). Morphology of (0001) c-plane GaN layer grown in the same growth run shows the evidence of 3D growth (b).

Another important process necessary for high-quality growth at low temperature on (0001) plane, which has also been a subject of intensive theoretical studies, is the adatom diffusivity. Perhaps the most insightful were recent results of modelling based on the density-functional theory, which pointed towards the existence of a very efficient lateral diffusion channel for nitrogen adatoms on semiconductor surface just below the thin metallic film [8]. Surprisingly small activation energies, for this so-called adlayer enhanced lateral diffusion (AELD), facilitate high quality step-flow epitaxy at low temperatures. This enhanced surface mobility of nitrogen adatoms, coupled with the reduction in the kink formation energy discussed earlier, promotes step-flow growth at temperatures far below those needed for "classical" group V-rich conditions.

The importance of the two mechanisms described above, i.e. kink activity and adatoms mobility, is well illustrated by recent results demonstrating that layer-by-layer growth of GaN is possible in PAMBE under "classical" N-rich conditions, *in the absence of a metallic layer on the surface*, provided that growth temperature is high enough (around 780°C) to cause concurrent decomposition of GaN in layer by latter process [33]. Such growth conditions provide very useful alternative for PAMBE growth of high quality GaN without the danger of Ga droplet formation.

10.4 Growth of InGaN

The growth of efficient, high In content InGaN quantum wells (QWs), required for green LEDs and LDs, continues to be a subject of intensive study [34–37]. Considerable technological hurdles encountered during the growth of high In content layers result in the rapid decrease of performance of light emitters with increasing emission wavelength. Up to now, green LDs have been demonstrated in nitride-based structures only by MOVPE [38, 39]. The melting point for InN crystal is much lower than for GaN — recent experimental data indicate that it can be around 1800°C [23]. Therefore for the InGaN growth, low temperatures are essential to avoid InGaN decomposition. As discussed earlier, PAMBE relies

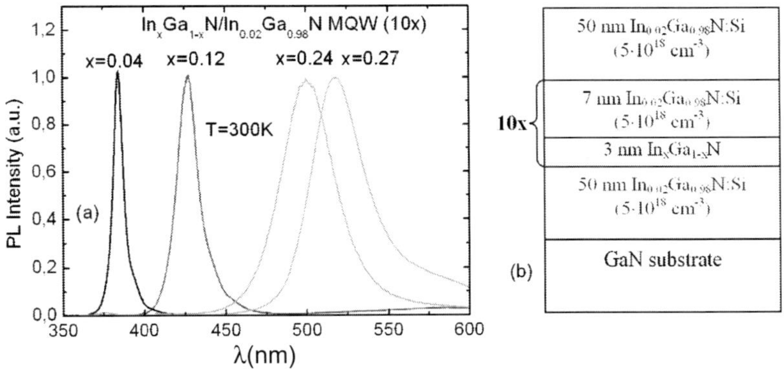

Figure 10.5. (a) Photoluminescence from InGaN/InGaN MQWs as a function of In content. (b) The MQWs have well and barrier thickness 3 and 7 nm, respectively.

on a very different growth mechanism than MOVPE, one enabling deposition of device-quality nitride structures at considerably lower temperatures (by 200–300°C compared to MOVPE). This makes PAMBE technology particularly suitable for the growth of strained, high In content structures required for green emitters.

In order to move from violet to the green region, the amount of In in the InGaN QWs should be increased from 8% to about 25%. Figure 10.5 shows the photoluminescence (PL) data from InGaN/InGaN MQWs grown by PAMBE for different In contents spanning the emission range from 380 to 525 nm.

We described already the growth conditions of atomically flat GaN layers in PAMBE. The metal-rich growth regime, similarly to growth of GaN, is beneficial also for high quality InGaN structures by PAMBE. There are only two differences: (1) the excess of indium is used (instead of Ga), in order to create thin liquid In layer on a grown surface and (2) to incorporate any amount of In into the grown InGaN layer, gallium flux should be lower than nitrogen flux. In this section, we will investigate in detail the limits of In incorporation into $In_xGa_{1-x}N$ layers during PAMBE process, namely the influence of (i) the growth temperature, T_G, (ii) the gallium flux, Φ_{Ga}, and (iii) the active nitrogen flux, Φ_N [40–46].

Lowering of T_G has been used extensively to increase the In content x of $In_xGa_{1-x}N$ layers. Indeed, at sufficiently low growth temperature and fixed Φ_N, the In content is limited only by Φ_{Ga}. This is because Ga efficiently displaces In from InN bond as a result of considerable gain in the bond energy. However, for high In content, the temperature necessary to uphold this simple relation is too low to ensure high-quality epitaxy. With increasing growth temperature, another effect becomes very important, namely thermal cracking of relatively weak In–N bonds. The rate of this process is influenced by a number of factors such as growth temperature, presence of metallic layer on the surface, surface orientation, and whether the grown InGaN layer is strained or relaxed. In general though, we expect a thermally activated rate of In loss from the layer. For fixed growth temperature, this decomposition rate will be faster with increasing effective layer composition [41, 47]. This will result in less In being incorporated during the steady state growth conditions, *decreasing the effective growth rate*, and leading to In content lower than expected.

10.4.1 *Growth Model of InGaN*

We describe the growth of $In_xGa_{1-x}N$ by a simple phenomenological model. We assume that *all Ga* is incorporated into the layer at growth temperature range we use for InGaN growth. For In rich regime, the growth rate, V_{GR}, is defined by the Gallium flux Ga, Φ_{Ga}, and effective indium flux, Φ^*_{In}, which is incorporated to $In_xGa_{1-x}N$ layer.

$$V_{GR} = \Phi_{Ga} + \Phi^*_{In} \tag{10.1}$$

where

$$\Phi^*_{In} = \Phi_N - \Phi_{Ga} - InN^{loss} \tag{10.2}$$

The term InN^{loss} in Eq. 10.2 describes the nitrogen losses due to the thermal decomposition of In–N bonds. Since *all Ga atoms* are incorporated into the $In_xGa_{1-x}N$ layer, $\Phi^*_{In}/\Phi_{Ga} = x/(1-x)$ and $V_{GR} = \Phi_{Ga}/(1-x)$. The InN^{loss} depends on growth temperature, T_G, and indium content x [41],

$$InN^{Loss} = x \cdot C_A \cdot \exp\left(-\frac{E_A}{kT_G}\right) \tag{10.3}$$

where C_A is a constant, E_A is given by $E_A = x \cdot E_{InN} + (1-x) \cdot E_{GaN}$, where E_{InN} and E_{GaN} are activation energies of InN and

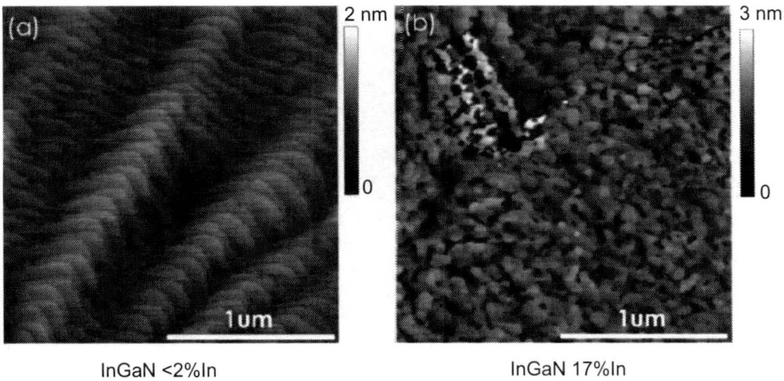

Figure 10.6. The morphology of InGaN layers on 0.2° miscut substrates (a) for 2% of In and (b) for 17% of In content, respectively.

GaN decomposition, respectively [48, 49]. The model described in Ref. [41] does not include very important microscopic factor related to the role which Ga atoms play for indium incorporation. In this work, we propose a new approach where not only thermal stability of InGaN is important but also kinetic and diffusion limited nature of the growth should be taken into account [50]. As we already mentioned, growth of nitrides is performed on GaN substrates with defined miscut angle (see Figs. 10.2b and 10.6). As an example, we show in Fig. 10.7a the GaN substrate with the miscut angle Θ, which defines width L of atomic terrace. We will demonstrate that Ga atoms, which are attached to the atomic kinks, block In-N bonds from decomposition. This process increases indium incorporation into the InGaN layer. Therefore, the *rate of Ga atoms*, υ_{Ga}, reaching atomic kink and the number of atomic kinks are essential for InN^{loss} [50]. The υ_{Ga} depends on the Ga flux, Φ_{Ga} (atoms/cm^{2}sec), and the surface fraction from which Ga adatoms can reach individual atomic kink, P_k,

$$\upsilon_{Ga} = \Phi_{Ga} \cdot P_k \tag{10.4}$$

For the case shown in Figure 10.7a, such fraction P_k is given by $L \cdot d_k$, where L is atomic terrace width, $L = c/(2\tan(\Theta))$, d_k is the mean distance between kinks on atomic edge, equal to $(a/2)\exp(E_A/12kT)$ [21], with a and c as GaN lattice constants. The kinetic process, where Ga atoms, which are attached to In-N

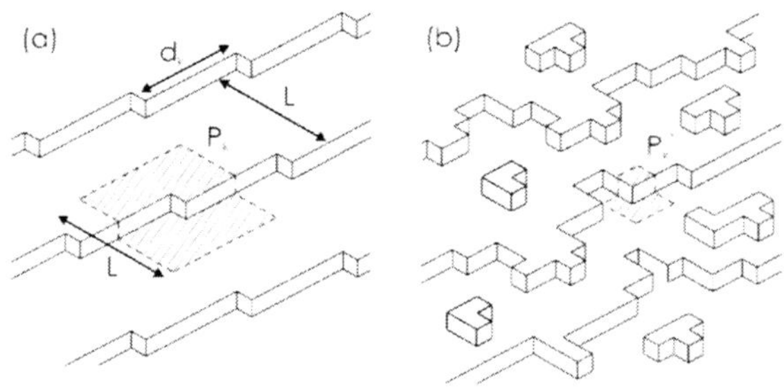

Figure 10.7. The surface fraction from which Ga adatom can reach one atomic kink for GaN growth (kinetic-limited growth) (a) and for InGaN (diffusion-limited growth) (b).

kinks, block In–N bonds from decomposition, is taken therefore into account in pre-exponent constant C_A for InN^{loss}. We can rewrite Eq. 10.3 by substituting

$$C_A = \frac{\Theta}{d_k \Phi_{Ga}} C \quad (10.5)$$

This picture, however, is only valid when the diffusion length of the Ga adatoms is higher than atomic terrace width L (kinetic-limited regime). For InGaN growth, we observe that such a scenario does work, only for low In content, namely 1–2% (see AFM image — Fig. 10.6a). For higher In content, the morphology of InGaN show instability, new islands, and a lot of new atomic kinks, as demonstrated on Fig. 10.6b. This indicates that diffusion length of Ga is smaller than the width of atomic terraces L (see Fig. 10.7b). We can define such process as diffusion-limited growth. The transition from kinetic- to diffusion- limited growth has been reported for instance for GaN growth in Ga- and N-rich regime [51], or during freezing of water where morphology of ice crystals grown in diffusion- and kinetic-regimes was studied [52]. The change of the AFM patterns for low and high In content InGaN was also reported for MOVPE growth [53].

As it can be seen from the AFM images presented in Fig. 10.6b, for growth of high In content structures the *number of atomic kinks increase rapidly*. In addition, the new islands emerge on atomic terraces. Figure 10.7 illustrates the surface fraction per one kink on initial GaN substrate (Fig. 10.7a) and during InGaN growth

(Fig. 10.7b). Therefore, for high content InGaN growth, the area per one kink, P_k^* *is smaller* than expected from miscut of the substrate $P_k^* << P_k$ and the rate $\upsilon_{Ga}^* = \Phi_{Ga} \cdot P_k^*$ is lower, which in turn *reduces indium incorporation*. Practically, for calculations of υ_{Ga}^*, the increase of the kink density for InGaN can be introduced as a growth on substrate with higher effective miscut angle equal to $\Theta + \Theta_0$. Taking into account peculiarities of InGaN growth, the diffusion- and kinetic-limited regimes, the InN^{Loss} is given by the following equation:

$$InN^{Loss} = C \cdot x \cdot \frac{(\Theta + \Theta_0)}{\Phi_{Ga}} \cdot \exp\left(-\frac{E_A}{kT_G}\right) \qquad (10.6)$$

The parameter Θ_0 can be used to define transition between kinetic- ($\Theta_0 = 0$) and diffusion-limited ($\Theta_0 > 0$) growth regimes.

10.4.2 *InGaN Growth: Experimental*

In this section, we will give experimental evidence that the growth of InGaN is influenced by (a) substrate temperature, T_G, (b) substrate miscut, Θ, (c) Ga flux, Φ_{Ga}, and (d) N flux, Φ_N. For the analysis, we have grown 40 nm thick InGaN layers. The In content was determined by XRD reciprocal space mapping in order to take into account possible lattice relaxation for the highest In content samples. The growth temperature was determined by a pyrometer.

The result for the dependence of the In content in InGaN on the growth temperature in PAMBE is shown in Fig. 10.8 (dots). The set of InGaN layers were grown on substrates with the miscut of 0.6° and constant Ga and N fluxes. Substrate temperature plays a dominant role in InN decomposition rate, so for relatively low nitrogen fluxes in PAMBE system, we have to reduce growth temperature to about 550–650°C in order to obtain high indium content InGaN layers. The results can be explained with assumption that during the growth the decomposition of InGaN takes place — as indicated by the arrow on Fig. 10.8. This decomposition is stronger for higher T_G. The solid line presents results of modelling (Eq. 10.6). Since we keep constant miscut Θ and Φ_{Ga}, the same curve can be also calculated using model from Ref. [41].

Situation becomes more complicated when we want to study the In content dependence on the miscut angle Θ. In Fig. 10.9, we

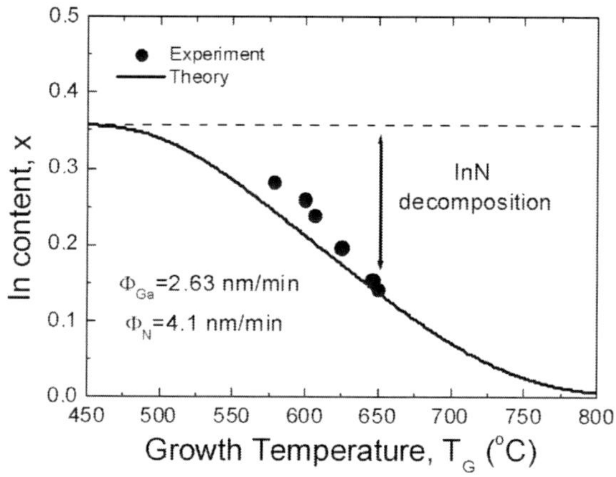

Figure 10.8. The dependence of the In content in InGaN layers as a function of the growth temperature. The experimental data (dots) were obtained from XRD measurements on substrate with miscut 0.6°.

present results of InGaN growth for Θ in a range up to 1.2°. We have grown layers using N flux equal to 5.8 nm/min and two different Ga fluxes of 2 nm/min and 4.3 nm/min, respectively. The dependence of the In content as a function of the miscut angle Θ demonstrates clearly that the Ga adatoms play important role in the InGaN growth, by "blocking" the In-N bonds. We can expect lower InN^{loss} for lower miscut angle Θ, since we have higher υ_{Ga} for constant Ga flux. Model presented in Ref. [41] does not account for such case — giving constant indium content as a function of Θ.

The solid line in Fig. 10.9 shows theoretical modelling according to Eq. 10.6 with $\Theta_0 = 1.4$ and $\Theta_0 = 0°$. As we mentioned already, the use of "effective" miscut $\Theta + \Theta_0$ (with $\Theta_0 = 1.4°$) is due to the fact that for high In content InGaN layers we observe increase of the kink density. As an example, AFM images are presented in Fig. 10.6 for low and high In content InGaN layers. For the ideal case, where the number of atomic kinks remains constant (like in Fig. 10.7a), the experimental data should be well fitted with $\Theta_0 = 0$, where much stronger In content variation vs. miscut Θ is expected (see Fig. 10.9).

The important consequence of the fact that the indium incorporation depends on the Ga attachment rate, υ^*_{Ga}, is growth of InGaN using very low Ga fluxes. In Fig. 10.10, we present the data for

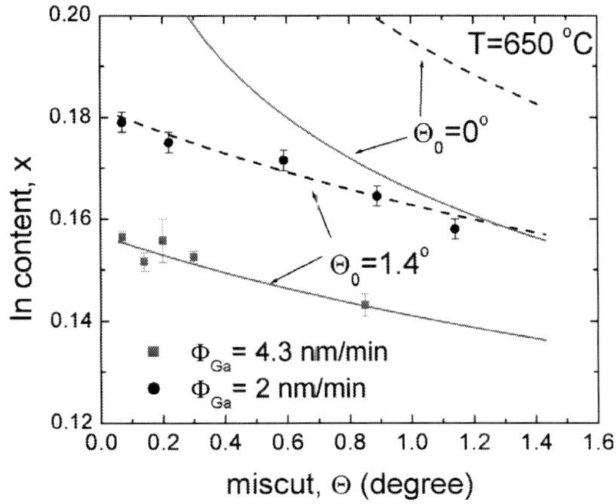

Figure 10.9. In content as a function of the miscut angle, Θ. Squares and dots — experimental data for Ga flux 4.3 nm/min and 2 nm/min, respectively. Solid and dashed lines — theoretical fitting for these Ga fluxes for two values of Θ_0; $\Theta_0 = 1.4$ and $\Theta_0 = 0°$, respectively.

InGaN growth for constant temperature 650°C as a function of the Φ_{Ga}. When we reduce the Φ_{Ga}, the InN decomposition takes place, as indicated by an arrow in Fig. 10.10, leading to less effective In incorporation. Initially, for high Ga fluxes comparable with N fluxes, such a process is well described by the approach presented in Ref. [41] (see dashed line in Fig. 10.10), but the most striking effect is for very low Φ_{Ga}, where we observe *decrease of the indium content.* Such an effect is a manifestation of the fact that smaller υ^*_{Ga} cannot stabilize many In-N bonds. The experiment using different Φ_{Ga} demonstrated also that the growth rate of InGaN is equal to $\Phi_{Ga}/(1-x)$. For $\Phi_{Ga} < 5.2$ nm/min, decomposition of InGaN is so high that *not all nitrogen* is used for growth.

The last part of the experimental survey concerns the role of nitrogen in the InGaN growth. We believe that increase of the Φ_N is beneficial for quality of the InGaN layers. In Fig. 10.11 the indium content x as a function of the nitrogen flux Φ_N is plotted. For fixed temperature, by increasing the Φ_N we can increase indium content from 12% up to 20%, with all other growth parameters fixed.

The Φ_N was calibrated by the growth of GaN in gallium rich regime. The increase of N flux from 6.5 nm/min to 13 nm/min was

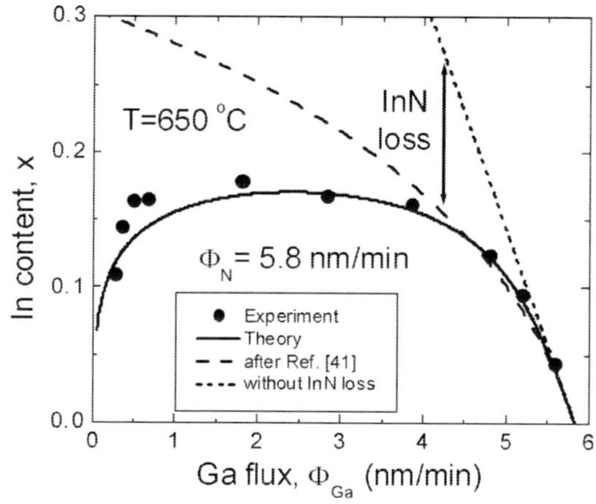

Figure 10.10. The dependence of the In content as a function of Ga flux.

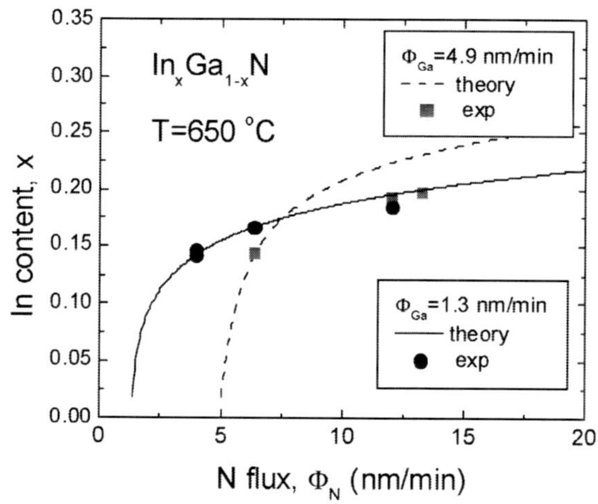

Figure 10.11. The dependence of the In content as a function of N flux, Φ_N.

realized by second RF plasma source mounted on MBE system. We will show later that increase of the nitrogen flux generated by RF plasma source was essential to achieve green lasing.

In this section, we demonstrated experimentally that the indium content of InGaN layers depend not only on the growth temperature but also on gallium and nitrogen fluxes, which gives an additional freedom to optimize InGaN-based structures. The results were described by phenomenological model, which takes into account that Ga attached to atomic kink (via nitrogen atom) block efficiently decomposition of In–N bonds. We would like to stress here that all experiments presented in this section were well fitted making use of Eq. 10.6 with the same set of parameters; $E_{InN} = 1.92$ eV [48], $E_{GaN} = 3.6$ eV [49], $C = 1.44 \times 10^{21}$ (nm/min)^2, $\Theta_0 = 1.4^\circ$.

10.5 Light-Emitting Diodes

Optimizing quantum efficiency of LEDs is an effective strategy when preparing for more demanding LDs. The LEDs were grown on 0.5° miscut GaN HVPE substrates. A 100 nm $In_{0.02}Ga_{0.98}N$:Si buffer layer was followed by 3 nm wells and 10 nm barriers, the 14 nm $In_{0.02}Al_{0.16}Ga_{0.82}N$:Mg electron blocking layer (EBL) and 200 nm $In_{0.02}Ga_{0.98}N$:Mg layer. The barriers were Si-doped at 2×10^{18} cm^{-3}, while the EBL and upper $In_{0.02}Ga_{0.98}N$:Mg layers were Mg-doped at 10^{20} cm^{-3} [37]. The LED structures were grown at the growth conditions established for MQWs. The light-current(L–I) characteristic of one of our LEDs (obtained on a non-processed wafer structure) is presented in Fig. 10.12. The insets to this figure show electroluminescence (EL) spectrum. The light was collected from the bottom side of the substrate (passing through GaN substrate) and measured by a silicon detector attached to the bottom side of the wafer. The output power for this LED exceeded 3 mW at 100 mA for 450 nm. The non-linear behaviour of L–I characteristic for high currents is mainly related to the heating of the indium p-type top contact. The optical power curves for LEDs grown by PAMBE measured at 20 mA for several wavelengths spanning from blue-violet to green are shown in Fig. 10.13. The inset to Fig. 10.13 shows details of the sample — detector configuration for EL experiments. We have obtained high optical output power of 1.5 mW at 20 mA for blue-violet LED. This power is reduced to 0.1–0.2 mW for the green LEDs. The reduction of the EL emission for the LEDs grown by PAMBE is similar to that observed on structures grown by MOVPE technique [36]. The possible explanation of the EL reduction for longer wave-

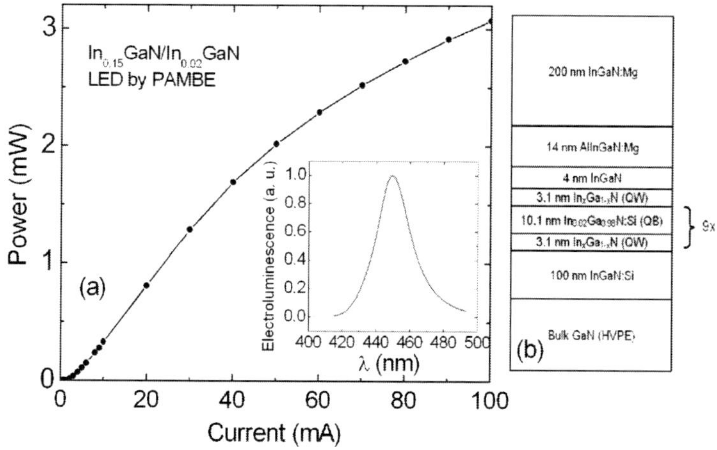

Figure 10.12. The output power of LED grown by PAMBE as a function of current (a) and the LED structure (b). The inset to Fig. 10.12a shows the electroluminescence spectrum.

lengths is related to deterioration of the structural quality of the QWs with increased indium composition, which results in indium content fluctuations and possible creation of dislocations. Other likely explanation is a presence of large built-in electric fields for the layers grown along the polar c-axis. The strength of these fields increases drastically for high indium content QWs, leading to the reduction of oscillator strength and laser gain [54]. These issues can be effectively circumvented by switching to non-polar or semi-polar substrates [55, 56]. Indeed, it is widely believed that low dislocation density non-polar or semi-polar GaN substrates can provide the key to successful fabrication of green LDs [39, 57, 58]. Another candidate for the process which cause observed quantum efficiency drop with increasing emission wavelength could be non-radiative recombination related to N vacancies (V_N). Such correlations have been observed for MOVPE growth of LDs [59]. Indeed, high In content InGaN layers require high In precursor overpressure during the growth. For insufficient ammonia supply, the concentration of V_N related recombination centres may increase drastically in the grown film. More studies, e.g. positron annihilation experiments, should elucidate the question whether V_N are present in PAMBE high In content layers.

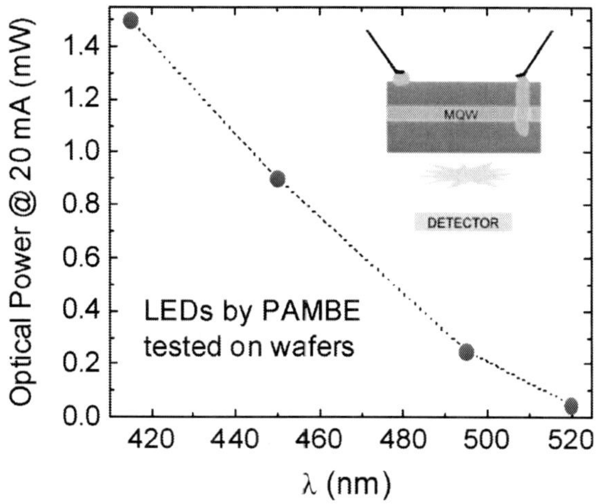

Figure 10.13. The optical power output for LEDs at current 20 mA as a function of the wavelength λ.

Finally, one should consider the acceleration of Auger recombination processes with narrowing the bandgap of the well material, which would lead to the decrease in the quantum efficiency for radiative recombination, particularly at high injection currents [60]. However, importance of Auger recombination for InGaN-based devices is still very controversial, with the most recent fully microscopic many body calculations showing that Auger processes are unlikely to play a role for this material [61].

10.6 Optical Pumped Lasers at 410–500 nm

As we mentioned for LEDs, the most important factors which reduce quantum efficiency are structural quality and non-radiative recombination centres. Since our growth is under metal rich conditions, the *N* vacancies are good candidates for such centres. For low excitation regime in LEDs, the indium composition fluctuations effectively localize carriers/excitons, making dislocations to play a rather limited role in recombination processes. At high carrier concentrations, potential fluctuations, present in quantum well planes, are screened/filled by free carriers. Then, dislocations become effective centres of non-radiative recombination. According to this concept, achievement of the stimulated emission, which occurs at high carrier densities, requires the use of material with the lowest possible density of dislocations.

The most important factor for achieving long wavelength lasing are therefore improvements of the quality of InGaN QWs. As we demonstrated, the increase of the indium content in InGaN layers can be realized by (a) reduction of the growth temperature or (b) increase in the nitrogen flux. The reduction of the growth temperature is effective in increasing the indium content, however, at low growth temperatures the optical quality of the InGaN is dramatically deteriorated — QWs efficiency is decreased, as demonstrated for LEDs.

A new insight into the quality of QWs can be gained from the time resolve photoluminescence (TRPL) experiments [47, 62]. In Fig. 10.14, we show PL intensity and PL decay time spectra measured by TRPL experiments at T = 300 K. The TRPL experiments provide information on the dynamics of excitons. The observed exciton decay time, τ_{EXC} is a function of radiative (τ_R) and non-radiative (τ_N) recombination times ($\tau_{EXC}^{1} = \tau_R^{-1} + \tau_N^{-1}$). At room temperature τ_{EXC} is mainly governed by τ_N. As we pointed out, by decreasing the growth temperature we can shift PL from 420 nm (see Fig. 10.14a) to 520 nm (Fig. 10.14b) for constant nitrogen flux 4.1 nm/min. However, PL decay times (dots in Fig. 10.14b) strongly increase for longer wavelengths, the effect which was not

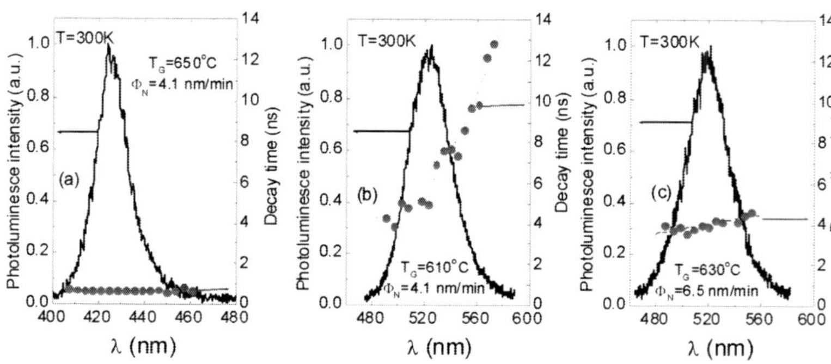

Figure 10.14. Photoluminescence intensity and decay time of PL for blue (a), green (b), and (c) emission.

present for sample shown in Fig. 10.14a with PL at 420 nm. Such an increase of τ_{EXC} as a function of λ indicates that for longer λ excitons are localized (e.g. by indium content fluctuations) which *effectively reduces* the influence of non-radiative recombination channels. Therefore the increase of τ_{EXC} is indicative of presence of localizations. When we increased the nitrogen flux, we were able to grow InGaN QWs with the same In content (around 24%) at higher growth temperature. The TRPL measurements show flat wavelength dependence for τ_{EXC} (see Fig. 10.14c) for QWs grown at such conditions. This indicates that such QWs are good gain media, much as the considerably lower indium content sample (with In content of 12%) presented in Fig. 10.14a. Therefore for efficient QWs at green spectral region, we tried to keep the growth temperature as high as possible increasing the nitrogen flux.

From cathodoluminescence (CL) experiments, we also found that QWs grown at higher nitrogen flux and higher growth temperature are more homogenous. It is also likely that we have fewer *N* vacancies in that case. The increase of the absolute value of τ_{EXC} for sample shown in Fig. 10.14c in comparison to Fig. 10.14a is related to the strong built-in electric field for high In content QWs, which results in spatial separation between holes and electrons, in consequence decreasing the oscillator strength of the related optical transitions. The laser structures for optical pumping were grown on HVPE bulk substrates with dislocation density 10^6–10^7 cm^{-2}. In order to keep the optical modes confined inside the waveguide, we have grown 0.5 μm $Al_{0.07}Ga_{0.93}N$ claddings. The active region consists of one or three 3 nm $In_xGa_{1-x}N$ QWs with $In_{0.06}Ga_{0.94}N$ barriers. The lasing was obtained from cleaved 0.7–1.4 mm structures (Fig. 10.15). The Nd:YAG aperture size was 0.25 mm.

The maximum of the nitrogen flux for one RF plasma source initially mounted on our MBE system was 6.5 nm/min. This was sufficient to grow structure which show lasing up to 475 nm at room temperature. Using two RF plasma sources, with nitrogen flux of 13 nm/min, we were able to grow structures, which show lasing at 501 nm at room temperature. In Fig. 10.16, we demonstrate lasing spectra from optically pumped structures made by PAMBE in the range from 410 to 501 nm.

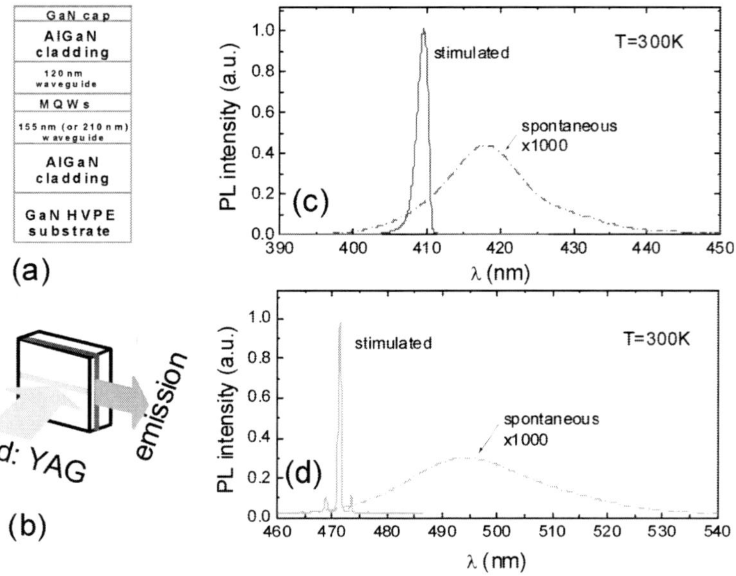

Figure 10.15. The structures for optical lasing (a). The experimental geometry with Nd:YAG pumping laser (b). The lasing spectra for 410 nm (c) and 470 nm (d).

We have found that lasing at long wavelengths can be realized when the electric field in QWs is not fully screened at high carrier injection. We have measured the lasing threshold for the different resonator length, *d*, for structures cleaved from the same wafer (Figs. 10.16b and 10.17). The mirrors were not coated. For longer resonators, the threshold power was lower, as can be expected due to the lower impact of mirror losses (we assume that all mirror are identical). Interestingly, for lower threshold power, i.e. for lower injected carrier density, the lasing wavelength is longer. This can be explained by the fact that effectiveness of screening by the injected carriers is increasing only up to lasing threshold. Above the lasing threshold, all additional electrons and holes injected into the system recombine radiatively, so the quasi-Fermi levels remain unchanged. Therefore one can conclude that for InAlGaN laser grown on polar (0001) c-plane, low light losses promote not only the low lasing threshold but also the longer wavelength of lasing.

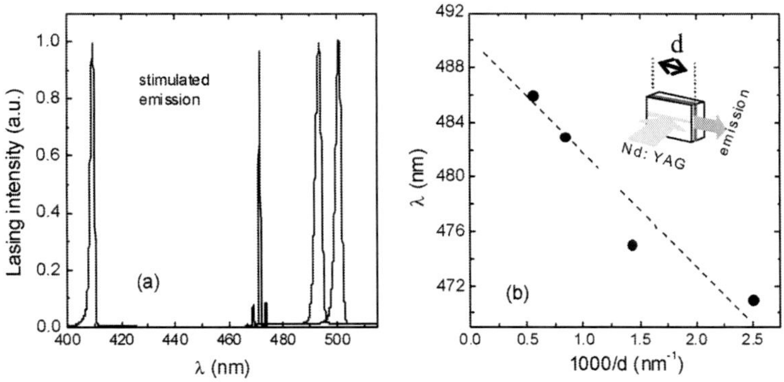

Figure 10.16. The lasing spectra for structures grown by PAMBE (a) and the lasing wavelength as function of resonator length, *d*, for structures from one wafer (b).

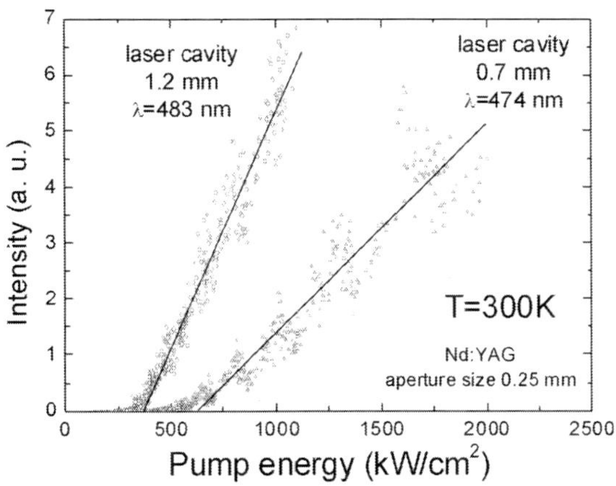

Figure 10.17. The optical lasing for two laser cavity lengths 0.7 and 1.2 mm.

10.7 Laser Diodes

Laser diode structures were grown on (0001) Ga-polarity, conductive GaN substrates [18, 19]. Two types of GaN substrates were used: (i) low dislocation density, below 10^4 cm^{-2}, high pressure-grown and (ii) HVPE with dislocation density of $10^6 - 10^7$ cm^{-2} For the lasers 405–420 nm, we used structure shown in Fig. 10.18. The 40

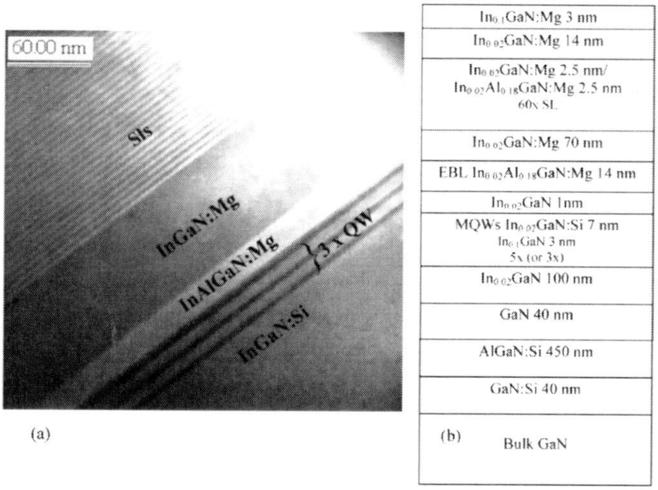

Figure 10.18. Transmission Electron Microscope image of the active region of a PAMBE LD (a) structure details (b).

nm GaN:Si buffer layer and 450 nm $Al_{0.08}Ga_{0.92}N$:Si cladding were grown under Ga- rich conditions at 720°C. The bottom waveguide, MQWs, EBL, top waveguide, top cladding, and contact layer were grown under In-rich conditions at 650°C. The active region consisted of three 3 nm $In_{0.1}Ga_{0.9}N$ wells with 7 nm $In_{0.02}Ga_{0.98}N$ barriers.

The devices were processed as ridge-waveguide, oxide-isolated lasers. The mesa structures were etched to a depth of 0.3 μm. The 20 μm-wide and 500 μm-long stripes were used as laser resonators. The oxidized Ni/Au ohmic contacts were deposited on the top surface of the devices, and Ti/Au contacts were deposited on the backside of the highly conducting n-GaN substrate crystal.

The cleaved laser mirror facets were coated with symmetrically reflecting mirrors. Figure 10.19a shows the Light-Current-Voltage ($L-I-V$) characteristics of the CW LDs with lasing threshold current density and voltage of 5.5 kA/cm^2 and 5.7 V, respectively. Lasing was observed up to 60 mW of optical output power (30 mW per facet) at a wavelength of 411 nm as indicated in Fig. 10.19b [19]. This confirms that growth of high quality layers by PAMBE has been achieved — the smooth interfaces required for LDs can be obtained as shown in Fig. 10.18a.

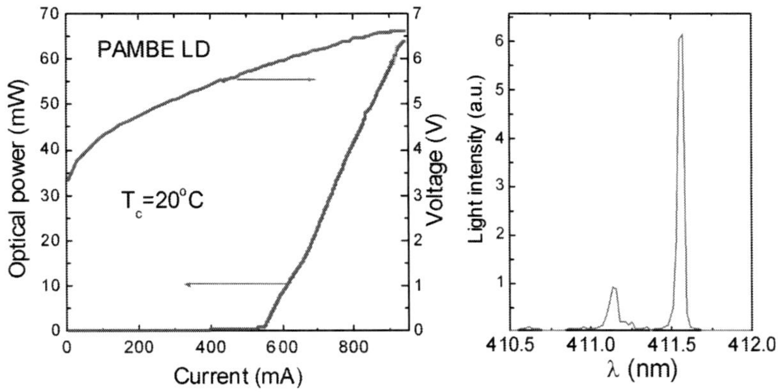

Figure 10.19. (a) L–I–V characteristics of PAMBE-grown cw LD; (b) lasing spectrum.

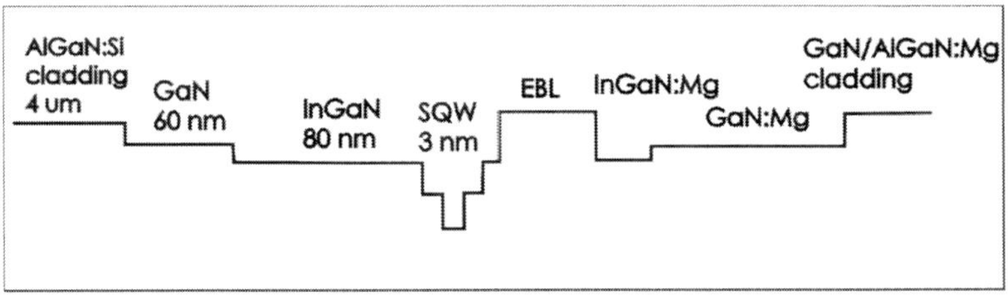

Figure 10.20. Structure details of active region of SQW LD.

The optically pumped stimulated emission at 500 nm was a good starting point for true blue and green LDs. It is worth to stress here that not only efficient QWs are essential for LDs but also such design of the LDs ensuring confinement of the optical modes inside laser waveguide. For the LDs operating at wavelengths longer than 430 nm, we found that the crucial factor for achieving lasing is optimisation of the claddings and the waveguide. The details of single quantum well (SQW) LDs structure lasing at 455 nm are presented in Fig. 10.20.

In order to reduce penetration of the optical modes into the HVPE substrate (which has rather low electron concentration — $5 \times 10^{17} cm^{-3}$), we have grown special claddings comprising of 2 μm heavily doped GaN:Si at the level of 7×10^{19} cm^{-3} and 4 μm GaN/$Al_{0.05}Ga_{0.95}N$ supperlattice. The heavily doped GaN:Si having

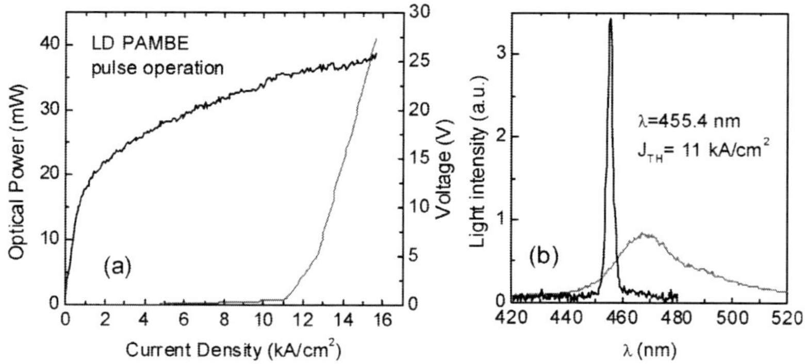

Figure 10.21. (a) L–I–V characteristics of PAMBE-grown blue SQW LD operating in pulse mode; (b) lasing spectrum.

lattice constant close to GaN acts as plasmonic cladding [63]. For further localisation of the optical mode inside waveguide and reduce loses in the gold p-type contact we used also "optical confinement layer", OCL, — 80 nm of $In_{0.07}Ga_{0.93}N$ (see Fig. 10.20).

Figure 10.21 demonstrate the L–I–V and spectral characteristics for blue LD operating in pulse mode at 455 nm. The lasing threshold current density is 11 kA/cm^2 (for uncoated mirrors). The thickness of the InGaN OCL and bottom cladding are essential to achieve efficient LDs. We can compare two LDs design: (a) with 80 nm $In_{0.07}Ga_{0.93}N$ OCL and plasmonic claddings with 4 um GaN/$Al_{0.05}Ga_{0.95}N$ superlattice and (b) with 40 nm $In_{0.07}Ga_{0.93}N$ OCL and more typical (used for 405 nm LDs) 2 μm $Al_{0.05}Ga_{0.95}N$. As presented in Fig. 10.22, the improved OCL and bottom cladding reduced the lasing threshold current and increased the laser slope efficiency. As it comes out from theoretical calculations, the new bottom cladding and waveguide with OCL helped significantly (i) to decrease absorption of the optical modes in upper gold contact and (ii) to reduce penetration of the optical modes to the GaN HVPE substrate. The other interesting feature of SQW LDs grown by PAMBE is a relatively small blue shift from spontaneous (at 465 nm) to lasing emission (at 455 nm) — Fig. 10.21b. This is achieved by employing staggered SQWs which reduce piezoelectric fields inside SQW.

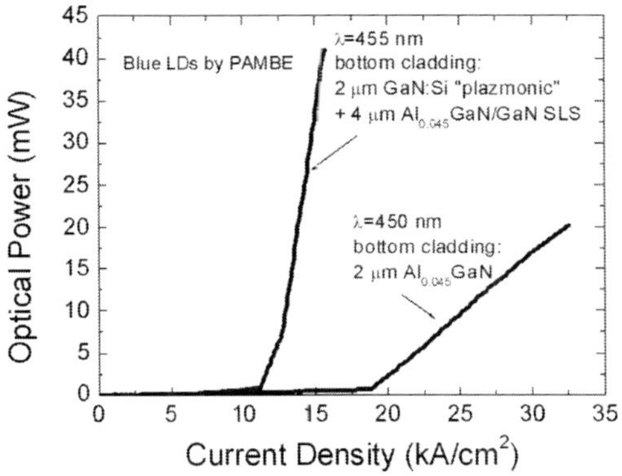

Figure 10.22. L–I characteristics for two blue LDs with different bottom claddings.

10.8 Summary

In this work, we demonstrated recent progress in PAMBE technology for growth of nitride-based Light Emission Diodes and LDs. We described the conditions for high quality growth at low growth temperatures. We showed systematic experimental study and proposed a theoretical model for InGaN growth by PAMBE. The key for the demonstrated improvement of the optical quality of QWs was the new understanding of the role of nitrogen overpressure and the growth temperature on the growth kinetics. We presented optically pumped structures operating from 410 nm to 501 nm and LDs in the spectral range from 410 to 455 nm. As can be seen from presented here results, there is no obvious barrier in extending MBE-grown LDs spectral range into longer wavelengths. Thus the long-term prospects for PAMBE in this area appear to be as good as, or better, than prospects for MOVPE. For nitride-based optoelectronic technology, this is a very new situation.

Acknowledgement

This work was supported partially by the Polish Ministry of Science and Higher Education Grant No. IT 13426 and DARPA VIGIL project. Author would like to thank M. Siekacz, H. Turski, M. Sawicka, G.

Cywinski, A. Feduniewicz-Zmuda for their contribution into PAMBE growth; P. Perlin, P. Wisniewski, W. Purgal for device processing; S. Grzanka, A. Khachapuridze, J. Smalc, M. Albrecht, M. Kryśko for device characterization; B. Lucznik, I. Grzegoy for bulk substrates; Z. Wasilewski, T. Suski, and S. Porowski for numerous stimulating discussions.

References

1. S. Nakamura, G. Fasol, and S. J. Pearton, *The Blue Laser Diode: The Complete Story*, 2 ed. (Springer-Verlag, 2000).
2. M. Kneissl, D. P. Bour, C. G. V. d. Walle, L. T. Romano, J. E. Northrup, R. M. Wood, *et al.*, *Appl. Phys. Lett.*, **75**, (1999) 581.
3. N. Grandjean, B. Damilano, and J. Massies, *J. Phys.: Condens. Matter*, **13**, (2001) 6945.
4. H. Morkoc, *J. Mater. Sci., Mater. Electron.*, **12**, (2001) 677.
5. T. D. Moustakas, E. Iliopoulos, A. V. Sampath, H. M. Ng, D. Doppalapudi, M. Misra, *et al.*, *J. Cryst. Growth*, **227**, (2001) 13.
6. M. Kauer, S. E. Hooper, V. Bousquet, K. Johnson, C. Zellweger, J. M. Barnes, *et al.*, *Electron. Lett.*, **41**, (2005) 23.
7. H. Riechert, R. Averbeck, A. Graber, M. Schienle, U. Strauss, and H. Tews, *Mater. Res. Soc. Symp. Proc.*, **449**, (1997) 149.
8. J. Neugebauer, T. K. Zywietz, M. Scheffler, J. E. Northrup, H. Chen, and R. M. Feenstra, *Phys. Rev. Lett.*, **90**, (2003) 056101.
9. J. E. Northrup, J. Neugebauer, R. M. Feenstra, and A. R. Smith, *Phys. Rev. B*, **61**, (2000) 9932.
10. C. R. Elsass, I. P. Smorchkova, B. Heying, E. Haus, P. Fini, K. Maranowski, *et al.*, *Appl. Phys. Lett.*, **74**, (1999) 3528.
11. B. Heying, R. Averbeck, L. F. Chen, E. Haus, H. Riechert, and J. S. Speck, *J. Appl. Phys.*, **88**, (2000) 1855.
12. B. Heying, E. J. Tarsa, C. R. Elsass, P. Fini, S. P. DenBaars, and J. S. Speck, *J. Appl. Phys.*, **85**, (1999) 6470.
13. C. Adelmann, J. Brault, D. Jalabert, P. Gentile, H. Mariette, G. Mula, *et al.*, *J. Appl. Phys.*, **91**, (2002) 9638.
14. M. J. Manfra, K. W. Baldwin, A. M. Sergent, K. W. West, R. J. Molnar, and J. Caissie, *Appl. Phys. Lett.*, **85**, (2004) 5394.

15. C. Skierbiszewski, K. Dybko, W. Knap, M. Siekacz, W. Krupczynski, G. Nowak, *et al.*, *Appl. Phys. Lett.*, **86**, (2005) 102106.
16. C. Skierbiszewski, Z. R. Wasilewski, M. Siekacz, A. Feduniewicz, B. Pastuszka, I. Grzegory, *et al.*, *Phys. Status Solidi A*, **201**, (2004) 320.
17. P. Waltereit, H. Sato, C. Poblenz, D. S. Green, J. S. Brown, M. McLaurin, *et al.*, *Appl. Phys. Lett.*, **84**, (2004) 2748.
18. C. Skierbiszewski, Z. R. Wasilewski, M. Siekacz, A. Feduniewicz, P. Perlin, P. Wisniewski, *et al.*, *Appl. Phys. Lett.*, **86**, (2005) 011114.
19. C. Skierbiszewski, P. Wisniewski, M. Siekacz, P. Perlin, A. Feduniewicz-Zmuda, G. Nowak, *et al.*, *Appl. Phys. Lett.*, **88**, (2006) 221108.
20. I. Grzegory and S. Porowski, *Thin Solid Films*, **367**, (2000) 281.
21. W. K. Burton, N. Cabrera, and F. C. Frank, *Philos. Trans. R. Soc. London*, **243**, (1951) 299.
22. A. Ishizaka and Y. Murata, *J. Phys.: Condens. Matter*, **6**, (1994) L693.
23. H. Saitoh, W. Utsumi, H. Kaneko, and K. Aoki, *J. Cryst. Growth*, **300**, (2007) 26.
24. J. A. Van Vechten, *Phys. Rev. B*, **7**, (1973) 1479.
25. J. Karpinski, J. Jun, and S. Porowski, *J. Cryst. Growth*, **66**, (1984) 1.
26. J. Karpinski and S. Porowski, *J. Cryst. Growth*, **66**, (1984) 11.
27. I. Grzegory, S. Krukowski, J. Jun, M. Bockowski, M. Wroblewski, and S. Porowski, in *AIP Conference Proceedings*, 309 ed. (1994), p. 565.
28. R. C. Schoonmaker, A. Buhl, and J. Lemley, *J. Phys. Chem.*, **69**, (1965) 3455.
29. R. Groh, G. Gerey, L. Bartha, and J. I. Pankove, *Phys. Status Solidi A*, **26**, (1974) 353.
30. N. Newman, in *Semiconductors and Semimetals*, **50**, (1998), p. 55.
31. M. Sawicka, H. Turski, M. Siekacz, J. Smalc-Koziorowska, M. Krysko, I. Dziecielewski, *et al.*, *Phys. Rev.*, **B83**, (2011) 245434.
32. D. A. Stocker, E. F. Schubert, and J. M. Redwing, *Appl. Phys. Lett.*, **73**, (1998) 2654.
33. G. Koblmuller, F. Wu, T. Mates, J. S. Speck, S. Fernandez-Garrido, and E. Calleja, *Appl. Phys. Lett.*, **91**, (2007) 221905.
34. X. Q. Shen, K. Furuta, N. Nakamura, H. Matsuhata, M. Shimizu, and H. Okumura, *J. Cryst. Growth*, **301–302**, (2007) 404.
35. T. Jang, O. H. Nam, K. H. Ha, S. N. Lee, J. K. Son, H. Y. Ryu, *et al.*, in *Recent achievements of AlInGaN based laser diodes in blue and green wavelength*, San Jose, CA, USA, 2007 (SPIE), p. 64730X.
36. C. Wetzel and T. Detchprom, *MRS Internet J. Nitride Semicond. Res.*, **10**, (2005) 2.

37. C. Skierbiszewski, Z. R.Wasilewski, I. Grzegory, S. Porowski, *J. Cryst. Growth*, **311**, (2009) 1632.
38. T. Miyoshi, S. Masui, T. Okada, T. Yanamoto, T. Kozaki, S. Nagahama, *et al.*, *Appl. Phys. Exp.*, **2**, (2009) 062201.
39. Y. Yoshizumi, M. Adachi, Y. Enya, T. Kyono, S. Tokuyama, T. Sumitomo, *et al.*, *Appl. Phys. Exp.*, **2**, (2009) 092101.
40. C. Adelmann, R. Langer, G. Feuillet, and B. Daudin, *Appl. Phys. Lett.*, **75**, (1999) 3518.
41. R. Averbeck and H. Riechert, *Phys. Status Solidi A*, **176**, (1999) 301.
42. T. Bottcher, S. Einfeldt, V. Kirchner, S. Figge, H. Heinke, D. Hommel, *et al.*, *Appl. Phys. Lett.*, **73**, (1998) 3232.
43. H. Chen, A. R. Smith, R. M. Feenstra, D. W. Greve, and J. E. Northrup, *MRS Internet J. Nitride Semicond. Res.*, **4S1**, (2000) G9.5.
44. M. Siekacz, A. Feduniewicz-Zmuda, G. Cywinski, M. Krysko, I. Grzegory, S. Krukowski, *et al.*, *J. Cryst. Growth*, **310**, (2008) 3983.
45. D. F. Storm, *J. Appl. Phys.*, **89**, (2001) 2452.
46. C. Skierbiszewski, *Solid State Phenomena*, **140**, (2008) 17.
47. M. Siekacz, M. Sawicka, H. Turski, G. Cywiński, A. Khachapuridze, P. Perlin, T. Suski, M. Boćkowski, J. Smalc-Koziorowska, M. Kryśko, R. Kudrawiec, M. Syperek, J. Misiewicz, Z. Wasilewski, S. Porowski and C. Skierbiszewski, *J. Appl. Phys.*, **110**, (2011) 063110.
48. C. S. Gallinat, G. Koblmüller, J. S. Brown, and J. S. Speck, *J. Appl. Phys.*, **102**, (2007) 064907.
49. N. Grandjean, J. Massies, F. Semond, S. Yu. Karpov, and R. A. Talalaev, *Appl. Phys. Lett.*, **74**, (1999) 1855.
50. H. Turski, M. Siekacz, M. Sawicka, G. Cywinski, A. Feduniewicz-Zmuda, J. Smalc-Koziorowska, *et al.*, *J. Vac. Sci. Technol.*, **B29**, (2011) 03C136.
51. M. H. Xie, M. Gong, E. K. Y. Pang, H. S. Wu, and S. Y. Tong, *Phys. Rev.*, **B74**, (2006) 085314.
52. A. A. Shibkov, M. A. Zheltov, A. A. Korolev, A. A. Kazakov, and A. A. Leonov, *J. Cryst. Growth*, **285**, (2005) 215–227.
53. D. D. Koleske, S. R. Lee, G. Thaler, M. H. Crawford, M. E. Coltrin, and K. C. Cross, *Appl. Phys. Lett.*, **97**, (2010) 071901.
54. K. Kojima, M. Funato, Y. Kawakami, Y. Narukawa, and T. Mukai, *Solid State Commun.*, **140**, (2006) 182.
55. A. E. Romanov, T. J. Baker, S. Nakamura, J. S. Speck, and E. J. U. Group, *J. Appl. Phys.*, **100**, (2006) 023522.

56. H. Teisseyre, C. Skierbiszewski, A. Khachapuridze, A. Feduniewicz-Zmuda, M. Siekacz, B. Lucznik, *et al.*, *Appl. Phys. Lett.*, **90**, (2007) 081104.

57. K. Okamoto, T. Tanaka, and M. Kubota, *Appl. Phys. Exp.*, **1**, (2008) 072201.

58. A. Tyagi, H. Zhong, R. B. Chung, D. F. Feezell, M. Saito, K. Fujito, *et al.*, *Jpn. J. Appl. Phys.*, **46**, (2007) L444.

59. K. Yanashima, S. Hashimoto, T. Hino, K. Funato, T. Kobayashi, K. Naganuma, *et al.*, *J. Electronic Materials*, **28**, (1999) 287.

60. Y. C. Shen, G. O. Mueller, S. Watanabe, N. F. Gardner, A. Munkholm, and M. R. Krames, *Appl. Phys. Lett.*, **91**, (2007) 141101.

61. J. Hader, J. V. Moloney, B. Pasenow, S. W. Koch, M. Sabathil, N. Linder, *et al.*, *Appl. Phys. Lett.*, **92**, (2008) 261103.

62. M. Gladysiewicz, R. Kudrawiec, J. Misiewicz, M. Siekacz, G. Cywinski, C. Skierbiszewski, *Physica Status Solidi (c)*, **8**, (2011) 2282.

63. P. Perlin, K. Holc, M. Sarzyñski, W. Scheibenzuber, L. Marona, R. Czernecki, *et al.*, *Appl. Phys. Lett.*, **95**, (2009) 261108.

Index

Color Insert

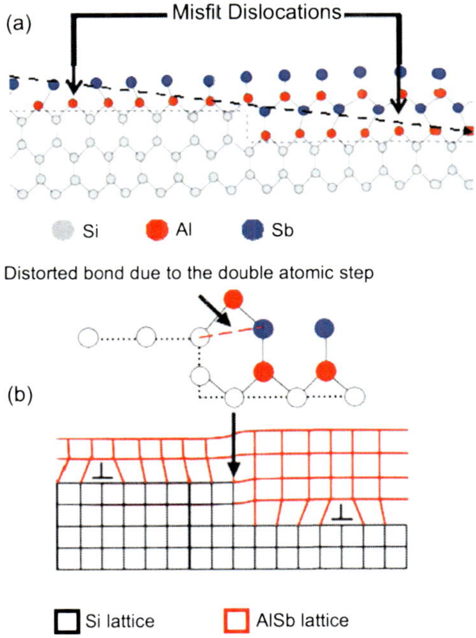

Figure 1.16

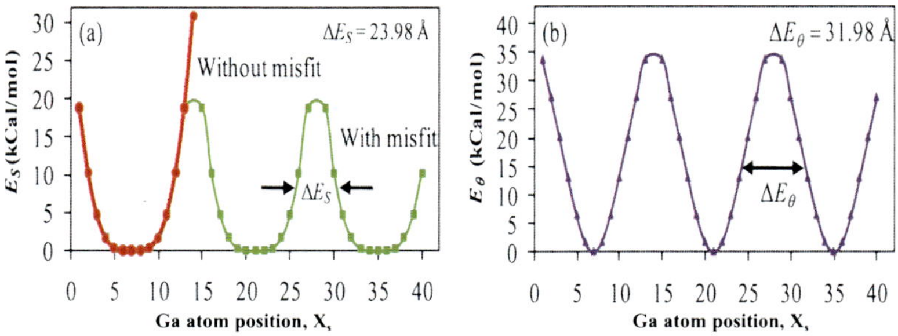

Figure 1.19

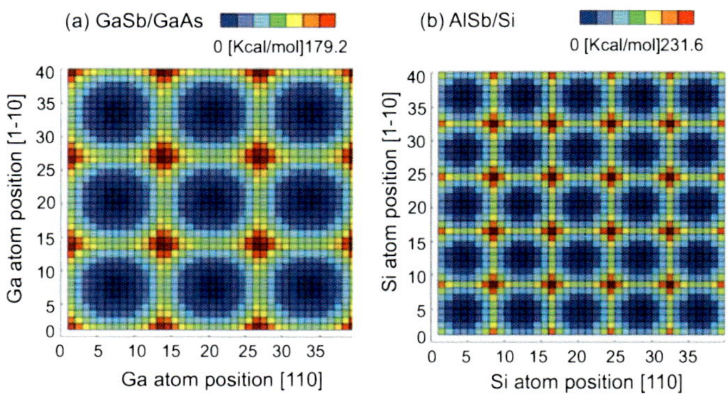

Figure 1.20

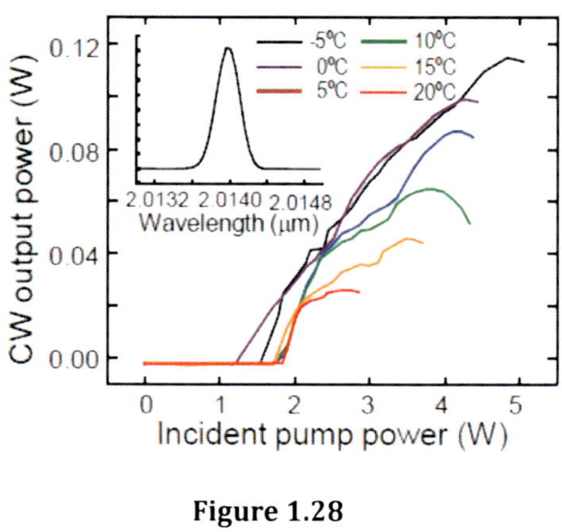

Figure 1.28

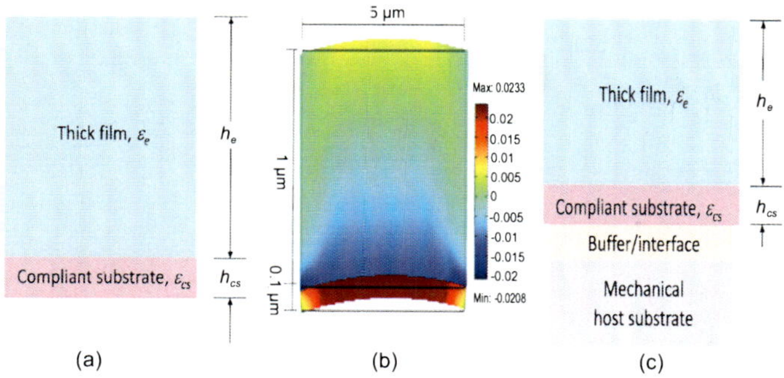

Figure 2.1

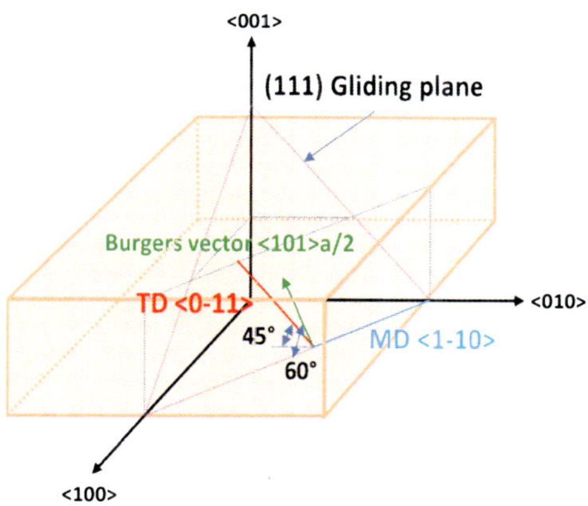

Figure 3.3

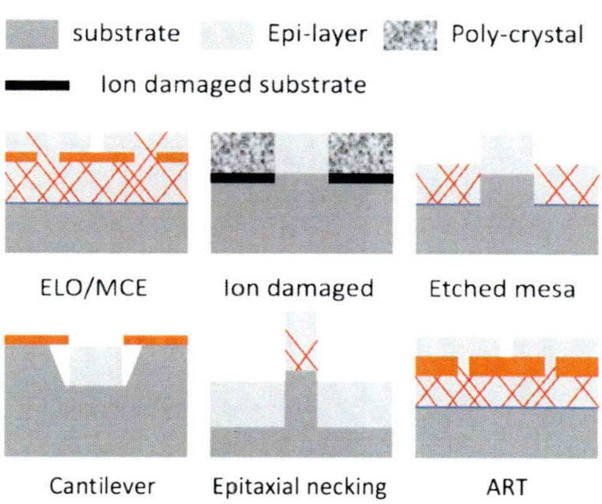

Figure 3.8

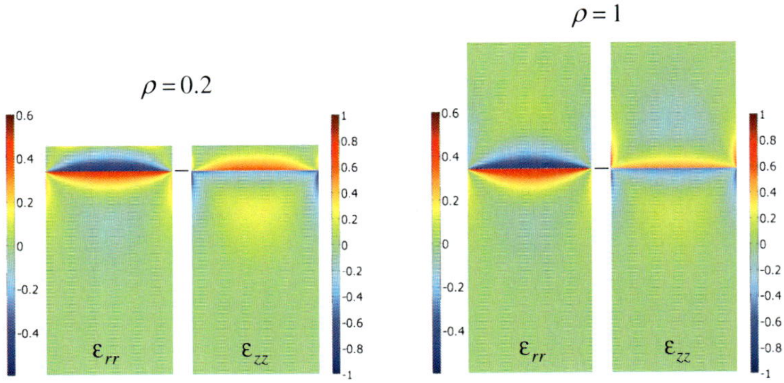

Figure 5.3

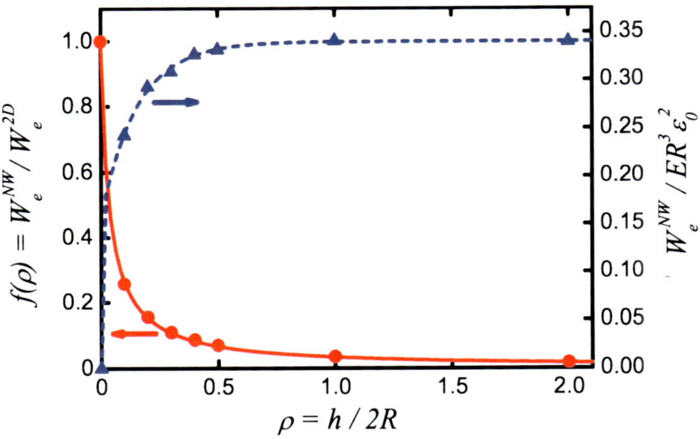

Figure 5.4

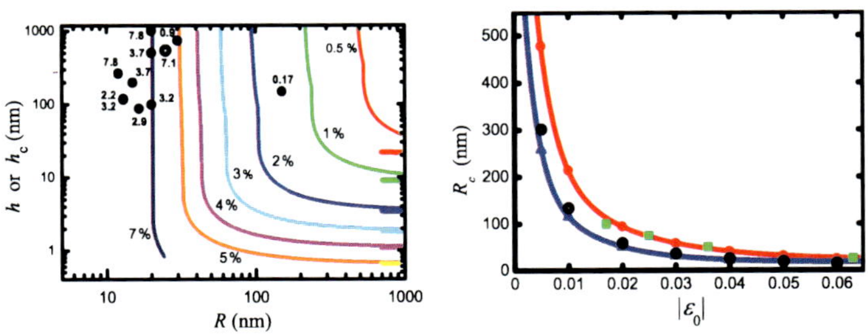

(a) (b)

Figure 5.5